Patterned Random
Matrices

Patterned Random Matrices

Arup Bose
Indian Statistical Institute
Kolkata, West Bengal, India

CRC Press
Taylor & Francis Group
Boca Raton London New York

CRC Press is an imprint of the
Taylor & Francis Group, an **informa** business

This book is on the spectrum of random matrices. The most heartwarming naturally occurring spectrum is the rainbow. In a double rainbow, the colors of the second rainbow are inverted.

CRC Press
Taylor & Francis Group
6000 Broken Sound Parkway NW, Suite 300
Boca Raton, FL 33487-2742

© 2018 by Taylor & Francis Group, LLC
CRC Press is an imprint of Taylor & Francis Group, an Informa business

No claim to original U.S. Government works

Printed on acid-free paper
Version Date: 20180430

International Standard Book Number-13: 978-1-138-59146-2 (Hardback)

Visit the Taylor & Francis Web site at
http://www.taylorandfrancis.com

and the CRC Press Web site at
http://www.crcpress.com

Contents

To the memory of my parents,

Satyendara Nath Basu,
who had a glint in his eyes at the prospect, but did not live to see it

and

Uma Basu,
who had the same glint, though unable to express it.

Preface

Approximately a dozen years ago, I was undeniably ignorant of the subject of random matrices. Then as I tried to answer a question posed to me by J.K. Ghosh, I was inextricably led to large-dimensional random matrices (LDRM).

I have been fortunate. A steady stream of wonderfully prepared students knocked, looking for a one- or two-semester master's project/dissertation outside their regular curriculum. I sold them LDRM. In chronological order of their appearance in my life, these students are: Sourav Chatterjee, Arnab Sen, Anirban Basak, Sanchayan Sen, Sayan Banerjee, Riddhipratim Basu and Shirshendu Ganguly. I also have had three wonderful collaborators, Sreela Gangopadhyay, Joydip Mitra and Rajat Subhra Hazra, and a dedicated doctoral student, Koushik Saha. But for a hiatus of three years in between when yours truly doubled up as a punching bag as the Unit Head, it has been a bustling and enjoyable journey, though not bereft of its trials and tribulations. The forays with my students are the sole props on which this book stands.

Random matrices is a many faceted sculpture offering myriad viewing angles. It has significant ramifications in other areas of science. There are several books on random matrices—by mathematicians, physicists and probabilists. In contrast to the vast array of topics these books address, the goal of the current one is modest. It attempts to provide an easy initiation into the field of LDRM. The focus is on investigating the existence and properties of the limiting spectral distribution of different patterned random matrices as the dimension becomes large. The main ingredients are the old fashioned simplistic method of moments with rudimentary combinatorics for support. Some easy matrix results come in handy too. By stretching the moment arguments, we also have a brush with the intriguing but difficult concepts of joint convergence of sequences of random matrices and their ramifications. If the reader finds this book to be a convenient stepping stone to other difficult, interesting and meaty issues in random matrices, I shall be content.

Enrichment came from many academic exchanges. I am grateful to Greg Andersen, Zhidong Bai, Wlodek Bryc, Amir Dembo, Alice Guionnet, Teifeng Jiang, Manjunath Krishnapur, Baiqi Miao, Steven J. Miller, Arunava Mukherjea, Tamer Oraby, Dimitris Politis, Mohshen Pourahmadi, Dimitri Shlyakhtenko, Jack Silverstein, Alexander Soshnikov, Roland Speicher, Ofer Zeitouni and Shurong Zheng.

Kolkata, March 16, 2013/Cincinnati, February 14, 2018   Arup Bose

Introduction

A matrix whose elements are random variables will be called a *random matrix*. The 2×2 contingency table, common in statistics, is perhaps the earliest and simplest example of a random matrix. The famous Wishart distribution (Wishart (1928)[107]) arose as the distribution of the *sample variance-covariance matrix* when the observations are from a normal distribution.

Then for a long period there does not seem to have been any activity in random matrices. The matrix which changed this and in particular which has propelled random matrices as an independent and major area of study is the *Wigner matrix*. Wigner (1955)[105] introduced this matrix while discussing statistical models for heavy nuclei atoms in physics.

Links were gradually established between random matrices and various branches of sciences which include areas as diverse as classical analysis, multivariate statistics, number theory, operator algebra, signal processing and wireless communication. Random matrices have also appeared in substantial ways in other branches of sciences such as high-dimensional data analysis, communication theory, dynamical systems, finance and diffusion processes.

The random matrix literature is now vast and rapidly growing. The most important papers on random matrix theory in physics from the early period are collected in Porter (1965)[86]. The book by Mehta (2004)[78] has been a standard source for a long time, especially for physicists. The book by Forrester (2010)[56] also has a heavy physics flavor. Books with pronounced probabilistic flavor are Bai and Silverstein (2010)[8], Anderson, Guionnet and Zeitouni (2010)[1] and Tao (2012)[100]. Couillet and Debbah (2011)[45] stress the use of random matrices in wireless communication. Other advanced mathematical books on random matrices include Deift (2000)[48], Deift and Gioev (2000)[50], Deift, Baik and Suidan (2016)[49] and Erdős and Yau (2017)[53]. The survey paper of Bai (1999)[5] has spurred a lot of activity in probabilistic aspects of random matrices. Our focus is *patterned random matrices*.

Any matrix (with real or complex entries) is a finite-dimensional linear operator and hence the study of random matrices is the study of these *random operators*. Any such operator is described by its *eigenvalues* (the *spectrum*) and *eigenvectors*. Hence, the primary focus of random matrix theory is to understand the behavior of these objects from different angles. Incidentally, problems on the eigenvectors are notoriously difficult and are not the subject matter of this book. The literature in this area is rather meagre.

A typical setup is where we have a sequence of square matrices whose dimension increases to ∞. These are called *large-dimensional random matrices*

(LDRM). The aspects that are usually investigated are the behavior of the *bulk distribution, extreme eigenvalues, spacings* between the eigenvalues and *linear statistics* of the eigenvalues. We concentrate on the bulk distribution.

Suppose A_n is an $n \times n$ matrix with eigenvalues $\lambda_1, \ldots, \lambda_n \in \mathbb{R}$ (or $\mathbb{C}$). Its empirical spectral measure on $\mathbb{R}$ (or $\mathbb{C}$) is given by

$$\mu_n = \frac{1}{n} \sum_{i=1}^{n} \delta_{\lambda_i},$$

where δ_x is the Dirac delta measure at x. The probability distribution function on $\mathbb{R}$ (or $\mathbb{C} = \mathbb{R}^2$) corresponding to μ_n, is known as the *Empirical Spectral Distribution* (ESD) of A_n. We will denote it by F^{A_n}. Thus,

$$F^{A_n}(x, y) = n^{-1} \sum_{i=1}^{n} \mathbb{I}\{\mathcal{R}(\lambda_i) \leq x, \ \mathcal{I}(\lambda_i) \leq y\}, \ x, y \in \mathbb{R}.$$

If the entries A_n are random, then F^{A_n} is a *random distribution function*. Further, its expectation $\mathrm{E}[F^{A_n}(x, y)]$ is a non-random distribution function and is called the *expected ESD*.

Let $\{A_n\}_{n=1}^{\infty}$ be a sequence of square matrices with the corresponding ESD $\{F^{A_n}\}_{n=1}^{\infty}$. Its *Limiting Spectral Distribution (LSD)* is the weak limit, if it exists, *in probability* or *almost surely*. For real symmetric or Hermitian matrices, all eigenvalues are real and hence the ESD and LSD are defined on the real line. Roughly speaking, any LSD is said to be *universal* if it does not change when the underlying distribution of the entries is changed, while holding the variances and the dependence structure fixed. In this book, only matrices which have some sort of independent structure are studied.

One of the most useful methods of studying the spectral distribution is enumerative combinatorics. This approach will be taken in this book, in an accessible and unified manner. As will be seen, it is ideally suited for matrices which are symmetric or Hermitian. Though it is not of direct use for non-symmetric/non-Hermitian matrices, it still has its value while dealing with them. Non-symmetric matrices (except for the Circulant-type matrices) are not dealt with in this book, but their singular values will be discussed. Other sophisticated tools are available to deal with non-symmetric situations, for example, Pastur's fundamental technique of *Stieltjes transforms* (which shall be touched upon) and the theory of *free probability* (which, as shall be seen, is important in the study of joint convergence of several matrices). The method of *normal approximation* shall be encountered in the context of Circulant-type matrices.

Often the matrices have specific *patterns* which arise from different modelling considerations—for example, from problems in computer science, mathematics, physics and statistics. The problem of proving the existence of LSD and deriving its properties for different *patterned matrices* has drawn significant attention in the literature. The study of this issue is one of the primary

aims of this book. Bose and Sen (2008)[29] initiated a unified approach based
on the work of Bryc, Dembo and Jiang (2006)[34] to prove the existence of
LSD of different LDRM and also to handle dependent entries. This method is
termed the *volume method* and is based on the method of moments.

Suppose, to begin with, the entries of the matrix come from an independent
identically distributed *input* sequence with mean zero and variance one and are
placed in some pattern in the matrix. One of the fundamental conditions that
is imposed on the pattern is *Property B: the maximum number of times any
entry is repeated in a row remains uniformly bounded across all rows as $n \to$
∞.* This property in particular is satisfied for the five well-known matrices,
namely Wigner, Toeplitz, Hankel, Reverse Circulant and Symmetric Circulant.
Under Property B and mild growth conditions on how many variables and
with what frequency they appear in the matrix, the sequence of ESD is tight
almost surely, and all subsequential limits have moments bounded by Gaussian
moments. If there is weak convergence, then it is almost sure, and the limit is
symmetric and universal, as long as the input satisfies some very mild moment
conditions and is independent. This is the content of Chapter 1. Specific matrix
models are discussed in the subsequent chapters.

Wigner matrix. The two most significant matrices whose LSD have been
extensively studied are the Wigner and the sample covariance matrices. The
Wigner matrix $W_n^{(s)}$ (with scaling) is a symmetric matrix and is defined as

$$W_n^{(s)} = \frac{1}{\sqrt{n}} \begin{bmatrix} x_{11} & x_{12} & x_{13} & \cdots & x_{1(n-1)} & x_{1n} \\ x_{12} & x_{22} & x_{23} & \cdots & x_{2(n-1)} & x_{2n} \\ & & \vdots & & & \\ x_{1n} & x_{2n} & x_{3n} & \cdots & x_{(n-1)n} & x_{nn} \end{bmatrix}.$$

Wigner (1958)[106] assumed the entries $\{x_{ij}\}$ to be i.i.d. real (standard) Gaussian. Suppose $\lambda_1/\sqrt{n} \geq \cdots \geq \lambda_n/\sqrt{n}$ are the eigenvalues. Then the joint
distribution is given by

$$f(\lambda_1, \ldots, \lambda_n) = \frac{\exp\left(-\sum_{i=1}^{n} \lambda_i^2/2\right)}{2^{n/2} \prod_{i=1}^{n} \Gamma\left(\frac{p+1-i}{2}\right)} \prod_{i<j}^{n} (\lambda_i - \lambda_j).$$

Wigner proved that the expected ESD of $W_n^{(s)}$ tends to the so-called *semi-
circle law* which has the density function

$$p_W(s) = \begin{cases} \frac{1}{2\pi}\sqrt{4-s^2} & \text{if } |s| \leq 2, \\ 0 & \text{otherwise.} \end{cases}$$

Several extensions of this result are available. The weakest known condition under which convergence to the semi-circle law holds was given by Pastur (1972)[81] who assumed that the entries x_{ij}'s are independent with zero mean and unit variance, and they satisfy the following Lindeberg-type condition which is now known as *Pastur's condition*:

$$\lim_{n\to\infty} \frac{1}{\delta^2 n^2} \sum_{i,j=1}^{n} \mathrm{E}[x_{ij}^2 \mathbb{I}_{(|x_{ij}|>\delta\sqrt{n})}] = 0, \quad \text{for any} \ \ \delta > 0.$$

The Wigner matrix will appear prominently in Chapters 2, 5, 9 and 10. Different extensions of the basic semi-circle LSD result are available in the literature. In particular, Banerjee and Bose (2010) [9] showed that the semi-circle law continues to hold even when we move away from the Wigner pattern to some extent. This is discussed in Chapter 5.

Sample covariance (S) matrix. Suppose that $\{x_{ij};\ i,j = 1,2,\ldots\}$ is a double array of i.i.d. real random variables with mean zero and variance 1. Write $x_k = (x_{1k}, x_{2k}, \ldots, x_{pk})'$ and $X = (x_1, x_2, \ldots, x_n)$. The sample covariance matrix is usually taken as $\frac{1}{n}\sum_{k=1}^{n}(x_k - \bar{x})(x_k - \bar{x})'$. However, in spectral analysis of LDRM, the *sample covariance matrix* is simply defined as

$$S = \frac{1}{n}\sum_{k=1}^{n} x_k x_k' = \frac{1}{n}XX'.$$

The first success in finding the LSD of S was made by Marčenko and Pastur (1967)[73]. There has been much subsequent work on this matrix. The basic LSD result can be described as follows:

Suppose $\{x_{ij}\}$ have finite fourth moment. Then as $p \to \infty$ and $p/n \to y \in (0, \infty)$, the ESD of S converges weakly almost surely. The limit, denoted by $\mathcal{L}_{MPy}$, is known as the *Marčenko-Pastur law* and can be described as follows: It has a positive mass $1 - \frac{1}{y}$ at the origin if $y > 1$. Elsewhere, it has the density:

$$p_{MPy}(x) = \begin{cases} \frac{1}{2\pi xy}\sqrt{(b-x)(x-a)} & \text{if } a \leq x \leq b, \\ \\ 0 & \text{otherwise} \end{cases}$$

where $a = a(y) = (1 - \sqrt{y})^2$ and $b = b(y) = (1 + \sqrt{y})^2$.

Now suppose $p \to \infty$ and $p/n \to 0$. Then under the same conditions, the LSD of $\sqrt{\frac{n}{p}}(XX'/n - I_p)$ is the semi-circle law. Here, I_p is the identity matrix of order p. The S matrix and extensions to matrices of the form XX' where X is a more general patterned matrix, will be dealt with in Chapter 3.

Toeplitz, Hankel and related matrices. Any matrix T_n of the form $((t_{i-j}))_{1\leq i,j\leq n}$ is called a *Toeplitz* matrix. For simplicity, suppose that T_n is symmetric (that is, $t_k = t_{-k}$ for all k) and that $\sum_{k=-\infty}^{\infty} |t_k|^2 < \infty$. The infinite-dimensional Toeplitz and Hankel matrices are important operators from l_2 to

l_2 (l_2 is the space of all square summable sequences). From Szegö's theory of *Toeplitz operators*, the LSD of T_n exists. Define, in the L_2 sense,

$$f(x) = \sum_{k=-\infty}^{\infty} t_k \exp\left(-2\pi i x k\right), \ x \in (0, \ 1].$$

Then the LSD is $f(U)$, where U is uniformly distributed on $(0, \ 1]$.

The random Toeplitz and Hankel matrices, T_n and H_n, with scaling are:

$$T_n = n^{-1/2} \begin{bmatrix} x_0 & x_1 & x_2 & \cdots & x_{n-2} & x_{n-1} \\ x_1 & x_0 & x_1 & \cdots & x_{n-3} & x_{n-2} \\ x_2 & x_1 & x_0 & \cdots & x_{n-4} & x_{n-3} \\ & & & \vdots & & \\ x_{n-1} & x_{n-2} & x_{n-3} & \cdots & x_1 & x_0 \end{bmatrix},$$

$$H_n = n^{-1/2} \begin{bmatrix} x_2 & x_3 & x_4 & \cdots & x_n & x_{n+1} \\ x_3 & x_4 & x_5 & \cdots & x_{n+1} & x_{n+2} \\ x_4 & x_5 & x_6 & \cdots & x_{n+2} & x_{n+3} \\ & & & \vdots & & \\ x_{n+1} & x_{n+2} & x_{n+3} & \cdots & x_{2n-1} & x_{2n} \end{bmatrix}.$$

Bai (1999)[5] posed the question of existence of the LSD of T_n and H_n. When the entries are i.i.d. with finite variance, the existence of the LSD was proved by Bryc, Dembo and Jiang (2006)[34]. Hammond and Miller (2005)[63] also proved the existence of the LSD of T_n. These results are proven in Chapter 2. The Toeplitz LSD is known to have a bounded density and the Hankel LSD is known to be not unimodal. Both have unbounded support and their $(2k)$th moments can be calculated by computing volumes of certain subsets of the hypercube in dimension $(k+1)$. No other properties of these LSDs are currently known. *Balanced* versions of the Toeplitz and Hankel matrices was investigated in Basak and Bose (2010)[12] and their results are presented in Chapter 6. The sample autocovariance matrix for a stationary time series is also a Toeplitz matrix. This matrix is discussed in Chapter 11.

k-Circulant matrices. For positive integers k and n, the $n \times n$ *k-Circulant* matrix with scaling is defined as

$$A_{k,n} = \frac{1}{\sqrt{n}} \begin{bmatrix} x_0 & x_1 & x_2 & \cdots & x_{n-2} & x_{n-1} \\ x_{n-k} & x_{n-k+1} & x_{n-k+2} & \cdots & x_{n-k-2} & x_{n-k-1} \\ x_{n-2k} & x_{n-2k+1} & x_{n-2k+2} & \cdots & x_{n-2k-2} & x_{n-2k-1} \\ & & & \vdots & & \\ x_k & x_{k+1} & x_{k+2} & \cdots & x_{k-2} & x_{k-1} \end{bmatrix}_{n \times n}.$$

Note that the subscripts of the entries are to be read modulo n. For $1 \leq j < n - 1$, the $(j+1)$-th row of the matrix is obtained by giving its j-th row a right circular shift by k positions (equivalently, k mod n positions). Specific choices of k yield the following matrices which are important and useful in their own right.

(i) If $k = 1$, then we obtain the following *Circulant matrix* with scaling:

$$C_n = \frac{1}{\sqrt{n}} \begin{bmatrix} x_0 & x_1 & x_2 & \cdots & x_{n-2} & x_{n-1} \\ x_{n-1} & x_0 & x_1 & \cdots & x_{n-3} & x_{n-2} \\ x_{n-2} & x_{n-1} & x_0 & \cdots & x_{n-4} & x_{n-3} \\ & & & \vdots & & \\ x_1 & x_2 & x_3 & \cdots & x_{n-1} & x_0 \end{bmatrix}_{n \times n}.$$

The eigenvalues of the Circulant matrix have a well-known closed form. They arise crucially in time series analysis via the *periodogram* which is a straightforward function of the eigenvalues of a Circulant matrix. Non-random Circulant matrices also play a crucial role in the study of non-random Toeplitz matrices. The LSD of C_n is complex Gaussian when the real or complex random variables $\{x_i\}$ are independent with suitable moment conditions. See for example Sen (2006)[92] and Meckes (2009)[76].

(ii) Imposition of the symmetry restriction on the Circulant, yields the *Symmetric Circulant matrix* (with scaling):

$$SC_n = \frac{1}{\sqrt{n}} \begin{bmatrix} x_0 & x_1 & x_2 & \cdots & x_2 & x_1 \\ x_1 & x_0 & x_1 & \cdots & x_3 & x_2 \\ x_2 & x_1 & x_0 & \cdots & x_2 & x_3 \\ & & & \vdots & & \\ x_1 & x_2 & x_3 & \cdots & x_1 & x_0 \end{bmatrix}_{n \times n}.$$

The LSD is then real Gaussian under reasonable conditions on the entries; see Bose and Sen (2008)[29]. The Gaussian LSD is rather an exception than a rule for the LSDs of patterned random matrices.

(iii) The choice of $k = n-1$ yields the *Reverse Circulant matrix* (with scaling):

$$RC_n^{(s)} = \frac{1}{\sqrt{n}} \begin{bmatrix} x_0 & x_1 & x_2 & \cdots & x_{n-2} & x_{n-1} \\ x_1 & x_2 & x_3 & \cdots & x_{n-1} & x_0 \\ x_2 & x_3 & x_4 & \cdots & x_0 & x_1 \\ & & & \vdots & & \\ x_{n-1} & x_0 & x_1 & \cdots & x_{n-3} & x_{n-2} \end{bmatrix}_{n \times n}.$$

This matrix has gained importance recently due to its connection with the concept of *half independence* which shall be briefly described later. Under similar conditions as described earlier, the LSD of RC_n has the density

$$f(x) = |x| \exp(-x^2), \quad -\infty < x < \infty.$$

The above two symmetric matrices are treated in Chapter 2.

There are other reasons to be interested in the general k-Circulant matrices. They, along with their block versions, appear in the area of multi-level supersaturated design of experiments and in time series analysis. The adjacency matrix of a De Bruijn graph is also a k-Circulant matrix.

One important aspect of these matrices is that there is a formula solution known for the eigenvalues. This eigenvalue formula exhibits the differing behavior of the eigenvalues depending on the common prime factors of n and k. This helps to a large extent in establishing properties of their eigenvalues.

Establishing the LSD for general k-Circulant matrices appears to be a difficult problem. Bose, Mitra and Sen (2012)[28] showed that if $\{x_i\}$ are i.i.d. $N(0,1)$, $k = n^{o(1)}$ (≥ 2) and $gcd(k,n) = 1$, then the LSD of $A_{k,n}$ is degenerate at zero, in probability. They also derived the following LSD. Suppose $\{x_i\}$ are i.i.d. with finite $(2 + \delta)$ moment. Let $\{E_i\}$ be i.i.d. $Exp(1)$, U_1 be uniformly distributed over $(2g)$th roots of unity, and U_2 be uniformly distributed over the unit circle where $\{U_i\}$, $\{E_i\}$ are mutually independent. Then the LSD of $A_{k,n}$ in probability is

(i) $U_1(\prod_{i=1}^{g} E_i)^{1/2g}$ if $k^g = -1 + sn$, $g \geq 1$ and $s = o(n^{1/3})$,

(ii) $U_2(\prod_{i=1}^{g} E_i)^{1/2g}$ if $k^g = 1 + sn$, $g \geq 2$ and

$$ s = \begin{cases} o(n) & \text{if } g \text{ is even} \\ o(n^{\frac{g+1}{g-1}}) & \text{if } g \text{ is odd.} \end{cases} $$

The k-Circulant matrices are dealt with in Chapter 4.

Triangular matrices. Triangular random matrices have gained importance recently in the context of DT operators. Dykema and Haagerup (2004)[52] were led to the consideration of the *asymmetric* matrix with scaling

$$ T_n = \frac{1}{\sqrt{n}} \begin{bmatrix} t_{1,1} & t_{1,2} & t_{1,3} & \cdots & t_{1,n-1} & t_{1,n} \\ 0 & t_{2,2} & t_{2,3} & \cdots & t_{2,n-1} & t_{2,n} \\ 0 & 0 & t_{3,3} & \cdots & t_{3,n-1} & t_{3,n} \\ & & & \vdots & & \\ 0 & 0 & 0 & \cdots & 0 & t_{n,n} \end{bmatrix} $$

where $(t_{i,j})_{1\leq i\leq j\leq n}$ are i.i.d. complex Gaussian random variables having mean 0 and variance 1. It does not seem to be easy to obtain the LSD of this nonsymmetric matrix. They obtained the LSD of $T_n'T_n$.

This motivates the study of the triangular versions of symmetric patterned matrices. This is done in Chapter 8. Though Property B is not satisfied since there are too many zero entries, it is easy to modify the methods of earlier chapters to deal with these. The problem of identification of the LSDs is non-trivial. For the Symmetric Triangular Wigner matrix, the LSD is universal, with interval $[-\sqrt{e}, \sqrt{e}]$, and is absolutely continuous with respect to the Lebesgue measure with density $|x|\psi(x^2)$ where ψ is the unique solution of

$$\psi\left(\frac{\sin v}{v}\exp(v\cot v)\right) = \frac{1}{\pi}\sin v\exp(-v\cot v).$$

The $(2k)$-th moment of this is given by $\frac{k^k}{(k+1)!}$. However, moment formulae or distributional properties of the LSD for other matrices are not known.

Joint convergence. So far, convergence of eigenvalues has been discussed for one matrix at a time. How does one give meaning to the joint convergence of several matrices in a non-trivial way? Note that matrix multiplication is a non-commutative operation. Thus, one needs to develop a theory of *non-commutative probability spaces*. A *non-commutative probability space* is a pair $(\mathcal{A}, \phi)$ where $\mathcal{A}$ is a unital complex algebra (with unity 1) and $\phi : \mathcal{A} \to \mathbb{C}$ is a linear functional that satisfies $\phi(1) = 1$.

For example, let $(X, \mathcal{B}, \mu)$ be a probability space and $\mathcal{A}$ be the space of $n \times n$ complex random matrices with entries that have finite moments of all order. Then ϕ equal to $\frac{1}{n}\mathrm{E}_\mu[\mathrm{Tr}(\cdot)]$ or $\frac{1}{n}[\mathrm{Tr}(\cdot)]$ both yield non-commutative probabilities.

For any non-commuting variables $x_1, \ldots, x_n$, let $\mathbb{C}\langle x_1, x_2, ..., x_n\rangle$ be the unital algebra of all complex polynomials in these variables. If $a_1, a_2, \ldots, a_n \in \mathcal{A}$ then their *joint distribution* $\mu_{\{a_i\}}$ is defined canonically by their *mixed moments* $\mu_{\{a_i\}}(x_{i_1} \cdots x_{i_m}) = \phi(a_{i_1} \cdots a_{i_m})$. That is,

$$\mu_{\{a_i\}}(P) = \phi(P(\{a_i\})) \quad \text{for} \quad P \in \mathbb{C}\langle x_1, x_2, ..., x_n\rangle.$$

Let $(\mathcal{A}_n, \phi_n)$, $n \geq 1$ and $(\mathcal{A}, \phi)$ be non-commutative probability spaces and let $\{a_i^n\}_{i\in J}$ be a sequence of subsets of $\mathcal{A}_n$, where J is any finite subset of $\mathbb{N}$. Then we say that $\{a_i^n\}_{i\in J}$ *converges in law* to $\{a_i\}_{i\in J} \subset \mathcal{A}$, if for all complex polynomials P,

$$\lim_{n\to\infty} \mu_{\{a_i^n\}_{i\in J}}(P) = \mu_{\{a_i\}_{i\in J}}(P).$$

Independent Wigner matrices converge jointly in the above sense, and various aspects of this joint convergence have been well studied. This joint convergence is tied to the idea of free independence developed by Voiculescu (1991)[102]. Suppose $\{\mathcal{A}_i\}_{i\in J} \subset \mathcal{A}$ are unital subalgebras. They are called *freely independent* or simply *free* if, $\phi(a_j) = 0$, $a_j \in \mathcal{A}_{i_j}$, and $i_j \neq i_{j+1}$ for all j implies $\phi(a_1 \cdots a_n) = 0$. The random variables (or elements of an algebra) $(a_1, a_2, \ldots)$ will be called free if the subalgebras generated by them are free. A well-known fact is that independent Wigner matrices are *asymptotically free*.

Bose, Hazra and Saha (2011)[26] proved some joint convergence results for other patterned matrices. Roughly speaking, if there is marginal convergence of a single patterned matrix, then there is joint convergence of its i.i.d.

copies. In particular, there is joint convergence of copies of any one of Toeplitz, Hankel, Symmetric Circulant and Reverse Circulant matrices. The Symmetric Circulant limits are (classically) independent. The Reverse Circulant limits exhibit *half independence*. To describe this, concepts of half commuting elements and symmetric monomials are essential. Let $\{a_i\}_{i \in J} \subset \mathcal{A}$. Say that they *half commute* if $a_i a_j a_k = a_k a_j a_i$, for all $i, j, k \in J$. Observe that, if $\{a_i\}_{i \in J}$ half commute, then a_i^2 commutes with a_j and a_j^2 for all $i, j \in J$. Suppose $\{a_i\}_{i \in J} \subset \mathcal{A}$. For any $k \geq 1$, and any $\{i_j\} \subset J$, let $a = a_{i_1} a_{i_2} \cdots a_{i_k}$ be an element of $\mathcal{A}$. For any $i \in J$, let $E_i(a)$ and $O_i(a)$ be, respectively, the number of times a_i has occurred in the even positions and in the odd positions in a. The monomial a is said to be *symmetric* (with respect to $\{a_i\}_{i \in J}$) if $E_i(a) = O_i(a)$ for all $i \in J$. Else it is said to be non-symmetric. Let $\{a_i\}_{i \in J}$ in $(\mathcal{A}, \phi)$ be half commuting. They are said to be half independent if (i) $\{a_i^2\}_{i \in J}$ are independent and (ii) whenever a is non-symmetric with respect to $\{a_i\}_{i \in J}$, we have $\phi(a) = 0$. The Toeplitz and Hankel limits do not seem to submit to any easy or explicit independence/dependence notions. This is the topic of Chapter 9.

Now suppose we have h different types of patterned matrices with p_j independent copies of type j. Do they converge jointly? Moreover, if there is joint convergence, then are the collection of Wigner matrices and the collection of other patterned matrices asymptotically free? Basu, Bose, Ganguly and Hazra (2012)[15] provided an affirmative answer to both of these questions. One interesting consequence is that any symmetric matrix polynomial of multiple copies of the above five matrices has an LSD. In particular, let A and B be any two of the Toeplitz, Hankel, Reverse Circulant and Symmetric Circulant matrices without scaling that are independent and satisfy suitable moment assumptions. Then the LSD for $\frac{A+B}{\sqrt{n}}$ exists in the almost sure sense, is symmetric, has unbounded support and does not depend on the distribution of the input sequences of A and B. All this is discussed in Chapter 10.

Sample autocovariance matrix. This matrix appears in time series analysis. Suppose $X = \{X_t\}$ is a *stationary* process with $\mathrm{E}(X_t) = 0$ and $\mathrm{E}(X_t^2) < \infty$. The *autocovariance function* $\gamma_X(\cdot)$ and the *autocovariance matrix* $\Sigma_n(X)$ of order n are defined as:

$$\gamma_X(k) = cov(X_0, X_k), \ \ k = 0, 1, \ldots \ \ \text{and} \ \ \Sigma_n(X) = ((\gamma_X(i-j)))_{1 \leq i,j \leq n}.$$

It may be noted that $\Sigma_n(X)$ is a Toeplitz matrix.

The *sample autocovariance matrix* is the usual *non-negative definite* estimate of $\Sigma_n(X)$ and equals

$$\Gamma_n(X) = ((\hat\gamma_X(i-j)))_{1 \leq i,j \leq n} \ \ \text{where} \ \ \hat\gamma_X(k) = n^{-1} \sum_{i=1}^{n-|k|} X_i X_{i+|k|}.$$

Chapter 11 is on $\Gamma_n(X)$ and its variations and is based on Basak, Bose and Sen (2013)[14]. The existence of their LSDs is shown when the time series is a linear process with reasonable restriction on the coefficients. The limit is universal in the sense that it does not depend on the distribution of the underlying driving i.i.d. sequence. Its support is unbounded. This limit does not coincide with the LSD of $\Sigma_n(X)$. However, it does so for a suitably tapered version of $\Gamma_n(X)$. For banded $\Gamma_n(X)$, the limit has unbounded support as long as the number of non-zero diagonals in proportion to the dimension of the matrix is bounded away from zero. If this ratio tends to zero, then the limit exists and again coincides with the LSD of $\Sigma_n(X)$. Finally, a naturally modified version of $\Gamma_n(X)$ which is not non-negative definite is also considered.

Chapters 1, 2, 3 (section on S matrix) and 9 are must-reads. Then one may pick and choose depending upon the interest.

About the Author

Arup Bose obtained his B.Stat., M.Stat. and Ph.D. degrees from the Indian Statistical Institute. He has been on the faculty at the Theoretical Statistics Mathematics Unit of the Institute in Kolkata, India, for more than twenty-five years. He has significant research contributions in the areas of statistics, probability, economics and econometrics. He is a Fellow of the Institute of Mathematical Statistics, and of all three national science academies of India. He is a recipient of the S.S. Bhatnagar Prize and the C.R. Rao Award.

1

A unified framework

1.1 Empirical and limiting spectral distribution

Suppose A_n is an $n \times n$ matrix. Let $\lambda_1, \ldots, \lambda_n$ denote its eigenvalues. When the eigenvalues are real, our convention will be to always write them in descending order. The empirical spectral measure μ_n of A_n is the measure given by

$$\mu_n = \frac{1}{n} \sum_{i=1}^{n} \delta_{\lambda_i}, \tag{1.1}$$

where δ_x is the Dirac delta measure at x. The probability distribution function corresponding to μ_n is known as the *Empirical Spectral Distribution* (ESD) of A_n. We will denote it by F^{A_n} and it is given by

$$F^{A_n}(x, y) = n^{-1} \sum_{i=1}^{n} \mathbb{I}\{\mathcal{R}(\lambda_i) \leq x, \ \mathcal{I}(\lambda_i) \leq y\}, \quad x, y \in \mathbb{R}.$$

This is also known as the *bulk* of the spectrum. If the entries are random, then the eigenvalues can be chosen in a measurable way. Then F^{A_n} is a measurable *random distribution function*. The ESD is a measure on $\mathbb{C}$. But if all the eigenvalues are real (for example if A_n is real symmetric), then the ESD is a measure on $\mathbb{R}$. Further, $\mathrm{E}[F^{A_n}(x, y)]$ is also a distribution function and is non-random. It is called the *expected ESD* (EESD). We need the following definition for convergence of random distribution functions.

Definition 1.1.1. A sequence of (measurable) random distribution functions $\{F_n(\cdot, \cdot)\}$ is said to converge (weakly) in probability if there exists a distribution function $F(\cdot, \cdot)$ such that for all continuity points (x, y) of F, $F_n(x, y) \to F(x, y)$ in probability. The sequence is said to converge (weakly) almost surely to F if, outside a null set, $F_n(x, y) \to F(x, y)$ for all continuity points (x, y) of F. $\diamond$

The above definition holds in a natural way if F is defined on $\mathbb{R}$.

Large-dimensional random matrices are matrices where the dimension n is very large or grows to ∞. Such matrices have arisen naturally in many areas of science and their behavior has received considerable attention.

One way to study their behavior is to study the nature of their ESD when the dimension tends to ∞. Let $\{A_n\}_{n=1}^{\infty}$ be a sequence of square matrices with the corresponding ESD $\{F^{A_n}\}_{n=1}^{\infty}$. The *Limiting Spectral Distribution* (LSD) of the sequence is defined as the weak limit of the sequence $\{F^{A_n}\}$, if it exists, *in probability* or *almost surely*. It may be noted that if the matrix is real symmetric or Hermitian, then all its eigenvalues are real and in that case the ESD is defined on the real line and so is the LSD, if it exists.

One of the most useful methods of studying the spectra was, and still is, enumerative combinatorics. This is the approach we will take but we will do so in a very accessible and unified manner. This approach, as we shall see, is ideally suited for matrices which are real symmetric or Hermitian. It is not of direct use for non-symmetric/non-Hermitian matrices (but still has its value while dealing with such matrices). Other sophisticated tools are available to help in these situations. Some of them are Pastur's fundamental technique of *Stieltjes transforms* (which we shall touch upon), *free probability* (which also we shall encounter) and the method of *normal approximation* which is specially useful for Circulant type matrices.

A tremendous amount of effort has gone into establishing the LSD of real symmetric matrices such as the Wigner, sample variance-covariance and the Symmetric Toeplitz. The arguments for each has been handled separately and often more than one proof is known.

We shall first place them in the common format of *patterned matrices* (also often called structured matrices). In this framework, the moment method is quite appropriate for studying the LSD. We shall demonstrate relatively short proofs for the LSD of common matrices and provide insight into the nature of different LSDs and their interrelations. The method is flexible enough to be applicable to matrices with appropriate dependent entries, banded matrices, and matrices of the form $A_p = \frac{1}{n}XX'$ where X is a $p \times n$ matrix with real entries and $p \to \infty$ with $n = n(p) \to \infty$ and $p/n \to y$ with $0 \leq y < \infty$.

This approach raises interesting questions about the class of patterns for which LSD exists and the nature of the possible limits. In many cases, the LSDs are not known in any explicit form and so deriving probabilistic properties of the limit are also interesting issues.

1.2   Moment method

We shall first deal with only real symmetric matrices and hence all eigenvalues are real. As discussed earlier, the moment method will be our choice for such matrices. We shall use the following notation for moments throughout this book. For any distribution F or any random variable X, $\beta_h(F)$ and $\beta_h(X)$, respectively, will denote their h-th moment.

The following easy result is well known in the literature of weak convergence of probability measures.

Lemma 1.2.1. Suppose $\{Y_n\}$ is a sequence of real-valued random variables with distributions $\{F_n\}$. Suppose that there exists some sequence $\{\beta_h\}$ such that as $n \to \infty$, $E[Y_n^h] = \int x^h dF_n(x) \to \beta_h$ for every positive integer h. Suppose that there is a unique distribution F whose moments are $\{\beta_h\}$. Then Y_n (or equivalently F_n) converges to F in distribution. ◇

Proof. Since all moments of $\{Y_n\}$ converge, it is a tight sequence and moreover, all powers of $\{Y_n\}$ are uniformly integrable. Consider any sub-sequence $\{n_k\}$ of $\{n\}$. By tightness, it has a further sub-sequence which converges in distribution, to say, G. By uniform integrability, all moments of G exist and

$$\int x^h dG(x) = \beta_n \ \forall h.$$

Now by our assumption, G is unique. This completes the proof. □

The existence of the unique distribution in the above lemma is guaranteed by the following results of Carleman (1926)[39] and Riesz (1923)[88].

Lemma 1.2.2. Let $\{\beta_k\}$ be the sequence of moments of the distribution function F. Then F is the unique distribution with these moments if any of the following two conditions are satisfied.

(a) Carleman's condition: $\displaystyle\sum_{h=1}^{\infty} \beta_{2h}^{-1/2h} = \infty.$

(b) Riesz's condition:

$$\liminf_{k \to \infty} \frac{1}{k} \beta_{2k}^{\frac{1}{2k}} < \infty. \tag{1.2}$$

◇

Proof. It is easy to see that Riesz's condition implies Carleman's condition and hence (a) yields a stronger result. However, we shall prove only (b) as its proof is much easier but the result is powerful enough for our purposes in this book. A proof of (a) can be found in Bai and Silverstein (2010)[8].

Let F and G be two distributions with common moments $\{\beta_k\}$. Let

$$f(t) = \int e^{itx} F(dx) \ \text{ and } \ g(t) = \int e^{itx} G(dx), \ i = \sqrt{-1}$$

denote the characteristic functions of F and G.

We need only show that $f(t) = g(t)$ for all $t \geq 0$. Since F and G have common moments, we have, for all $j = 0, 1, \ldots$,

$$f^{(j)}(0) = g^{(j)}(0) = i^j \beta_j.$$

Define

$$t_0 = \sup\{t \geq 0; g^{(j)}(s) = f^{(j)}(s), \ 0 \leq s \leq t, j \geq 1\}.$$

Then Lemma 1.2.2 will follow if $t_0 = \infty$ for all $j = 0, 1, \ldots$. Suppose, if possible, $t_0 < \infty$. Then appealing to continuity,

$$\int_{-\infty}^{\infty} x^j e^{it_0 x}[F(ds) - G(ds)] = 0.$$

By condition (1.2), there is a constant $M > 0$ such that

$$\beta_{2k} \leq (Mk)^{2k} \text{ for infinitely many } k.$$

Recall the inequality

$$|e^{ia} - 1 - ia \ldots - (ia)^k/k!| \leq |a|^{k+1}/(k+1)!, \quad \text{for all } k \geq 1. \qquad (1.3)$$

Choose $s \in (0, 1/(eM))$. Use the above inequality and the relation $k! > (k/e)^k$, to obtain that, for any fixed $j \geq 0$,

$$\left| f^{(j)}(t_0 + s) - g^{(j)}(t_0 + s) \right| = \left| \int_{-\infty}^{\infty} x^j e^{i(t_0+s)x}[F(dx) - G(dx)] \right|$$

$$= \left| \int_{-\infty}^{\infty} x^j e^{it_0 x} \left[e^{isx} - 1 - isx - \ldots - \frac{(isx)^{2k-j-1}}{(2k-j-1)!} \right] [F(dx) - G(dx)] \right|$$

$$\leq 2 \frac{s^{2k-j}\beta_{2k}}{(2k-j)!} \leq 2 \frac{(sMk)^{2k}}{s^j(2k-j)!}$$

$$\leq 2(esMk/(2k-j))^{2k}(2k/s)^j \to 0,$$

as $k \to \infty$ along those k such that $\beta_{2k} \leq (Mk)^{2k}$. The last inequality above follows from *Stirling's approximation* for factorial. This violates the definition of t_0. The proof of Lemma 1.2.2 is complete.   □

The following lemma will be useful to verify Riesz's condition (1.2).

Lemma 1.2.3. If for some $0 < \Delta < \infty$, $\{\beta_k\}$ satisfies

$$\beta_{2k} \leq \frac{(2k)!}{k!2^k} \Delta^k, \quad k = 0, 1, \ldots$$

then it satisfies Carleman's as well as Riesz's condition (1.2).   ◇

Proof. Recall that by Stirling's approximation, for some $C_1, C_2 > 0$,

$$C_1(2\pi)^{\frac{1}{2}} e^{-n} n^{n+\frac{1}{2}} \leq n! \leq C_2(2\pi)^{\frac{1}{2}} e^{-n} n^{n+\frac{1}{2}}, \ n \geq 1.$$

Using this for all large k, for some $C > 0$,

$$\frac{1}{k} (\beta_{2k})^{\frac{1}{2k}} \leq \frac{1}{k} \left[C \frac{e^{-2k}(2k)^{2k+\frac{1}{2}}}{2^k e^{-k} k^{k+\frac{1}{2}}} \right]^{\frac{1}{2k}} = C^{\frac{1}{2k}} e^{-1/2} 2^{\frac{1}{2}+\frac{1}{4k}} \frac{1}{k^{3/2}},$$

and hence

$$\lim_{k \to \infty} \frac{1}{k} \beta_{2k}^{\frac{1}{2k}} = 0. \qquad \qquad □$$

The usual central limit theorem (CLT), when we assume in addition that all moments are finite, may be established by the above approach.

Example 1.2.1. Consider the sequence of moments of a standard normal variable N,

$$\beta_{2k}(N) = \frac{(2k)!}{2^k k!} \quad \text{and} \quad \beta_{2k+1}(N) = 0 \quad \text{for all} \quad k = 0, 1, 2, \ldots. \qquad (1.4)$$

By Lemma 1.2.3, this sequence satisfies Riesz's condition. Now suppose $\{x_i\}$ are i.i.d. random variables with mean zero and variance one and all moments finite. Let $Y_n = n^{-1/2}(x_1 + x_2 + \cdots + x_n)$. By using binomial expansion and taking term-by-term expectations and then using elementary order calculations, it can be shown that

$$\mathrm{E}[Y_n^{2k+1}] \to 0 \quad \text{and} \quad \mathrm{E}[Y_n^{2k}] \to \beta_{2k}(N). \qquad (1.5)$$

Hence $Y_n \overset{\mathcal{D}}{\to} N$.   ▲

Now consider the ESD of A_n. It's h-th moment has the following nice form:

$$h\text{-th moment of the ESD of } A_n = \frac{1}{n}\sum_{i=1}^{n}\lambda_i^h = \frac{1}{n}\mathrm{Tr}(A_n^h) = \beta_h(A_n) \text{ (say)}$$

$$(1.6)$$

where Tr denotes the trace of a matrix. This is often known as the *trace-moment* formula.

Let E_{F_n} and E denote, respectively, the expectations with respect to the ESD F_n and the probability on the space where the entries of the random matrices are defined. Thus, Lemma 1.2.1 comes into force, except that now the moments are also random. Lemma 1.2.4 links convergence of moments of the ESD and LSD. Consider the following conditions:

(M1) For every $h \geq 1$, $\mathrm{E}[\beta_h(A_n)] \to \beta_h$.

(M2) $\mathrm{Var}[\beta_h(A_n)] \to 0$ for every $h \geq 1$.

(M4) $\displaystyle\sum_{n=1}^{\infty} \mathrm{E}[\beta_h(A_n) - \mathrm{E}(\beta_h(A_n))]^4 < \infty$ for every $h \geq 1$.

(R) The sequence $\{\beta_h\}$ satisfies Riesz's or Carleman's condition.

Note that (M4) implies (M2). The following lemma can be easily derived from Lemma 1.2.1 and the *Borel-Cantelli Lemma*. We omit its proof.

Lemma 1.2.4. (a) If (M1), (M2) and (R) hold, then $\{F^{A_n}\}$ converges in probability to F determined by $\{\beta_h\}$.

(b) If further (M4) holds, then the convergence in (a) is a.s.   ◇

We may mention that while showing these conditions, the trace-moment formula (1.6) replaces the binomial expansion that was used in Example 1.2.1. However, the calculation/estimation of the leading term and the bounding of the lower-order terms lead to combinatorial issues. We shall see that under suitable conditions, (M2), (M4) and (R) may be verified in a unified way for a class of matrices. However, (M1) must be verified on a case-by-case basis.

1.3   A metric for probability measures

It is well known that weak convergence of probability measures is metrizable. Our metric of choice will be a metric which is a special case of the class of Mallow's metric (also known as the Wasserstein metric or the Kantorovitch metric). It is defined as follows on the space of all probability distributions with finite second moment. Let F and G be two distribution functions with finite second moment. Then the W_2 distance between them is defined as

$$W_2(F,G) = \left[\inf_{(X \sim F, Y \sim G)} E[X - Y]^2 \right]^{\frac{1}{2}}. \tag{1.7}$$

Here $(X \sim F, Y \sim G)$ means that the joint distribution of (X, Y) is such that their marginal distributions are F and G. The following lemma links weak convergence and convergence in the above metric.

Lemma 1.3.1. The metric W_2 is complete and $W_2(F_n, F) \to 0$ if and only if $F_n \xrightarrow{\mathcal{D}} F$ and $\beta_2(F_n) \to \beta_2(F)$.

Proof. First suppose $W_2(F_n, F) \to 0$. By definition of W_2, we can get (X_n, Y_n) such that X_n has distribution F_n, Y_n has distribution F, and $E(X_n - Y_n)^2 \to 0$. Further, it can be so arranged that all these random variables are defined on the same probability space.

By Chebyshev's inequality, $X_n - Y_n \xrightarrow{P} 0$. Moreover, since all Y_n have the same distribution F, by applying Slutsky's theorem, $X_n \xrightarrow{\mathcal{D}} F$.

Now it remains to show that $E(X_n^2) \to E(X^2)$ where X has distribution F. For this, it is enough to show that $E(X_n^2) - E(Y_n^2) \to 0$.

Note that $E(X_n^2)$ is bounded. Hence,

$$[E(X_n^2) - E(Y_n^2)]^2 \le E(X_n - Y_n)^2) E(X_n + Y_n)^2$$
$$\le [E(X_n - Y_n)^2)][2 E(X_n)^2 + 2 E(Y_n^2)] \to 0.$$

Conversely, suppose $F_n \xrightarrow{\mathcal{D}} F$ and $\beta_2(F_n) \to \beta_2(F)$. To show that $W_2(F_n, F) \to 0$. First get $\{Y_n\}$, Y on the same probability space such that $Y_n \to Y$ almost surely, Y_n is distributed as F_n and Y is distributed as F. Fix a real number k large. Define $\bar{Y}_{n,k} = Y_n$ if $|Y_n| < k$, $\bar{Y}_{n,k} = k$ if $Y_n \ge k$ and $\bar{Y}_{n,k} = -k$ if $Y_n \le -k$. Define $\bar{Y}_k$ similarly.

Let $G_{n,k}$ and G_k denote the distributions of $\bar{Y}_{n,k}$ and $\bar{Y}_k$ respectively.
Note that as $n \to \infty$, $\bar{Y}_{n,k} \to \bar{Y}_k$ almost surely. By DCT, $E[\bar{Y}_{n,k} - \bar{Y}_k]^2 \to 0$
as $n \to \infty$. Hence $W_2(G_{n,k}, G_k) \to 0$ as $n \to \infty$.

Now

$$W_2(F_n, F) \leq W_2(F_n, G_{n,k}) + W_2(G_{n,k}, G_k) + W_2(G_k, F).$$

By DCT,

$$W_2^2(G_k, F) \leq E[\bar{Y}_k - Y]^2 \leq E[|Y|^2 \mathbb{I}(|Y| \geq k)] \to 0 \quad \text{as} \quad k \to \infty.$$

$$W_2^2(G_{n,k}, F_n) \leq E[\bar{Y}_{n,k} - Y_n]^2 \leq E[|Y_n|^2 \mathbb{I}(|Y_n| \geq k)].$$

Choose k (large) such that $\pm k$ is a continuity point of F. Note that this is
possible since the number of discontinuity points of F is countable. Now

$$E[|Y_n|^2 \mathbb{I}(|Y_n| \geq k)] + E[|Y_n|^2 \mathbb{I}(|Y_n| < k)] = E(Y_n^2) = E(Y^2).$$

By DCT, as $n \to \infty$ the second term on the left side of the equality converges
to $E[|Y|^2 \mathbb{I}(|Y| < k)]$. Hence the first term converges to $E[|Y|^2 \mathbb{I}(|Y| \geq k)]$. This
in turn converges to 0 as $k \to \infty$.

Hence combining all the above, we get

$$\limsup W_2(F_n, F) \to 0.$$

That W_2 is a metric and is complete is easy to prove. $\qquad\square$

An estimate of the metric distance W_2 between two ESD in terms of the
trace will be crucial to us. *In the following result and similar results later, we
may and will assume that the eigenvalues of any real symmetric matrix are
indexed in a descending order of value.*

Lemma 1.3.2. Suppose A, B are $n \times n$ symmetric real matrices with eigenvalues $\{\lambda_i(A)\}$ and $\{\lambda_i(B)\}$, both written in a descending order, $1 \leq i \leq n$.
Then

$$W_2^2(F^A, F^B) \leq \frac{1}{n} \sum_{i=1}^{n} (\lambda_i(A) - \lambda_i(B))^2 \leq \frac{1}{n} \operatorname{Tr}(A - B)^2. \tag{1.8}$$

$\diamond$

Proof. The first inequality follows by considering the joint distribution which
puts mass $1/n$ at $(\lambda_i(A), \lambda_i(B))$. Then the marginals are the two ESDs of A
and B. The second inequality follows from the *Hoffmann-Wielandt inequality*
(see Hoffman and Weilandt (1953)[66]). $\qquad\square$

1.4   Patterned matrices: A unified approach

A sequence or a bi-sequence of variables $\{x_i; i \geq 0\}$ or $\{x_{ij};\ i,j \geq 1\}$ will be called an *input sequence*. All matrices that we shall consider will be constructed out of such an input sequence in the following way.

Let $\mathbb{Z}$ be the set of all integers and let $\mathbb{Z}_+$ denote the set of all non-negative integers. Let

$$L_n : \{1,2,\ldots,n\}^2 \to \mathbb{Z}^d,\ n \geq 1,\ d = 1\ \text{or}\ 2 \tag{1.9}$$

be a sequence of functions. For notational convenience, we shall write $L_n = L$ and call it the *link* function. By abuse of notation we write $\mathbb{Z}_+^2$ as the common domain of $\{L_n\}$. If $L_{n+1}(i,j) = L_n(i,j)$ whenever $1 \leq i,j \leq n$, then the $\{L_n\}$ are said to be *nested*.

The matrices we consider will be of the form

$$A_n = ((x_{L(i,j)})). \tag{1.10}$$

Such matrices will be termed *patterned matrices*. If L is symmetric, that is $L(i,j) = L(j,i)$ for all i,j, then the matrix is symmetric. It is important to note that if we impose the condition that no element of the input sequence is degenerate, then matrices with zero entries in particular would be excluded from the current definition. Once we have enough experience, we shall see how we may allow zero entries to deal with band and triangular matrices.

Some well-known patterned matrices with appropriate link functions:

(i) Wigner matrix $W_n^{(s)}$. $L : \mathbb{Z}_+^2 \to \mathbb{Z}_+^2$, $L(i,j) = (\min(i,j), \max(i,j))$.

$$W_n^{(s)} = \begin{bmatrix} x_{11} & x_{12} & x_{13} & \cdots & x_{1(n-1)} & x_{1n} \\ x_{12} & x_{22} & x_{23} & \cdots & x_{2(n-1)} & x_{2n} \\ & & & \vdots & & \\ x_{1n} & x_{2n} & x_{3n} & \cdots & x_{(n-1)n} & x_{nn} \end{bmatrix}.$$

(ii) Symmetric Toeplitz matrix $T_n^{(s)}$. $L : \mathbb{Z}_+^2 \to \mathbb{Z}_+$, $L(i,j) = |i-j|$.

$$T_n^{(s)} = \begin{bmatrix} x_0 & x_1 & x_2 & \cdots & x_{n-2} & x_{n-1} \\ x_1 & x_0 & x_1 & \cdots & x_{n-3} & x_{n-2} \\ x_2 & x_1 & x_0 & \cdots & x_{n-4} & x_{n-3} \\ & & & \vdots & & \\ x_{n-1} & x_{n-2} & x_{n-3} & \cdots & x_1 & x_0 \end{bmatrix}.$$

(iii) Symmetric Hankel matrix $H_n^{(s)}$. $L : \mathbb{Z}_+^2 \to \mathbb{Z}_+$, $L(i,j) = i + j$.

$$H_n^{(s)} = \begin{bmatrix} x_2 & x_3 & x_4 & \cdots & x_n & x_{n+1} \\ x_3 & x_4 & x_5 & \cdots & x_{n+1} & x_{n+2} \\ x_4 & x_5 & x_6 & \cdots & x_{n+2} & x_{n+3} \\ & & \vdots & & & \\ x_{n+1} & x_{n+2} & x_{n+3} & \cdots & x_{2n-1} & x_{2n} \end{bmatrix}.$$

(iv) Reverse Circulant matrix $R_n^{(s)}$. $L : \mathbb{Z}_+^2 \to \mathbb{Z}_+$, $L(i,j) = (i+j-2) \mod n$.

$$R_n^{(s)} = \begin{bmatrix} x_0 & x_1 & x_2 & \cdots & x_{n-2} & x_{n-1} \\ x_1 & x_2 & x_3 & \cdots & x_{n-1} & x_0 \\ x_2 & x_3 & x_4 & \cdots & x_0 & x_1 \\ & & \vdots & & & \\ x_{n-1} & x_0 & x_1 & \cdots & x_{n-3} & x_{n-2} \end{bmatrix}.$$

(v) Symmetric Circulant matrix $C_n^{(s)}$.
$L : \mathbb{Z}_+^2 \to \mathbb{Z}$, $L(i,j) = n/2 - |n/2 - |i-j||$.

$$C_n^{(s)} = \begin{bmatrix} x_0 & x_1 & x_2 & \cdots & x_2 & x_1 \\ x_1 & x_0 & x_1 & \cdots & x_3 & x_2 \\ x_2 & x_1 & x_0 & \cdots & x_2 & x_3 \\ & & \vdots & & & \\ x_1 & x_2 & x_3 & \cdots & x_1 & x_0 \end{bmatrix}.$$

We often write it also as SC_n. The Symmetric Circulant is also a Doubly Symmetric Toeplitz matrix.

(vi) Doubly Symmetric Hankel matrix DH_n.
$L(i,j) = n/2 - |n/2 - (i+j) \mod n|$.

$$DH_n = \begin{bmatrix} x_0 & x_1 & x_2 & \cdots & x_3 & x_2 & x_1 \\ x_1 & x_2 & x_3 & \cdots & x_2 & x_1 & x_0 \\ x_2 & x_3 & x_4 & \cdots & x_1 & x_0 & x_1 \\ & & \vdots & & & & \\ x_2 & x_1 & x_0 & \cdots & x_5 & x_4 & x_3 \\ x_1 & x_0 & x_1 & \cdots & x_4 & x_3 & x_2 \end{bmatrix}.$$

(vii) Palindromic matrices PT_n and PH_n. For these symmetric matrices, the first row is a palindrome. PT_n is given below and PH_n is defined similarly.

$$PT_n = \begin{bmatrix} x_0 & x_1 & x_2 & \cdots & x_2 & x_1 & x_0 \\ x_1 & x_0 & x_1 & \cdots & x_3 & x_2 & x_1 \\ x_2 & x_1 & x_0 & \cdots & x_4 & x_3 & x_2 \\ & & \vdots & & & & \\ x_1 & x_2 & x_3 & \cdots & x_1 & x_0 & x_1 \\ x_0 & x_1 & x_2 & \cdots & x_2 & x_1 & x_0 \end{bmatrix}.$$

Symmetric matrices of the form XX' where X is a suitable non-symmetric matrix will be introduced later and analyzed once we have sufficient experience with the above symmetric matrices.

It is noteworthy that all the link functions above possess *Property B* (B for bounded) given below. This property will be very crucial to us. It assures that no element of the input sequence is used too many times in any row or column. We use the notation

$$\#A = \text{the number of elements in the set} A.$$

Definition 1.4.1. (Property B) The link L is said to satisfy *Property B* if

$$\Delta(L) = \sup_n \sup_{t\in\mathbb{Z}_+^d} \sup_{1\leq k\leq n} \#\{l : 1\leq l\leq n,\ L(k,l)=t\} < \infty. \qquad (1.11)$$

$\diamond$

1.4.1   Scaling

That $\sqrt{n}$ is the correct scaling in the CLT can be established by computing the variance. Similarly the matrices also need appropriate scaling for existence of their LSD. To understand what this scaling should be, assume that $\{x_i\}$ have mean zero and variance 1. Let F_n denote the ESD of $T_n^{(s)}$ and let X_n be the corresponding random variable. Then

$$\mathrm{E}_{F_n}(X_n) = \frac{1}{n}\sum_{i=1}^n \lambda_{i,n} = \frac{1}{n}\,\mathrm{Tr}(T_n^{(s)}) = x_0 \ \text{ and } \ \mathrm{E}[\mathrm{E}_{F_n}(X_n)] = 0,$$

$$\mathrm{E}_{F_n}(X_n^2) = \frac{1}{n}\sum_{i=1}^n \lambda_{i,n}^2 = \frac{1}{n}\,\mathrm{Tr}\left(T_n^{(s)^2}\right)$$
$$= \frac{1}{n}[nx_0^2 + 2(n-1)x_1^2 + \ldots + 2x_{n-1}^2] \ \text{ and } \ \mathrm{E}[\mathrm{E}_{F_n}(X_n^2)] = n.$$

Hence, the appropriate scaled matrix to consider is $n^{-1/2}T_n^{(s)}$. It may be noted that the above argument continues to hold for the other patterned matrices that we have introduced so far.

1.4.2   Reduction to bounded case

The following assumptions impose varied conditions on the moments of the input sequence. All of them shall be used in this book depending on the context.

Assumption I $\{x_i, x_{ij}\}$ are i.i.d. and uniformly bounded with mean 0 and variance 1.

Assumption II $\{x_i, x_{ij}\}$ are i.i.d. with mean zero and variance 1.

Assumption III $\{x_i, x_{ij}\}$ are independent with mean zero and variance 1 and with uniformly bounded moments of all order.

Note that if a sequence satisfies Assumption I then it satisfies Assumptions II and III. If the sequence satisfies Assumption III and in addition is identically distributed, then it satisfies Assumption II. As we shall see later, Assumption III is particularly suited to dealing with joint convergence of sequences of matrices.

Definition 1.4.2. (α_n, k_n). Let $\{A_n\}$ be a sequence of $n \times n$ random matrices with link function L_n. Then k_n (increasing to ∞) and α_n are defined by

$$\{L_n(i,j) : 1 \le i, j \le n\} \subset \{1, \ldots, k_n\}^d \text{ (smallest } k_n)(1.12)$$

$$\max_k \#\{(i,j) : L_n(i,j) = k, \ 1 \le i, j \le n\} = \alpha_n. \tag{1.13}$$

The number k_n^d is the total number of variables used in the matrix. For example, for Toeplitz and Hankel matrices, $d = 1$ and $k_n = n$. For the Wigner matrix $d = 2$ and $k_n = n$.

The number α_n is the maximum number of times a variable has appeared in the matrix. For Toeplitz and Hankel matrices, $\alpha_n = n$ and for the Wigner matrix $\alpha_n = 2$.

Lemma 1.4.1 is from Bose and Sen (2008)[29]. It shows to what extent it is sufficient to work with input sequences which are uniformly bounded.

Remark 1.4.1. The lemma below continues to hold also under Assumption III. Proof of this is left as an exercise. ◇

Lemma 1.4.1. Suppose A_n is a sequence of symmetric matrices with link function L that satisfies Property B. Suppose $k_n \to \infty$ and $k_n^d \alpha_n = O(n^2)$. If $\{F^{n^{-1/2} A_n}\}$ converges to a non-random F a.s. whenever the input sequence satisfies Assumption I, then the same limit holds almost surely if Assumption I is replaced by Assumption II. ◇

Proof. For simplicity, we will only consider the case when the input is a single sequence $\{x_0, x_1, x_2, \ldots\}$. That is, $d = 1$ in the definition of the link function. If $d > 1$, the proof works mutatis mutandis. We shall use the W_2 metric. For $t > 0$, let

$$\mu(t) = \mathrm{E}[x_0 \mathbb{I}(|x_0| > t)] = -\mathrm{E}[x_0 \mathbb{I}(|x_0| \le t)].$$
$$\sigma^2(t) = \mathrm{Var}(x_0 \mathbb{I}(|x_0| \le t)) = \mathrm{E}[x_0^2 \mathbb{I}(|x_0| \le t)] - \mu(t)^2.$$

Since $\mathrm{E}(x_0) = 0$ and $\mathrm{E}(x_0^2) = 1$, we have $\sigma^2(t) \le 1$ and as $t \to \infty$,

$$\mu(t) \to 0 \quad \text{and} \quad \sigma(t) \to 1.$$

Define random variables $\{x_i^*\}$ and $\{\bar{x}_i\}$ by

$$
\begin{aligned}
x_i^* &= \frac{x_i \mathbb{I}(|x_i| \le t) + \mu(t)}{\sigma(t)} = \frac{x_i - \bar{x}_i}{\sigma(t)}, & (1.14) \\
\bar{x}_i &= x_i \mathbb{I}(|x_i| > t) - \mu(t) = x_i - \sigma(t)x_i^*. & (1.15)
\end{aligned}
$$

It is easy to see that as $t \to \infty$,

$$\mathrm{E}(\bar{x}_0^2) = 1 - \sigma^2(t) - \mu(t)^2 \to 0. \tag{1.16}$$

Further, $\{x_i^*\}$ are i.i.d. bounded, mean zero and variance one random variables. Let A_n^* denote the matrix with the same link function L but input sequence $\{x_i^*\}_{i \ge 0}$.

By triangle inequality and (1.8),

$$
\begin{aligned}
W_2^2\big(F^{n^{-1/2}A_n}, F^{n^{-1/2}A_n^*}\big) &\le 2W_2^2\big(F^{n^{-1/2}A_n}, F^{n^{-1/2}\sigma(t)A_n^*}\big) \\
&\quad + 2W_2^2\big(F^{n^{-1/2}A_n^*}, F^{n^{-1/2}\sigma(t)A_n^*}\big) \\
&\le \frac{2}{n^2}\,\mathrm{Tr}(A_n - \sigma(t)A_n^*)^2 + \frac{2(1-\sigma(t))^2}{n^2}\,\mathrm{Tr}(A_n^*)^2.
\end{aligned}
$$

Consider the second term first. By the hypotheses on A_n, using the strong law of large numbers, we get (letting $a_{i,j,n}^*$ be the $(i,\ j)$-th element of A_n^*)

$$
\begin{aligned}
\frac{1}{n^2}\,\mathrm{Tr}(A_n^*)^2 &= \frac{1}{n^2} \sum_{1 \le i,j \le n} (a_{i,j,n}^*)^2 \\
&\le \frac{\alpha_n}{n^2} \sum_{h=0}^{k_n} x_h^{*2} \quad \text{(we have used (1.12) here)} \\
&\le \frac{C}{k_n} \sum_{h=0}^{k_n} x_h^{*2} \overset{a.s.}{\to} C\,\mathrm{E}(x_0^{*2}) = C.
\end{aligned}
$$

Similarly, for the first term (using 1.16), as $t \to \infty$,

$$\frac{1}{n^2}\,\mathrm{Tr}(A_n - \sigma(t)A_n^*)^2 \le \frac{C}{k_n} \sum_{h=0}^{k_n} \bar{x}_h^2 \overset{a.s.}{\to} C\,\mathrm{E}(\bar{x}_0^2) \to 0.$$

Thus, as $t \to \infty$,

$$\limsup_n W_2^2\big(F^{n^{-1/2}A_n}, F^{n^{-1/2}A_n^*}\big) \to 0 \quad \text{almost surely} \tag{1.17}$$

and the proof is complete. $\square$

1.4.3 Trace formula and circuits

In Example 1.2.1 we showed how the moment method can be used to prove the CLT. There the binomial expansion was used to show that certain terms dropped out of contention in the limit. We now proceed to develop a similar approach for the convergence of the ESD by using (1.6) which connects the moment of the ESD to the trace. However, unlike the CLT proof, here the issue of which terms contribute to the limit is much more complicated.

Let $A_n = ((x_{L(i,j)}))$. Using (1.6), the h-th moment of $F^{n^{-1/2}A_n}$ is given by the following *trace formula* or the *trace-moment* formula.

$$\frac{1}{n}\operatorname{Tr}\Big(\frac{A_n}{\sqrt{n}}\Big)^h = \frac{1}{n^{1+h/2}}\sum_{1 \le i_1, i_2, \ldots, i_h \le n} x_{L(i_1,i_2)} \cdots x_{L(i_{h-1},i_h)} x_{L(i_h,i_1)}. \quad (1.18)$$

We now proceed to develop a systematic way of keeping track of the terms in the trace formula. This is done by defining an appropriate equivalence relation between the different terms in the sum. The concepts we develop below will remain useful to us for the rest of this book.

Circuit: $\pi : \{0, 1, 2, \ldots, h\} \to \{1, 2, \ldots, n\}$ with $\pi(0) = \pi(h)$ is called a *circuit* of *length* $l(\pi) := h$. The dependence of a circuit on h and n will be suppressed. Clearly, the convergence in (M1), (M2) and (M4) may be written in terms of circuits. For example, assuming all appearing moments are finite,

(M1)

$$\mathrm{E}[\beta_h(n^{-1/2}A_n)] = \mathrm{E}[\frac{1}{n}\operatorname{Tr}\Big(\frac{A_n}{\sqrt{n}}\Big)^h] \quad (1.19)$$

$$= \frac{1}{n^{1+h/2}}\sum_{\pi:\ \pi \text{ circuit}} \mathrm{E}\, \mathrm{X}_\pi \to \beta_h \quad (1.20)$$

where

$$\mathrm{X}_\pi = x_{L(\pi(0),\pi(1))} x_{L(\pi(1),\pi(2))} \cdots x_{L(\pi(h-2),\pi(h-1))} x_{L(\pi(h-1),\pi(h))}.$$

Matched Circuit: For any π, any $L(\pi(i-1), \pi(i))$ is an *L-value*. If an L-value is repeated exactly e times, we say that the circuit has an *edge of order* e $(1 \le e \le h)$. We may visualize $(\pi(i-1), \pi(i))$ as an edge. If π has all $e \ge 2$, then it is called *L-matched* (in short *matched*). For any non-matched π, $\mathrm{E}[\mathrm{X}_\pi] = 0$ and hence only matched π are relevant in (1.19). If π has only order two edges, then it is called *pair-matched*.

To deal with (M2) or (M4), we need multiple circuits: suppose we have k circuits $\pi_1, \pi_2, \ldots, \pi_k$. They are *jointly-matched* if each L-value occurs at least twice across all circuits. They are *cross-matched* if each circuit has at least one L-value which occurs in at least one of the *other* circuits. Note that this implies that none of them are self-matched.

Equivalence of circuits: The following defines an equivalence relation between the circuits: π_1 and π_2 are *equivalent* if and only if their L-values respectively match at the same locations. That is, for all i, j,

$$L(\pi_1(i-1), \pi_1(i)) = L(\pi_1(j-1), \pi_1(j))$$
$$\Longleftrightarrow \quad L(\pi_2(i-1), \pi_2(i)) = L(\pi_2(j-1), \pi(j)). \qquad (1.21)$$

1.4.4 Words

Any equivalence class can be indexed by a partition of $\{1, 2, \ldots, h\}$. Each block of a given partition identifies the positions where the L-matches take place. We can label these partitions by *words* of length $l(w) = h$ of letters where the first occurrence of each letter is in alphabetical order. For example if $h = 5$ then the partition $\{\{1, 3, 5\}, \{2, 4\}\}$ is represented by the word *ababa*. This identifies all circuits π for which $L(\pi(0), \pi(1)) = L(\pi(2), \pi(3)) = L(\pi(4), \pi(5))$ and $L(\pi(1), \pi(2)) = L(\pi(3), \pi(4))$. Let $w[i]$ denote the i-th entry of w.

The class Π: The equivalence class corresponding to w will be denoted by

$$\Pi(w) = \{\pi : w[i] = w[j] \Leftrightarrow L(\pi(i-1), \pi(i)) = L(\pi(j-1), \pi(j))\}.$$

The number of partition blocks corresponding to w will be denoted by $|w|$. This is the same as the number of distinct letters in w. If $\pi \in \Pi(w)$, then clearly,

$$|w| = \#\{L(\pi(i-1), \pi(i)) : 1 \le i \le h\}.$$

By varying w, we obtain all the equivalence classes. It is important to note that for any fixed h, as $n \to \infty$, the number of words (equivalence classes) remains finite but the number of circuits in any given $\Pi(w)$ may grow indefinitely.

The above notions carry over to words. For instance the word *ababa* is matched. The word *abcadbaa* is non-matched with edges of order 1, 2 and 4 and the corresponding partition of $\{1, 2, \ldots, 8\}$ is $\{\{1, 4, 7, 8\}, \{2, 6\}, \{3\}, \{5\}\}$. A word is *pair-matched* if every letter appears exactly twice in that word. Let

$$\mathcal{W}_{2k}(2) = \{w : w \text{ is pair-matched of length } 2k\}. \qquad (1.22)$$

For technical reasons it becomes easier to deal with a class larger than Π. Let

$$\Pi^*(w) = \{\pi : w[i] = w[j] \Rightarrow L(\pi(i-1), \pi(i)) = L(\pi(j-1), \pi(j))\}$$

so that the implication in (1.21) is only one sided. Clearly $\Pi^*(w) \supset \Pi(w)$.

1.4.5 Vertices

Any i (or $\pi(i)$ by abuse of notation) will be called a *vertex*. It is *generating*, if either $i = 0$ or $w[i]$ is the *first* occurrence of a letter. Otherwise it is called *non-generating*. For example, if $w = abbcab$ then $\pi(0)$, $\pi(1)$, $\pi(2)$, $\pi(4)$ are generating and $\pi(3)$, $\pi(5)$, $\pi(6)$ are *non-generating*. When Property B holds, a circuit is completely determined, *up to finitely many choices*, by its generating vertices. The number of generating vertices is $|w| + 1$ and hence

$$\#\Pi(w) \leq \#\Pi^*(w) = O(n^{|w|+1}).$$

1.4.6 Pair-matched word

From the discussion after (1.19) in Section 1.4.3, as far as verifying (M1) is concerned, whenever the input sequence satisfies Assumption III, it is enough to consider matched circuits. The next lemma helps to reduce the number of terms in (1.19) and shows that we can further restrict attention to only pair-matched words. Lemma 1.4.2 is from Bose and Sen (2008)[29]. Relation (1.23) is proved in Bryc, Dembo and Jiang (2006)[34] for L-functions of the Toeplitz and Hankel matrices. A careful scrutiny of their proof reveals that the same argument works for the present case.

Let $\lfloor x \rfloor$ denote the integral part of x and let

$$N_{h,3+} = \#\{\pi : \pi \text{ is matched}, \ l(\pi) = h \text{ and at least one } e \geq 3\}.$$

Lemma 1.4.2. (a) If L satisfies Property B, then there is a constant C_h depending on L and h such that

$$N_{h,3+} \leq C_h n^{\lfloor (h+1)/2 \rfloor} \text{ and hence } n^{-(1+h/2)} N_{h,3+} \to 0 \text{ as } n \to \infty. \quad (1.23)$$

(b) Suppose $\{A_n\}$ is a sequence of $n \times n$ patterned random matrices where the input sequence $\{x_i\}$ or $\{x_{i,j}\}$ satisfies Assumption III and L satisfies Property B. Then

$$\text{if } h \text{ is odd}, \quad \lim_n E[\beta_h(n^{-1/2} A_n)] = \lim_n E\left[\frac{1}{n} \text{Tr}\left(\frac{A_n}{\sqrt{n}}\right)^h\right] = 0 \quad (1.24)$$

$$\text{and if } h = 2k, \quad \sum_{w \in \mathcal{W}_{2k}(2)} \lim_n \frac{1}{n^{1+k}} \# [\Pi^*(w) - \Pi(w)] = 0. \quad (1.25)$$

Further,

$$\begin{aligned}
\lim_n E[\beta_{2k}(n^{-1/2} A_n)] &= \sum_{w \in \mathcal{W}_{2k}(2)} \lim_n \frac{1}{n^{1+k}} \#\Pi(w) \\
&= \sum_{w \in \mathcal{W}_{2k}(2)} \lim_n \frac{1}{n^{1+k}} \#\Pi^*(w), \quad (1.26)
\end{aligned}$$

provided any of the last two limits above exist. $\diamond$

Proof. (a) Since the number of words is finite, it is enough to fix a word and argue. So let w be a word of length h with at least one edge of order greater or equal to three. Either $h = 2k$ or $h = 2k - 1$ for some k. In both cases the number of generating vertices in w, $|w|$, is less than or equal to $k - 1$. If we fix the generating vertices, by Property B, the maximum number of choices for the non-generating vertices is $\Delta(L)^t$ for some integer t. Hence,

$$\#\Pi(w) \le n\Delta(L)^t n^{k-1} = C_h n^{\lfloor (h+1)/2 \rfloor}.$$

Relation (1.23) is an immediate consequence.

(b) Relation (1.24) follows since every potentially contributing circuit must have at least one 3- or higher-order match and since all moments involved in the sum are uniformly bounded by some constant C_h say. Relation (1.25) follows immediately since if $\pi \in \Pi^*(w) - \Pi(w)$ then π must have an edge of order at least 4 and we can apply (1.23).

We now prove (1.26). From mean zero and independence assumption (provided the last limit below exists),

$$\lim \mathrm{E}[\beta_{2k}(n^{-1/2}A_n)] = \lim \frac{1}{n^{1+k}} \sum_{\pi \text{ circuit}} \mathrm{E}\,\mathrm{X}_\pi$$

$$= \sum_{w \text{ matched}} \lim \frac{1}{n^{1+k}} \sum_{\pi \in \Pi(w)} \mathrm{E}\,\mathrm{X}_\pi. \qquad (1.27)$$

By Holder's inequality and Assumption III, for some constant C_{2k},

$$\left| \sum_{\pi: \ \pi \in \Pi(w)} \mathrm{E}\,\mathrm{X}_\pi \right| \le \#\Pi(w) C_{2k}.$$

Therefore, from part (a), matched circuits which have edges of order three or more do not contribute to the limit in (1.26). So,

$$\lim E[\beta_{2k}(n^{-1/2}A_n)] = \sum_{w \in \mathcal{W}_{2k}(2)} \lim_n \frac{1}{n^{1+k}} \#\Pi(w)$$

$$= \sum_{w \in \mathcal{W}_{2k}(2)} \lim_n \frac{1}{n^{1+k}} \#\Pi^*(w), \qquad (1.28)$$

provided any of the last two limits above exist. This establishes (1.26). The proof is now complete. □

Define, for every k and for $w \in \mathcal{W}_{2k}(2)$,

$$p(w) = \lim_n \frac{1}{n^{1+k}} \#\Pi^*(w) = \lim_n \frac{1}{n^{1+k}} \#\Pi(w) \qquad (1.29)$$

whenever any one (and hence both) of the limits exists. For any fixed word, this limit will be positive and finite only if the number of elements in the set $\Pi(w)$ is of *exact* order n^{k+1}. From Lemma 1.4.2, it follows that then the limiting $(2k)$-th moment (provided the limit above exists) is the finite sum

$$\beta_{2k} = \sum_{w \in \mathcal{W}_{2k}(2)} p(w). \tag{1.30}$$

This would essentially establish the (M1) condition. We shall see later in Theorem 1.4.4 how Property B may be used to verify Carleman's/Riesz's condition for $\{\beta_{2k}\}$.

The next Lemma 1.4.3 helps to verify (M4). Again, relation (1.31) is proved in Bryc, Dembo and Jiang (2006)[34] for L-functions of Toeplitz and Hankel matrices. The same proof works for general L functions which satisfy Property B. We provide all the details of the proof. Let

$$Q_{h,4} = \#\{(\pi_1, \pi_2, \pi_3, \pi_4) : l(\pi_i) = h, 1 \leq i \leq 4 \text{ jointly- and cross-matched}\}.$$

Lemma 1.4.3. (a) If L satisfies Property B, then there exists a K, depending on L and h, such that,

$$Q_{h,4} \leq K n^{2h+2}. \tag{1.31}$$

(b) Suppose $\{A_n\}$ is a sequence of $n \times n$ random matrices where the input sequence $\{x_i\}$ or $\{x_{i,j}\}$ satisfies Assumption III and the link function L satisfies Property B. Then

$$E\left[\frac{1}{n} \operatorname{Tr}\left(\frac{A_n}{\sqrt{n}}\right)^h - E\frac{1}{n} \operatorname{Tr}\left(\frac{A_n}{\sqrt{n}}\right)^h\right]^4 = O(n^{-2}) \tag{1.32}$$

and hence (M4) holds. Moreover, $\beta_h(n^{-1/2}A_n) - E[\beta_h(n^{-1/2}A_n)] \to 0$ *a.s.* $\diamond$

Proof. (a) Consider all circuits $(\pi_1, \pi_2, \pi_3, \pi_4)$ of length h which are jointly-matched and cross-matched with respect to L. Consider all possible edges $(\pi_j(i-1), \pi_j(i))$, $1 \leq j \leq 4$ and $1 \leq i \leq h$. Since the circuits are jointly-matched and cross-matched, there are at most $2h$ distinct L-values in these $4h$ edges.

Fix an integer $u \leq 2h$. Note that the number of possible partitions of the $4h$ edges into u distinct groups of L-matched edges, with at least two edges in each partition block, is independent of n. So it is enough to establish the required upper bound when the number of quadruples of such circuits has u distinct L-values. Understandably, the argument becomes tougher as the value of u gets larger.

First assume $u \leq 2h - 2$. Then we count the total number of choices in the following way:

(i) The generating vertices $\pi_1(0), \pi_2(0), \pi_3(0)$ and $\pi_4(0)$ may be chosen in n^4 ways.

(ii) Now arrange the values $L(\pi_j(i-1), \pi_j(i))$, $1 \le j \le 4$, $1 \le i \le h$, from left to right, starting with π_1 and then π_2 and so on. Then the generating vertices $\pi_j(i)$, for which $L(\pi_j(i-1), \pi_j(i))$ is the first appearance of that L-value in this sequence, have at most n^u choices.

(iii) Having chosen these vertices, using Property B, the rest of the vertices in all the circuits, may be chosen from left to right, in at most $[\Delta(L)]^{4h}$ ways.

Since $u \le 2h - 2$, by using (i), (ii) and (iii), the total number of choices is bounded by

$$n^4 n^u [\Delta(L)]^{4h-u} \le C n^{2h+2}. \tag{1.33}$$

Now we consider the two remaining cases: $u = 2h - 1$ and $u = 2h$. In these cases we need to do a more careful counting.

First suppose $u = 2h - 1$. Then there are only two possibilities. Either

(i) two L-values are shared by three edges each and the rest are shared by exactly two edges (then the total number of edges equals $2 \times 3 + 2(2h-3) = 4h$) or

(ii) one L-value is shared by four edges and the rest are shared by exactly two edges (hence the total number of edges equals $4 + 2(2h - 2) = 4h$ as it should be).

Note that none of the circuits is self-matched and only the above two choices for partitions of L-values are possible. Thus, there is at least one circuit, say π_1, with an edge say $(\pi_1(i-1), \pi_1(i))$ whose L-value is not repeated in π_1. Leave aside this L-value from the set of u distinct L-values.

There remains $u - 1$ distinct L-values. As before, write their L-values in order, starting with those from π_1. Now counting as before, since there are $u - 1$ generating vertices, the total number of ways in which all these vertices and $\pi_j(0)$ may be chosen is

$$n^{u-1} n^4 = n^{2h+2}. \tag{1.34}$$

Note that the generating vertex $\pi_1(i)$ has *not* been counted above in the $u-1$ generating vertices. If we can now show that this vertex has finitely many choices, we would be done. Now, note that the L-value $L(\pi_1(i), \pi_1(i+1))$ appears elsewhere and hence has been chosen. Moreover, $\pi_1(i+1)$ has also been already chosen. Hence by Property B, this leaves only finitely many choices for $\pi_1(i)$ and we are done with this case.

Now assume $u = 2h$. Then each L-value is shared by exactly two edges. Now we seek *two* generating vertices which have finitely many choices.

As before, upon re-ordering the four circuits, we have an L-value that is assigned, as the first and only one, to exactly one edge, say $(\pi_1(i-1), \pi_1(i))$, of say π_1.

Now, we have the following three sub-cases:

(i) All the L-values of π_1 which appear exactly once in π_1 have the corresponding matches in only one other circuit, say π_2. Then π_3 has at least one singly occurring L-value, say $L(\pi_3(j-1), \pi_3(j))$, which does not appear in π_1 and, since the circuits are cross-matched, anywhere else in π_3 either. Re-order the circuits as $\pi_1, \pi_3, \pi_2, \pi_4$.

In the case $u = 2h - 1$, we had set aside one L-value. Now we set aside the *two* L-values $L(\pi_1(i-1), \pi_1(i))$ and $L(\pi_3(j-1), \pi_3(j))$. Then the number of choices for the generating vertices corresponding to the rest of the $u - 2$ edges is n^{u-2}.

Then, we repeat the dynamic construction of π_1 and π_3 as given earlier in the case $u = 2h - 1$ for π_1. Similar arguments now imply that the two generating vertices $\pi_1(i)$ and $\pi_3(j)$ obtained above have only finitely many choices and hence we get the required bound as $Cn^{u-2}n^4 = Cn^{2h+2}$.

(ii) π_1 has matches in only two other circuits, say π_2 and π_3. Then we re-order the circuits as $\pi_1, \pi_4, \pi_2, \pi_3$. Now clearly we have at least one L-value of π_1 which has not appeared in π_4 and elsewhere in π_1.

On the other hand, since the circuits are cross-matched, in π_4 there is at least one (singly occurring) L-value which appears in π_2 or π_3 and hence does not occur in π_1. So consider π_1 and π_4 and repeat the above argument, given in (i) for (π_1, π_3), with these two circuits and the chosen L-values.

(iii) π_1 has matches with all the three circuits, say π_2, π_3 and π_4.

Then we have two further sub-cases:

A. One of the three circuits (say π_2) has one L-value identical with one other L-value in a different circuit (say in π_3). Then again, there exists an L-value in π_2 along with one from π_1 that appear in π_4 and repeat the earlier argument, given in (i), on π_1 and π_2.

B. None of the three circuits have matches with any other. In this case, identify one L-value from π_2 that has appeared in π_1 (and hence not in π_3). Then identify one L-value from π_3 that has appeared in π_1 (hence not in π_2). Now re-order the circuits as π_2, π_3, π_1 and π_4 and repeat the earlier argument, given in (i), with the above two L-values (and the corresponding generating vertices) from the circuits π_2 and π_3.

This completes the proof of (a).

(b) We write the fourth moment as

$$\frac{1}{n^{2h+4}} \operatorname{E}[\operatorname{Tr} A_n^h - \operatorname{E}(\operatorname{Tr} A_n^h)]^4 = \frac{1}{n^{2h+4}} \sum_{\pi_1,\pi_2,\pi_3,\pi_4} \operatorname{E}[\prod_{i=1}^{4}(\operatorname{X}_{\pi_i} - \operatorname{E}\operatorname{X}_{\pi_i})].$$

If $(\pi_1, \pi_2, \pi_3, \pi_4)$ are not jointly-matched, then one of the circuits, say π_j, has an L-value which does not occur anywhere else. Also note that $\operatorname{E}\operatorname{X}_{\pi_j} = 0$. Hence, using independence,

$$\operatorname{E}[\prod_{i=1}^{4}(\operatorname{X}_{\pi_i} - \operatorname{E}\operatorname{X}_{\pi_i})] = \operatorname{E}[\operatorname{X}_{\pi_j} \prod_{i=1,i\neq j}^{4}(\operatorname{X}_{\pi_i} - \operatorname{E}\operatorname{X}_{\pi_i})] = 0.$$

So we need to consider only jointly-matched circuits. Now, if $(\pi_1, \pi_2, \pi_3, \pi_4)$ is jointly matched but is not cross-matched then one of the circuits, say π_j is only self-matched, that is, none of its L-values are shared with those of the other circuits. Then by independence,

$$\begin{aligned} \operatorname{E}[\prod_{i=1}^{4}(\operatorname{X}_{\pi_i} - \operatorname{E}\operatorname{X}_{\pi_i})] &= \operatorname{E}[(\operatorname{X}_{\pi_j} - \operatorname{E}\operatorname{X}_{\pi_j})] \operatorname{E}[\prod_{i=1,i\neq j}^{4}(\operatorname{X}_{\pi_i} - \operatorname{E}\operatorname{X}_{\pi_i})] \\ &= 0. \end{aligned}$$

Thus, we can focus on the circuits which are both jointly- and cross-matched. Since the input sequence satisfies Assumption III, $\operatorname{E}[\prod_{i=1}^{4}(\operatorname{X}_{\pi_i} - \operatorname{E}\operatorname{X}_{\pi_i})]$ is bounded uniformly across all possible circuits. Therefore by part (a),

$$\operatorname{E}\left[\frac{1}{n}\operatorname{Tr}\left(\frac{A_n}{\sqrt{n}}\right)^h - \operatorname{E}\frac{1}{n}\operatorname{Tr}\left(\frac{A_n}{\sqrt{n}}\right)^h\right]^4 \leq K\frac{n^{2h+2}}{n^4(n^{h/2})^4} = \operatorname{O}(n^{-2}), \qquad (1.35)$$

proving the lemma completely.   □

1.4.7 Sub-sequential limit

Theorem 1.4.4 is due to Bose and Sen (2008)[29]. This result shows that when Property B is satisfied, Carleman's or Riesz's condition is automatically guaranteed. Further, the moment convergence (1.19) is essentially equivalent to the existence of the LSD.

Theorem 1.4.4. Suppose $\{A_n = ((x_{L(i,j)}))_{i,j=1}^n\}$ is a sequence of $n \times n$ random matrices where the input sequence $\{x_i\}$ or $\{x_{i,j}\}$ satisfies Assumption III and L satisfies Property B. A non-random LSD G exists for $\{n^{-1/2}A_n\}$, if and only if for every h, almost surely,

$$\lim \beta_h(n^{-1/2}A_n) = \beta_h(G) \text{ say.} \qquad (1.36)$$

In that case,

(a) $\beta_h(G) = 0$ whenever h is odd,

(b) $\beta_{2k}(G) \leq \frac{(2k)!\Delta(L)^k}{k!2^k}$ and

(c) $G(x) + G(-x) = 1$ for all x.

Hence G has sub-Gaussian moments, satisfies Carleman's and Riesz's condition and represents a probability distribution function which is symmetric about 0.

Even when the limit in (1.36) does not exist, $\{F^{n^{-1/2}A_n}\}$ is tight almost surely and any sub-sequential limit S satisfies the properties of G listed in (a), (b) and (c) above.   ◇

Proof. For convenience, we write F_n for $F^{n^{-1/2}A_n}$. Let $\hat{F}_n = \mathrm{E}(F_n)$ be the expected ESD. Since Property B holds, we immediately have the following:

(i) (Odd moments) Lemma 1.4.2(b) implies $\mathrm{E}[\beta_{2k-1}(F_n)] \to 0$.

(ii) (Even moments) Fix a *matched* word w with $l(w) = 2k$. If $|w| = k$, then we have $(k+1)$ generating vertices. Having fixed the generating vertices arbitrarily, we have at most $\Delta(L)$ choices for each of the remaining k vertices. Thus,

$$\#\Pi^*(w) \leq \Delta(L)^k n^{k+1}. \tag{1.37}$$

On the other hand, if $|w| < k$ then the number of generating vertices is less than k, and so

$$\#\Pi^*(w) = O_k(n^k) \tag{1.38}$$

where the $O_k(\cdot)$ term may involve k. Combining (1.37) and (1.38), we get

$$\mathrm{E}[\beta_{2k}(F_n)] \leq \sum_{w \in \mathcal{W}_{2k}(2)} \frac{1}{n^{k+1}} \#\Pi^*(w) \leq \frac{(2k)!}{2^k k!}\Delta(L)^k + O_k(n^{-1}). \tag{1.39}$$

(iii) Lemma 1.4.3(b) implies that for each $h \geq 1$, $\beta_h(n^{-1/2}A_n) - \mathrm{E}[\beta_h(n^{-1/2}A_n)] \to 0$ *a.s.* Also using (i), we have for all $k \geq 1$,

$$\beta_{2k-1}(F_n) \overset{a.s.}{\to} 0$$

and using (ii), for any sub-sequential limit S,

$$\beta_{2k}(S) \leq \frac{(2k)!\Delta(L)^k}{k!2^k}.$$

Now suppose F_n converges to G a.s. Fix a continuity point x of G. Then $F_n(x) \overset{a.s.}{\to} G(x)$. Using DCT, we get $\hat{F}_n(x) \to G(x)$. Hence $\hat{F}_n$ converges to G. Now

$$\int x^{2k} d\hat{F}_n(x) = \mathrm{E}[\beta_{2k}(F_n)] \leq \frac{(2k)!\Delta(L)^k}{k!2^k} + O_k(n^{-1}) \quad \text{for all } k \geq 1.$$

As a consequence, if Y_n has distribution $\hat{F}_n$ then $\{Y_n^k\}$ is uniformly integrable for every k.

Since $\hat{F}_n$ converges to G, this implies that $\int x^{2k} d\hat{F}_n(x) \to \int x^{2k} dG(x)$ for all $k \geq 1$ and then by applying (iii) even moments of F_n converge to corresponding even moments of G a.s.

Conversely, suppose all moments $\beta_{2k}(n^{-1/2}A_n)$ converge almost surely. From (iii), $\mathrm{E}[\beta_{2k}(F_n)] \to \beta_{2k}(G)$. It now follows from (b), that

$$\beta_{2k}(G) = \int x^{2k} dG(x) \leq \frac{(2k)!\Delta(L)^k}{k!2^k} \quad \text{for all } k \geq 1.$$

Hence G satisfies Carleman's and Riesz's condition because the sequence $\{\frac{(2k)!\Delta(L)^k}{k!2^k}\}$ itself does so (see Lemma 1.2.3). The rest of the proof is now immediate by the moment method. The fact that $G(x) + G(-x) = 1$ is easy to establish, since the characteristic function of G has a power series expansion. We omit the details.   □

Remark 1.4.2. (a) Suppose A_n is a sequence of patterned matrices where Assumption I, II or III holds for the input sequence and Property B holds for the link function, and $p(w)$ exists for every w of length $2k$, for all $k \geq 1$. Suppose further that $k_n \to \infty$ and $\alpha_n k_n^d = O(n^2)$. Then from the discussion in this chapter, the LSD is *universal*, that is, the LSD does not depend on the underlying distribution of the input sequence.

(b) In all of the examples given so far, it turns out that we can find a set $\Pi^{**}(w) \subseteq \Pi^*(w)$ so that the following two statements hold.

(i)
$$\lim_n n^{-(k+1)} \left[\#\Pi^*(w) - \#\Pi^{**}(w) \right] = 0$$

(ii) for any $\pi \in \Pi^{**}(w)$ if the generating vertices are fixed then there is *at most one* choice for the other vertices.

It follows that
$$p(w) \leq 1 \quad \text{for every } w.$$

As a consequence,
$$\beta_{2k} \leq \frac{(2k)!}{2^k k!}.$$

Thus, the limit moments are dominated by the Gaussian moments. Detailed proof of these facts and more will be given in the next chapters.   ◇

1.5   Exercises

1. Work out all the details in the proof of Lemma 1.2.1.

2. Show that W_2 is indeed a metric and is complete.

3. Check that the equivalence relation defined between circuits is a genuine equivalence relation.

4. Show that
$$\#\mathcal{W}_{2k}(2) = \frac{(2k)!}{2^k k!}.$$

5. Under Assumption III, complete the moment method proof of the CLT outlined in Example 1.2.1. Then using the W_2 metric, establish the CLT under Assumption II.

6. Prove Lemma 1.4.1 under Assumption III.

7. Show that Riesz's condition implies Carleman's condition.

8. Show that the moments of any distribution with bounded support satisfies Carleman's/Riesz's condition.

9. The *moment generating function* (m.g.f.) of a real random variable X is defined as $M_X(t) = \mathrm{E}(e^{tX})$. Show that, if the m.g.f. is finite in a neighborhood of zero, then Riesz's condition is satisfied.

10. Suppose $\{\beta_{2k}\}$ satisfies Carleman's/Riesz's condition and $0 \le c_{2k} \le \beta_{2k}$ for all k. Then show that $\{c_{2k}\}$ also satisfies Carleman's/Riesz's condition.

11. Give a complete proof of Lemma 1.2.4.

12. Suppose X is a random variable with bounded support and all its odd moments are 0. Show that the distribution of X is symmetric about 0.

13. Suppose X is a random variable all of whose odd moments are 0. Suppose further that $\beta_{2k}(X) \le \frac{(2k)!\Delta^k}{k!2^k}$ for some $\Delta < \infty$. Show that the distribution of X is symmetric about 0.

14. Show that for all non-negative integers k, $k! > (k/e)^k$.

15. Show that for all non-negative integers k and all real numbers a,
$$|e^{ia} - 1 - ia \ldots - (ia)^k/k!| \le |a|^{k+1}/(k+1)!. \qquad (1.40)$$

16. Let F_n denote the ESD of any of the matrices $n^{-1/2}W_n^{(s)}$, $n^{-1/2}H_n^{(s)}$, $n^{-1/2}RC_n^{(s)}$, $n^{-1/2}SC_n^{(s)}$, $n^{-1/2}H_n^{(s)}$. Show that if the input sequence is i.i.d. with mean zero and variance one, then in each case, $E[\beta_2(F_n)] \to 1$.

17. Let F_n be as in the previous problem. Suppose the input sequence is i.i.d. with mean zero and variance one and finite fourth moment. Show that

(a) for the Wigner, Hankel and Reverse Circulant matrices, $E[\beta_4(F_n)] \to 2$;

(b) for the Toeplitz matrix $E[\beta_4(F_n)] \to 8/3$; and

(c) for the Symmetric Circulant matrix, $E[\beta_4(F_n)] \to 3$.

18. Read the proof of Carleman's criterion Lemma 1.2.2 from Bai and Silverstein (2010)[8].

19. Suppose A_n is a scaled real symmetric Wigner matrix whose input satisfies Assumption I or III. Show that for every integer $h \geq 1$,

$$\sum_{n=1}^{\infty} E[\beta_h(A_n) - E(\beta_h(A_n))]^2 < \infty.$$

20. Suppose A_n is a scaled real symmetric random matrix with link function that satisfies Property B and $\alpha_n k_n^d = O(n^2)$. Suppose further that the input sequence satisfies Assumption I or III. Show that for every h,

$$E[\beta_h(A_n) - E(\beta_h(A_n))]^2 = O(n^{-1})$$

and hence $\beta_h(A_n) - E(\beta_h(A_n)) \xrightarrow{P} 0$ for every h.

2

Common symmetric patterned matrices

In this chapter we shall put to use the results and concepts of the previous chapter to provide proofs of existence of the LSD of some real symmetric patterned matrices. These matrices are: Wigner, Toeplitz, Hankel, Reverse Circulant and the Symmetric Circulant. We shall see how certain types of words assume significance in the context of each matrix. It turns out that, for all of these matrices, $p(w) = 1$ for all w in one common subclass of words, which we call the *Catalan words*. Moreover, for the Wigner matrix, $p(w) = 0$ for all words outside this class and, for the Symmetric Circulant, $p(w) = 1$ for all pair-matched words. The LSD is bounded for the Wigner matrix but is unbounded for all the other matrices. There are many interesting questions that arise from the main LSD results and associated words. The LSD do not depend on the specific nature of the input sequence but on the pattern. This feature is termed *universality*. We can define it formally as follows. First recall the three main assumptions from the previous chapter.

Assumption I $\{x_i, x_{ij}\}$ are i.i.d. and uniformly bounded with mean 0 and variance 1.

Assumption II $\{x_i, x_{ij}\}$ are i.i.d. with mean zero and variance 1.

Assumption III $\{x_i, x_{ij}\}$ are independent with mean zero and variance 1 and with uniformly bounded moments of all order.

Definition 2.0.1. For a sequence of patterned matrices, we will say that the LSD is *universal* if the LSD is the same for all input sequences which satisfy Assumption I, II or III. ◇

2.1 Wigner matrix

The earliest appearance of a random matrix is in statistics. One can recall the two 2×2 contingency tables. The famous *Wishart distribution* arose as the distribution of the *sample variance-covariance matrix* when the observations are from a Gaussian distribution. See Wishart (1928)[107]. We shall deal with this matrix in the next chapter.

After a long period of inactivity in this area, Wigner (1955)[105] introduced a random matrix while discussing statistical models for heavy nuclei atoms in physics. He considered the *real symmetric matrix* all of whose entries are independent and identically distributed. This matrix and its variations are now collectively known as the *Wigner matrices*. More recently, its triangular version has shown up in significant ways in operator theory as the so-called *DT* operator. As described earlier, the symmetric (unscaled) Wigner matrix $W_n^{(s)}$ equals

$$
W_n^{(s)} = \begin{bmatrix}
x_{11} & x_{12} & x_{13} & \cdots & x_{1(n-1)} & x_{1n} \\
x_{12} & x_{22} & x_{23} & \cdots & x_{2(n-1)} & x_{2n} \\
& & \vdots & & & \\
x_{1n} & x_{2n} & x_{3n} & \cdots & x_{(n-1)n} & x_{nn}
\end{bmatrix}.
$$

Suppose that the entries of the Wigner matrix are real normal with mean zero, and variances 1 and 1/2, respectively, for the entries on and above the diagonal. Then the joint distribution of its eigenvalues can be calculated explicitly. If $\lambda_1 \geq \cdots \geq \lambda_n$ are the eigenvalues, then it is not difficult to prove (see Mehta (1967)[77]) that the joint density of the scaled ordered eigenvalues $\{x_i = \frac{\lambda_i}{\sqrt{n}}\}$ is:

$$
f(x_1, \ldots, x_n) = \frac{\exp\left(-\sum_{i=1}^{n} x_i^2/2\right)}{2^{n/2} \prod_{i=1}^{n} \Gamma\left(\frac{p+1-i}{2}\right)} \prod_{i<j}^{n} (x_i - x_j). \tag{2.1}
$$

In general, if the entries are not normal, the distribution of the eigenvalues cannot be found in a closed form.

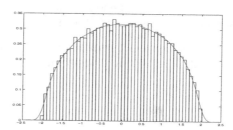

FIGURE 2.1
Smoothed ESD of $n^{-1/2}W_n^{(s)}$ for $n = 400$, 15 replications. Input is normalized i.i.d. symmetric Bernoulli.

Figure 2.1 shows the simulated ESD of $n^{-1/2}W_n^{(s)}$ with i.i.d. standardized Bernoulli entries as the input sequence. It turns out that this simulation exhibits a general phenomenon; under broad conditions on the entries, its LSD is the (standard) *semi-circle law* given below in (2.3). The law of the positive part of a semi-circle random variable is known as the *quarter-circle law* (see Girko and Repin (1995)[59]).

Historically, Wigner (1955)[105] assumed the entries to be i.i.d. real Gaussian and established the convergence of the EESD $E[F^{n^{-1/2}W_n^{(s)}}(\cdot)]$ to the semi-circle law. Wigner (1958)[106] also noted that the semi-circle law is the LSD of much more general symmetric matrix models where the entries on and above the diagonal are independent and the entries have symmetric distribution function with variance σ^2 for the non-diagonal entries and $2\sigma^2$ for the diagonal ones and all higher moments are uniformly bounded. This claim generated interest amongst researchers to relax the conditions on entries of the matrix to the maximum possible extent.

Assuming the existence of finite moments of all orders, Grenander (1963, pages 179 and 209)[61] established the convergence of the ESD in probability. Arnold (1967)[4] obtained the almost sure convergence of the ESD under the finiteness of the fourth moment of the entries.

It appears that the weakest known condition under which the convergence to the semi-circle law holds was given by Pastur (1972)[81]. He assumed that $\{x_{ij}\}$ are independent with zero mean and unit variance, and they satisfy the following Lindeberg-type condition, known as *Pastur's condition*:

$$\lim_{n \to \infty} \frac{1}{\delta^2 n^2} \sum_{i,j=1}^{n} E\left[|x_{ij}|^2 x_{ij}^2 \mathbb{I}_{(|x_{ij}|>\delta\sqrt{n})}\right] = 0, \quad \text{for any} \ \delta > 0. \tag{2.2}$$

For a detailed exposition, see Bai (1999)[5] or Khorunzhy, Khoruzhenko and Pastur (1996)[70].

All the earlier proofs use the Moment Method. Recently, Chatterjee (2005)[40] proved Wigner's semi-circle law under Pastur's condition by an ingenious method. The Stieltjes transform method was used in Bai (1999)[5]. Incidentally, the method of Stieltjes transform works well for matrices whose last row/column is independent of the rest of the matrix—a feature absent in the Toeplitz and Hankel matrices for instance. Bai (1999)[5] also provided two extensions of the matrix model of Wigner. Let W_n be an $n \times n$ Hermitian matrix whose entries above the diagonal are i.i.d. complex random variables with variance 1 and whose diagonal entries are i.i.d. real random variables (without any moment requirement), or let W_n be an $n \times n$ Hermitian whose entries above the diagonal are independent complex random variables with a common mean 0 and variance 1 and which satisfy the Lindeberg-type condition (2.2). It was shown in Bai (1999)[5] that, under either assumption, as $n \to \infty$ the ESD of $n^{-1/2}W_n$ converges weakly to the semi-circle law almost surely.

While we do provide a proof that the semi-circle law is indeed the LSD of a Wigner matrix, we do not aim for the greatest possible generality. In fact we work under Assumption I, II or III mentioned in Chapter 1. In Chapter

5, we will explore, to what extent the semi-circle law remains the LSD of non-Wigner random matrices.

Recall that the link function of the Wigner matrix is

$$L(i, j) = (\min(i, j), \max(i, j)).$$

It is easy to see that this link function satisfies Property B (see Definition 1.4.1) with $\Delta = 1$.

Also recall the definitions of k_n and α_n from Chapter 1. Note that for the Wigner link function,

$$d = 2, \ k_n = n \ \text{ and } \ \alpha_n = 2.$$

Thus, $k_n \to \infty$ and $k_n^d \alpha_n = O(n^2)$. As a result, Lemma 1.4.1 applies.

2.1.1 Semi-circle law, non-crossing partitions, Catalan words

Definition 2.1.1. (*Semi-circle law*) The (standard) *semi-circle law* $\mathcal{L}_W$, has the density function

$$p_W(s) = \begin{cases} \frac{1}{2\pi}\sqrt{4 - s^2} & \text{if } |s| \leq 2, \\ 0 & \text{otherwise.} \end{cases} \tag{2.3}$$

We shall denote a random variable with this density by W. As we shall see, this law arises as the LSD of $n^{-1/2}W_n^{(s)}$. Its moments are given in the next lemma.

Lemma 2.1.1. For $k = 0, 1, 2, \ldots$, we have

$$\begin{aligned} \beta_{2k+1}(W) &= 0, \\ \beta_{2k}(W) &= \frac{1}{k+1}\binom{2k}{k} = \frac{(2k)!}{k!(k+1)!}. \end{aligned}$$

$\diamond$

Proof. Since the semi-circle distribution is symmetric about 0, we have $\beta_{2k+1}(W) = 0$. Also,

$$\begin{aligned} \beta_{2k}(W) &= \frac{1}{2\pi}\int_{-2}^{2} x^{2k}\sqrt{4 - x^2}dx \\ &= \frac{1}{\pi}\int_{0}^{2} x^{2k}\sqrt{4 - x^2}dx \\ &= \frac{2^{2k+1}}{\pi}\int_{0}^{1} y^{k-1/2}(1-y)^{1/2}dy \quad \text{(by setting } x = 2\sqrt{y}) \\ &= \frac{2^{2k+1}}{\pi}\frac{\Gamma(k+1/2)\Gamma(3/2)}{\Gamma(k+2)} = \frac{1}{k+1}\binom{2k}{k} = \frac{(2k)!}{k!(k+1)!}. \end{aligned}$$

$\square$

This sequence of numbers is connected to the count of the so-called *non-crossing partitions*. They also play a crucial role in *free probability* which we shall encounter in Chapters 9 and 10.

Definition 2.1.2. (*Non-crossing pair-partitions*) Consider the set of integers $\{1, 2, \ldots, 2k\}$. Arrange them on a circle sequentially. Consider any *pair-partition* of this set and draw an edge between the two points of each partition. Then the partition is said to be *non-crossing* if none of the edges cross each other. The set of all non-crossing pair-partitions will be denoted by $\mathcal{NC}_{2k}$. ◇

A pictorial illustration is given in Figure 2.2.

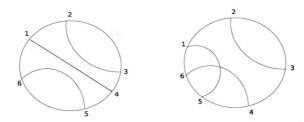

FIGURE 2.2
Non-crossing and crossing pair-partitions.

Equivalently, in our language of words, the following sub-class of the pair-matched words is intimately tied to the LSD of the Wigner matrix.

Definition 2.1.3. (*Catalan word*) A pair-matched word $w \in \mathcal{W}_{2k}(2)$ is called a Catalan word if

(i) it has at least one *double* letter (a letter repeated immediately) and

(ii) the reduced word after removing all double letters also has property (i) and sequential removal all double letters ultimately leads to the empty word.

The set of all Catalan words of length $2k$ will be denoted by $\mathcal{C}_{2k}$. ◇

For example, $abba, aabbcc, abccbdda$ are Catalan words but $abab$ and $abcddcab$ are not. Count of the number of Catalan words is given below. A more elaborate counting of Catalan words with a given number of generating vertices at the odd/even positions is presented in Lemma 3.2.4 of Chapter 3.

Lemma 2.1.2. (a)
$$\#\mathcal{C}_{2k} = \frac{(2k)!}{(k+1)!k!}.$$

(b) The number of non-crossing pair-partitions of $\{1, \ldots, 2k\}$ equals $\frac{(2k)!}{(k+1)!k!}$.

(c) $\mathcal{NC}_{2k}$ and $\mathcal{C}_{2k}$ are in bijection. ◇

Proof. A proof of (b) is available in Chapter 2 of Anderson, Guionnet and Zeitouni (2009)[1]. A proof of (a) is available in Bose and Sen (2008)[29].

We now sketch a proof of (a). For any Catalan word, mark the first and second occurrences of a letter by $+1$ and -1 respectively. For example, *abba* and *abccbdda* are represented, respectively, by $(1, 1, -1, -1)$ and $(1, 1, 1, -1, -1, 1, -1, -1)$. Consider the last $+1$. It will be followed by a -1. Pair these two. Then move to the previous $+1$ and pair it with the next -1 available. Continue the process till all $+1$ and -1 are exhausted. This provides a bijection between the Catalan words of length $2k$ and sequences $\{u_l\}_{1 \leq l \leq 2k}$ which satisfy: each $u_l = \pm 1$, $S_l = \sum_{j=1}^{l} u_j \geq 0 \ \forall \ l \geq 1$ and $S_{2k} = 0$. By the *reflection principle* (recall the well-known ballot problem), the total number of such paths is easily seen to be $\frac{(2k)!}{(k+1)!k!}$. We omit the details.

It is easy to see how each Catalan word leads to a non-crossing pair-partition and vice versa. For example the Catalan word *abbcca* corresponds to the non-crossing partition $\{\{1, 6\}, \{2, 3\}, \{4, 5\}\}$. Thus, proofs of (b) and (c) are then immediate. $\qquad \square$

2.1.2   LSD

Theorem 2.1.3. Let $W_n^{(s)}$ be the $n \times n$ Wigner matrix with the entries $\{x_{ij} : 1 \leq i \leq j, \ j \geq 1\}$ which satisfies Assumption I, II or III. Then $\{F^{n^{-1/2} W_n^{(s)}}\}$ converges weakly almost surely to the semi-circle law $\mathcal{L}_W$ given in (2.3). In particular the LSD is universal. $\qquad \diamond$

Proof. By Lemmata 1.4.2, 1.4.3 and 2.1.2 it is enough to show that for every $w \in \mathcal{W}_{2k}(2)$,

$$\lim \frac{1}{n^{1+k}} \#\Pi^*(w) = \begin{cases} 0 & \text{if } w \notin \mathcal{C}_{2k} \\ 1 & \text{if } w \in \mathcal{C}_{2k}. \end{cases} \qquad (2.4)$$

Note that if $\pi \in \Pi^*(w), w[i] = w[j] \Rightarrow L(\pi(i-1), \pi(i)) = L(\pi(j-1), \pi(j))$. Then

$$(\pi(i-1), \ \pi(i)) = \begin{cases} (\pi(j-1), \ \pi(j)) & \text{(constraint (C1)) \ or} \\ (\pi(j), \ \pi(j-1)) & \text{(constraint (C2)).} \end{cases} \qquad (2.5)$$

Incidentally, we will use constraints (C1) and (C2) in the later chapters also.

Now, for any $w \in \mathcal{W}_{2k}(2)$, there are k such constraints. Since each constraint is either (C1) or (C2), there are at most 2^k choices in all. Let λ be a typical choice of k constraints and $\Pi_\lambda^*(w)$ be the subset of $\Pi^*(w)$ corresponding to λ and so,

$$\Pi^*(w) = \bigcup_\lambda \Pi_\lambda^*(w) \quad \text{(a disjoint union)}, \qquad (2.6)$$

where the union extends over all 2^k possible choices. Some of the sets in the union may be vacuous.

Fix w and λ. We now count the number of elements in $\Pi_\lambda^*(w)$. For $\pi \in \Pi_\lambda^*(w)$, consider the graph with vertices $\pi(0), \pi(1), \ldots, \pi(2k)$. By abuse of notation, $\pi(i)$ thus denotes both a vertex and its numerical value. Vertices *within* the following pairs are connected with a single edge:

(i) The pairs $(\pi(i-1),\ \pi(j-1))$ and $(\pi(i),\ \pi(j))$ if $w[i] = w[j]$ yields constraint (C1). This implies $\pi(i-1) = \pi(j-1)$ and $\pi(i) = \pi(j)$.

(ii) The pairs $(\pi(i'-1),\ \pi(j'))$ and $(\pi(i'),\ \pi(j'-1))$ if $w[i'] = w[j']$ yields constraint (C2). This implies $\pi(i'-1) = \pi(j')$ and $\pi(i') = \pi(j'-1)$.

(iii) The pair $(\pi(0),\ \pi(2k))$, ensuring that π is indeed a circuit.

So, the graph has a total of $(2k+1)$ edges. These may include both loops and double edges. For example if $L(\pi(1), \pi(2)) = L(\pi(2), \pi(3)) = L(\pi(3), \pi(4))$, then we have a double edge between $\pi(2)$ and $\pi(4)$, and a loop between $\pi(3)$ and $\pi(3)$. In any connected component, the numerical values of the vertices are of course equal. Observe that the maximum number of vertices in the circuit π whose numerical value can be chosen freely is the same as the number of *connected components* in the graph.

Note that the number of generating vertices for π equals $(k+1)$. Consider these vertices. Then all other vertices are connected to one or more of these. Hence the number of connected components in λ is bounded by $(k+1)$.

Claim. λ has exactly $(k+1)$ connected components if and only if $w \in \mathcal{W}_{2k}(2)$ *and* all constraints are (C2).

To see this, first suppose λ is such that the graph has $(k+1)$ connected components. By the pigeon hole principle, there exists a vertex, say $\pi(i)$, which is connected to itself. Note that this is possible if and only if $w[i] = w[i+1]$ and (C2) is satisfied. But this implies that $w[i]w[i+1]$ is a double letter in w.

We remove this double letter and consider the reduced word w' of length $2(k-1)$. We claim that the reduced word still has a double letter. To show this, in the original graph, coalesce the vertices $\pi(i-1)$ and $\pi(i+1)$. Delete the vertex $\pi(i)$ and remove the (C2) constraint edges $(\pi(i-1),\ \pi(i+1))$ and $(\pi(i),\ \pi(i))$ but retain all the other earlier edges. For example, any other edge that might have existed earlier between $\pi(i-1)$, $\pi(i+1)$ is now a loop. This gives a new graph with $2k+1-2 = 2(k-1)+1$ vertices and has k connected components. Proceeding as before, there must exist a self-edge implying a double letter yy in w'. Proceeding inductively, after k steps, we are left with just a single vertex with a loop. In other words, $w \in \mathcal{W}_{2k}(2)$ and all constraints are (C2).

Conversely, it is easy to verify that if $w \in \mathcal{W}_{2k}(2)$ and all constraints in λ are (C2), then the number of connected components in the graph is indeed $(k+1)$. To see this, essentially retrace the steps given above. First identify a double letter (the last new letter is followed by itself). This gives a (C2) constraint. Remove it and proceed inductively. For example, coalesced vertices will fall in the same connected component. We omit the details.

We denote by $\lambda = \lambda_0$ the case when $w \in \mathcal{W}_{2k}(2)$ and all constraints are (C2). Observe that in this case, $w \in \mathcal{C}_{2k}$. Then clearly

$$\#\Pi^*_{\lambda_0}(w) = n^{k+1}. \tag{2.7}$$

On the other hand if $w \in \mathcal{W}_{2k}(2)$ and $\lambda \neq \lambda_0$, then the corresponding graph has at most k connected components and hence $\#\Pi^*_\lambda(w) \leq n^k$. This implies

$$\frac{1}{n^{k+1}} \# \left(\bigcup \Pi^*_{\lambda \neq \lambda_0}(w) \right) \to 0. \tag{2.8}$$

Combining the above facts we get

$$\lim \frac{1}{n^{1+k}} \#\Pi^*(w) = \begin{cases} 1 & \text{if } w \in \mathcal{C}_{2k}, \\ \\ 0 & \text{otherwise.} \end{cases} \tag{2.9}$$

This completes the proof of Theorem 2.1.2. $\qquad\qquad\qquad\qquad\qquad$ □

2.2   Toeplitz and Hankel matrices

Any matrix T_n of the form $((t_{i-j}))_{1 \leq i, j \leq n}$ is called a *Toeplitz* matrix. For simplicity suppose that T_n is symmetric (that is $t_k = t_{-k}$ for all k) and that $\sum_{k=-\infty}^{\infty} |t_k| < \infty$. Likewise the Hankel matrix H_n is of the form $((t_{i+j}))_{1 \leq i, j \leq n}$.

Let l_2 be the space of all square summable sequences. Then the infinite-dimensional Toeplitz and Hankel matrices are operators from l_2 to l_2.

From Szego's theory of *Toeplitz operators* (see for example Böttcher and Silberman (1998)[31]), it is known that the LSD of T_n exists and has the following description. Let f be the Fourier function corresponding to $\{t_k\}$:

$$f(x) = \sum_{k=-\infty}^{\infty} t_k \exp\left(-2\pi i x k\right), \; x \in (0, \, 1] \; \text{ in the } \; L_2 \; \text{ sense.}$$

Then the LSD is (the distribution of) $f(U)$ where U is uniformly distributed on $(0, \, 1]$.

Toeplitz matrices also appear in time series analysis, in shift-invariant linear filtering and in many aspects of combinatorics and harmonic analysis. Suppose $X = \{X_t\}$ is a *stationary* process with $\mathrm{E}(X_t) = 0$ and $\mathrm{E}(X_t^2) < \infty$. The *autocovariance function* $\gamma_X(\cdot)$ and the *autocovariance matrix* $\Sigma_n(X)$ of order n are defined as:

$$\gamma_X(k) = cov(X_0, X_k), \; k = 0, 1, \ldots \; \text{ and } \; \Sigma_n(X) = ((\gamma_X(i-j)))_{1 \leq i, j \leq n}.$$

It may be noted that $\Sigma_n(X)$ is a Toeplitz matrix.

The *sample autocovariance matrix* is the usual *non-negative definite* estimate of $\Sigma_n(X)$ and equals

$$\Gamma_n(X) = ((\hat{\gamma}_X(i-j)))_{1 \le i,j \le n} \ \text{ where } \ \hat{\gamma}_X(k) = n^{-1} \sum_{i=1}^{n-|k|} X_i X_{i+|k|}. \quad (2.10)$$

$\Gamma_n(X)$ is a random Toeplitz matrix with a triangular input sequence. These two matrices appear frequently in time series analysis.

In this chapter we will deal with the simplest random symmetric Toeplitz matrix $T_n^{(s)}$ with an i.i.d. input sequence which is defined as

$$T_n^{(s)} = \begin{bmatrix} x_0 & x_1 & x_2 & \cdots & x_{n-2} & x_{n-1} \\ x_1 & x_0 & x_1 & \cdots & x_{n-3} & x_{n-2} \\ x_2 & x_1 & x_0 & \cdots & x_{n-4} & x_{n-3} \\ & & \vdots & & & \\ x_{n-1} & x_{n-2} & x_{n-3} & \cdots & x_1 & x_0 \end{bmatrix}.$$

Figure 2.3 gives the simulated ESD of $n^{-1/2} T_n^{(s)}$. Note that the estimated

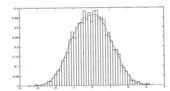 

FIGURE 2.3
Histogram and smoothed ESD of $T_n^{(s)}$ for $n = 400$, 15 replications. Standardized triangular($-1, 1$) (*left*) and Bernoulli(0.5) (*right*) input.

density looks very much like the normal density. However, as we shall see shortly, the LSD is not Gaussian. The question of the existence of the LSD was raised in Bai (1999)[5]. The existence of LSD was established by Bryc, Dembo and Jiang (2006)[34] and Hammond and Miller (2005)[63] when the input sequence is i.i.d. with finite variance. Sen and Virag (2011)[93] proved that the LSD is absolutely continuous with respect to the Lebesgue measure and has a bounded density.

Extensions of the above results when the input sequence is an appropriate *linear process* were done in Bose and Sen (2008)[29] but we shall not discuss those results here. Basak, Bose and Sen (2011)[14] proved the existence of the LSD of $\Gamma_n(X)$. We shall see the details in Chapter 11. The LSD of

band Toeplitz and Hankel matrices have been discussed in Basak and Bose
(2010)[13], Kargin (2009)[69] and Liu and Wang (2009)[71]. We study these
banded matrices in Chapter 7. *Balanced* Hankel and Toeplitz matrices are
discussed in Chapter 6.

The symmetric Hankel matrix $H_n^{(s)}$ is defined as

$$
H_n^{(s)} = \begin{bmatrix}
x_0 & x_1 & x_2 & \cdots & x_{n-3} & x_{n-2} \\
x_1 & x_2 & x_3 & \cdots & x_{n-2} & x_{n-1} \\
x_2 & x_3 & x_4 & \cdots & x_{n-1} & x_n \\
& & & \vdots & & \\
x_{n-1} & x_n & x_{n+1} & \cdots & x_{2n-3} & x_{2n-2}
\end{bmatrix}.
$$

Figure 2.4 provides the simulated ESD of $n^{-1/2}H_n^{(s)}$.

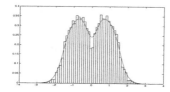 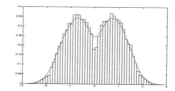

FIGURE 2.4
Histogram and smoothed ESD of the Hankel matrix for $n = 400$, 15 replications. Input is
$N(0,1)$ (*left*) and Exponential(1) (*right*).

Bryc, Dembo and Jiang (2006)[34] proved the existence of the LSD of
$n^{-1/2}H_n^{(s)}$ when the input sequence is i.i.d. with finite variance. The Hankel
LSD is evidently *bimodal*, and is absolutely continuous with respect to the
Lebesgue measure but there are no proofs for these yet.

2.2.1   Toeplitz matrix

Consider the symmetric Toeplitz matrix with i.i.d. input. It has the link function $L : \mathbb{Z}_+^2 \to \mathbb{Z}_+$ where $L(i,j) = |i-j|$. Moreover, $d = 1$, $k_n = n$ and $\alpha_n = n$
so that $k_n \alpha_n = n^2$ and hence Lemma 1.4.1 applies.

The following theorem was proved by Bryc, Dembo and Jiang (2006)[34]
and Hammond and Miller (2005)[63] for i.i.d. input.

Theorem 2.2.1. *If* $\{x_i\}$ *satisfies Assumption I, II or III, then a.s.,*
$\{F^{n^{-1/2}T_n^{(s)}}\}$ *converges to, say* $\mathcal{L}_T$, *which is symmetric about 0. The limit
distribution is universal and its* $(2k)$-*th moment can be expressed as a sum of
volumes of some subsets of the unit hypercube in* $k+1$ *dimension.*   ◇

Proof. Recall that we are concerned with only pair-matched words. Define the *slopes* as

$$s(i) = \pi(i) - \pi(i-1), \ 1 \le i \le 2k. \tag{2.11}$$

Now a match of the i-th and j-th vertices occurs if and only if

$$|s(i)| = |s(j)|.$$

This gives rise to two possibilities (both may hold simultaneously):

$$\text{either} \ \ s(i) + s(j) = 0 \ \ \text{or} \ \ s(i) - s(j) = 0. \tag{2.12}$$

The following lemma shows that the number of pair-matches where there is at least one pair of slopes with $s(i) - s(j) = 0$ is negligible. *This idea of slope and its properties will come in handy also in later chapters.*

Lemma 2.2.2. Let N be the number of pair-matched circuits of length $2k$ such that for at least one pair of matched vertices i, j, $s(i) - s(j) = 0$. Then, as $n \to \infty$, $n^{-(k+1)} N = \mathrm{O}(n^{-1}) \to 0$. ◇

Proof. Since $\#\mathcal{W}_{2k}(2) < \infty$, we are free to argue by fixing any pair-matched word. Then one way to count all possible circuits, that is all possible values of $\pi(i), 0 \le i \le 2k$, is to count the number of possible values of the $(2k+1)$ variables, $\{\pi(0), s(i), 1 \le s(i) \le 2k\}$. For each possible choice, $\pi(i)$ for all i are determined uniquely, one by one, from left to right.

Note that (2.12), when considered over all pairs i, j, gives rise to k linear constraints. Since there is at least one pair of slopes with $s(i) = s(j)$ (so they have the same sign), the circuit condition

$$0 = \pi(0) - \pi(2k) = \sum_{i=1}^{2k} s(i) \tag{2.13}$$

is *not* satisfied automatically and is an *additional* linear constraint. Thus, we have $(k+1)$ linear constraints on $\{\pi(0), s(i), 1 \le s(i) \le 2k\}$, that is, on $\{\pi(0), \pi(i), 1 \le i \le 2k\}$, and so the total number of circuits is at most of the order $\mathrm{O}(n^{(2k+1-(k+1))}) = \mathrm{O}(n^k)$. This proves the lemma. □

Now we get back to the proof of Theorem 2.2.1. Since the L-function satisfies Property B (see Definition 1.4.1) and Lemmata 1.4.1 and 1.4.3 are satisfied, it is enough to show that $\lim_{n \to \infty} \frac{1}{n^{1+k}} \#\Pi^*(w)$ exists for all $w \in \mathcal{W}_{2k}(2)$. Using Lemma 2.2.2, this limit, if it exists, is equal to $\lim_{n \to \infty} \frac{1}{n^{1+k}} \#\Pi^{**}(w)$, where

$$\Pi^{**}(w) = \{\pi : w[i] = w[j] \Rightarrow s(i) + s(j) = 0\}.$$

Let

$$v_i = \pi(i)/n \ \ \text{and} \ \ U_n = \{0, 1/n, 2/n, \ldots, (n-1)/n\}. \tag{2.14}$$

The number of elements in $\Pi^{**}(w)$ is the number of $\{v_i, 0 \leq i \leq 2k\}$ which satisfy

$$v_0 = v_{2k}, v_i \in U_n \text{ and } v_{i-1} - v_i + v_{j-1} - v_j = 0 \text{ if } w[i] = w[j]. \quad (2.15)$$

Recall that the set of generating vertices is given by

$$S = \{0\} \cup \{\min(i,j) : w[i] = w[j], i \neq j\} \quad (2.16)$$

and $\#S = k+1$. If $\{v_i\}$ satisfy the k equations in (2.15), then each v_i is a unique linear combination of $\{v_j\}$ where $j \in S$ and $j \leq i$. Let

$$v_S = \{v_i : i \in S\}. \quad (2.17)$$

Then we may write

$$v_i = L_i^T(v_S), \quad i = 0, 1, \ldots, 2k \quad (2.18)$$

where $L_i^T(\cdot)$ are linear functions of the vertices in v_S and depend on the word w. Clearly,

$$L_i^T(v_S) = v_i \text{ if } i \in S. \quad (2.19)$$

Also, summing the k equations $s(i) + s(j) = 0$ implies $L_{2k}^T(v_S) = v_0$. So

$$\#\Pi^{**}(w) = \#\{v_S : L_i^T(v_S) \in U_n \text{ for all } i = 0, 1, \ldots, 2k\}. \quad (2.20)$$

Since $\frac{1}{n^{1+k}} \#\Pi^{**}(w)$ is nothing but the $(k+1)$-dimensional Riemann sum for the function $\mathbb{I}(0 \leq L_i^T(v_S) \leq 1, \forall i \notin S \cup \{2k\})$ over $[0,1]^{k+1}$,

$$\lim_{n\to\infty} \frac{\#\Pi^{**}(w)}{n^{1+k}} = \underbrace{\int_0^1 \cdots \int_0^1}_{k+1} \mathbb{I}\left(0 \leq L_i^T(x) \leq 1, \forall i \notin S \cup \{2k\}\right) dx := p_T(w)$$

$$(2.21)$$

where $dx = \prod_{i \in S} dx_i$ denotes the $(k+1)$-dimensional Lebesgue measure. Clearly,

$$\beta_{2k}(\mathcal{L}_T) = \sum_{w \in \mathcal{W}_{2k}(2)} p_T(w). \qquad \square$$

Note that the above result does not provide the explicit value of $p_T(w)$ but tells us how to obtain it by computing certain integrals. More information on $p_T(w)$ and the limit distribution is given in Section 2.5.

2.2.2 Hankel matrix

Note that the link function for $H_n^{(s)}$ with i.i.d. input is $L(i,j) = i+j$. Moreover, $d = 1$, $k_n = 2n$ and $\alpha_n \leq n$, so that $k_n \alpha_n = O(n^2)$. Thus, Lemma 1.4.1 is applicable.

Theorem 2.2.3. If $\{x_i\}$ satisfies Assumption I, II or III, then almost surely $\{F^{n^{-1/2}H_n^{(s)}}\}$ converges to say, $\mathcal{L}_H$, which is symmetric about 0. The distribution is universal and its moments may be expressed as sums of volumes of certain subsets of the unit hypercubes. $\diamond$

Proof. If the LSD exists, let H denote a random variable which is distributed as this LSD. Then we know that provided the limit below exists,

$$\beta_{2k}(H) = \sum_{w \in \mathcal{W}_{2k}(2)} \lim_{n \to \infty} \frac{1}{n^{1+k}} \# \Pi^*(w) \tag{2.22}$$

where

$$\Pi^*(w) = \{\pi : w[i] = w[j] \Rightarrow \pi(i-1) + \pi(i) = \pi(j-1) + \pi(j)\}. \tag{2.23}$$

We can proceed as in the proof of the previous theorem to write v_i as a linear combination $L_i^H(v_S)$ of generating vertices for all $i \notin S$. As before, realizing $n^{-(k+1)}\#\Pi^*(w)$ as a Riemann sum we may conclude that it converges to the expression given below. We emphasize that unlike in the Toeplitz case, we do not automatically have $L_{2k}^H(v_S) = v_{2k}$ for every word w. Hence this restriction appears as an "additional" constraint. Combining all this, we have

$$\beta_{2k}(H) = \sum_{w \in \mathcal{W}_{2k}(2)} p_H(w)$$

where

$$p_H(w) = \underbrace{\int_0^1 \int_0^1 \cdots \int_0^1}_{k+1} \mathbb{I}(0 \leq L_i^H(x) \leq 1, \ \forall \ i \notin S \cup \{2k\})\mathbb{I}(x_0 = L_{2k}^H(x))dx$$

$$\tag{2.24}$$

where, as in the Toeplitz case, $dx = \prod_{i \in S} dx_i$ is the $(k+1)$-dimensional Lebesgue measure. $\square$

Of course, for some words, $L_{2k}^H(x) = x_0$ is a genuine extra linear constraint, and in such cases $p_H(w) = 0$. More information on $p_H(w)$ and the LSD is given in Section 2.5.

2.3 Reverse Circulant matrix

The $n \times n$ k-Circulant matrices will be studied in detail in Chapter 4. In these matrices after writing the first row, the next rows are sequentially written as the (right) circular shift by k positions of the previous row. The Reverse

Circulant matrix is a special case of the k-Circulant matrix—the only k-Circulant which is in general symmetric, and involves a right shift by $k = n-1$ positions. It is defined as

$$R_n^{(s)} = \begin{bmatrix} x_0 & x_1 & x_2 & \cdots & x_{n-2} & x_{n-1} \\ x_1 & x_2 & x_3 & \cdots & x_{n-1} & x_0 \\ x_2 & x_3 & x_4 & \cdots & x_0 & x_1 \\ & & & \vdots & & \\ x_{n-1} & x_0 & x_1 & \cdots & x_{n-3} & x_{n-2} \end{bmatrix}.$$

It has gained importance recently due to its connection with the concept of *half independence* which we shall briefly encounter in Chapters 9 and 10.

Note that the upper half of the anti-diagonal of this matrix is identical to the corresponding portion of the Hankel matrix. Figure 2.5 shows the simulated ESD of $n^{-1/2}R_n^{(s)}$.

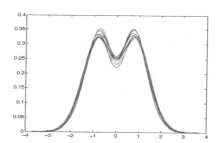

FIGURE 2.5
Smoothed ESD of $R_n^{(s)}$ for $n = 400$, 15 replications. Input is i.i.d. $N(0, 1)$. (Color figure appears following page 120)

The eigenvalues of $n^{-1/2}RC_n^{(s)}$ can be obtained as a special case of the formula for general k-Circulant matrices discussed in Chapter 4. See Bose and Mitra (2002)[27], for a direct derivation that uses *elementary transformations*. This formula is at the heart of obtaining the LSD in a closed form. We use the following notion from time series analysis.

Definition 2.3.1. The *periodogram* of any finite sequence $\{x_i\}, 0 \le i \le n-1$ is defined by

$$I_n(\omega_k) = \frac{1}{n}\Big|\sum_{t=0}^{n-1} x_t e^{-it\omega_k}\Big|^2, \quad k = 0, 1, \ldots, n-1, \tag{2.25}$$

where $\omega_k = 2\pi k/n$ are the *Fourier frequencies* and $i = \sqrt{-1}$.   ◇

The eigenvalues of $n^{-1/2} R_n^{(s)}$ are as follows:

$$
\begin{cases}
\lambda_0 & = n^{-1/2} \sum_{t=0}^{n-1} x_t, \\
\lambda_{n/2} & = n^{-1/2} \sum_{t=0}^{n-1} (-1)^t x_t, \quad \text{if } n \text{ is even,} \\
\lambda_k = -\lambda_{n-k} & = \sqrt{I_n(\omega_k)}, \ 1 \le k \le [\frac{n-1}{2}].
\end{cases}
$$

The form of these eigenvalues leads us *heuristically* to the LSD as follows. Suppose for the moment that $\{x_i\}$ is an i.i.d. sequence of standard Gaussian variables. As far as computing the LSD is concerned, we can ignore the two eigenvalues λ_0 and $\lambda_{n/2}$ and then consider only the positive eigenvalues. Let

$$
a_{t,n} = \sum_{l=0}^{n-1} x_l \cos\left(\frac{2\pi tl}{n}\right), \quad b_{t,n} = \sum_{l=0}^{n-1} x_l \sin\left(\frac{2\pi tl}{n}\right). \tag{2.26}
$$

Then for every n, $n^{-1/2} a_{t,n}, n^{-1/2} b_{t,n}$, $1 \le t \le (n/2) - 1$ are i.i.d. Gaussian with mean zero and variance $1/2$. Thus, $2I_n(\omega_k)$ are all independent chi-squared variables with two degrees of freedom. Since the empirical distribution F_n of i.i.d. variables with distribution F, converges to F itself, the ESD converges to the symmetrized distribution, say $\mathcal{L}_R$, of the symmetrized square root of $\sqrt{\chi_2^2/2}$ where χ_2^2 is a chi-squared random variable with two degrees of freedom. Let R denote such a random variable. Then its distribution has density and moments as

$$
\begin{aligned}
f_R(x) &= |x| \exp(-x^2), \quad -\infty < x < \infty, \tag{2.27} \\
\beta_{2k}(\mathcal{L}_R) &= k!, \quad \beta_{2k+1}(\mathcal{L}_R) = 0, \quad k = 0, 1, \dots. \tag{2.28}
\end{aligned}
$$

When the input sequence is not necessarily Gaussian but has a finite third moment, Bose and Mitra (2002)[27] first established the LSD of $n^{-1/2} R_n^{(s)}$ by using a normal approximation technique. On the other hand, clearly, this matrix is closely related to the Hankel matrix. Its link function is $L : \mathbb{Z}_+^2 \to \mathbb{Z}_+$ where $L(i, j) = (i + j - 2) \mod n$. Moreover $d = 1$, $k_n = n$ and $\alpha_n = n$ and hence $k_n \alpha_n = n^2$. Thus, Lemma 1.4.1 is applicable. As a consequence, its LSD may be established by arguments similar but simpler than those given earlier for the Hankel matrix. These arguments are due to Bose and Sen (2008)[29]. We shall take this approach below.

Just as the Catalan words play a crucial role for Wigner matrices, the following words play a crucial role for Reverse Circulant matrices.

Definition 2.3.2. (*Symmetric word*). A pair-matched word is *Symmetric* if each letter occurs once each in an odd and an even position. We shall denote the set of Symmetric words of length $2k$ by $\mathcal{S}_{2k}$. ◇

For example, $w = aabb$ is a Symmetric word while $w = abab$ is not. All Catalan words are Symmetric. A simple counting argument leads to the following lemma.

Lemma 2.3.1. The size of the set $\mathcal{S}_{2k}$ is $k!$.   ◇

Proof. The first letter is a. Choose the other (even) place for a from any of the k available positions. From left to right, move to the next empty slot and place the next letter b there. This position may be odd or even. Depending on that, there are remaining $(k-1)$ even or odd positions to place the other b. Proceeding in this manner we exhaust all choices and clearly the total number of choices is $k!$.   □

Theorem 2.3.2. If $\{x_i\}$ satisfies Assumption I, II or III, then almost surely, $\{F^{n-1/2}R_n^{(s)}\}$ converges to $\mathcal{L}_R$.   ◇

Proof. As in the previous proofs, without any loss, we can work under Assumption I and it is enough to show that

$$\lim_n E[\beta_{2k}(n^{-1/2}RC_n^{(s)})] = \sum_{w\in\mathcal{W}_{2k}(2)} \lim_n \frac{1}{n^{k+1}} \#\Pi^*(w) = k!.$$

Now due to Lemma 2.3.1, it is enough to verify the following two statements:

(i) If $w \in \mathcal{W}_{2k}(2) \cap \mathcal{S}_{2k}^c$ then $\lim_{n\to\infty} \frac{1}{n^{k+1}} \#\Pi^*(w) = 0$.

(ii) If $w \in \mathcal{S}_{2k}$ then for every choice of the generating vertices there is exactly one choice for the non-generating vertices and hence $\lim_{n\to\infty} \frac{1}{n^{k+1}} \#\Pi^*(w) = 1$.

Proof of (i) We use the notation from the proof of Theorem 2.2.1. Also let

$$t_i = v_i + v_{i-1}. \tag{2.29}$$

Note that the vertices i, j match if and only if

$$(\pi(i-1) + \pi(i) - 2) \mod n = (\pi(j-1) + \pi(j) - 2) \mod n.$$
$$\Leftrightarrow t_i - t_j = 0, 1 \text{ or } -1.$$

Since w is pair-matched, let $\{(i_s, j_s), 1 \leq s \leq k\}$ be such that $w[i_s] = w[j_s]$ and $j_s, 1 \leq s \leq k$, is in ascending order and $j_k = 2k$. So, $\#\Pi^*(w)$ equals

$$\sum_{(r_i, 1\leq i\leq k)\in\{0,\pm1\}^k} \#\Big\{(v_0, v_1, \ldots, v_{2k}) : v_0 = v_{2k}, \ v_i \in U_n, t_{i_s} - t_{j_s} = r_s\Big\}. \tag{2.30}$$

Let $r = (r_1, \ldots, r_k)$ be a sequence in $\{0, \pm1\}^k$ and let $v_S = \{v_i : i \in S\}$. Observe that $v_i = L_i^H(v_S) + a_i^{(r)}, i \notin S$ for some integer $a_i^{(r)}$. Arguing as in the Hankel case, we easily reach the following equality (compare with (2.24)), $\lim_{n\to\infty} \frac{1}{n^{k+1}} \#\Pi^*(w) =$

$$\sum_{r\in\{0,\pm1\}^k} \underbrace{\int_0^1 \cdots \int_0^1}_{k+1} \mathbb{I}(0 \leq L_i^H(v_S) + a_i^{(r)} \leq 1, i \notin S \cup \{2k\})$$

$$\times \mathbb{I}(v_0 = L_{2k}^H(v_S) + a_{2k}^{(r)})dv_S. \tag{2.31}$$

Now assume $\lim \frac{1}{n^{k+1}} \Pi^*(w) \neq 0$. Then one of the terms in the above sum must be non-zero. For the integral to be non-zero, we must have

$$v_0 = L_{2k}^H(v_S) + a_{2k}^{(r)}. \tag{2.32}$$

Now $(t_{i_s} - t_{j_s} - r_s) = 0$ for all $s, 1 \leq s \leq k$. Hence trivially, for any choice of $\{\alpha_s\}$,

$$v_{2k} = v_{2k} + \sum_{s=1}^{k} \alpha_s(t_{i_s} - t_{j_s} - r_s). \tag{2.33}$$

Let us choose integers $\{\alpha_s\}$ as follows: let $\alpha_k = 1$. Having fixed $\alpha_k, \alpha_{k-1}, \ldots, \alpha_{s+1}$, we choose α_s as follows:

(a) if $j_s + 1 \in \{i_m, j_m\}$ for some $m > s$, then set $\alpha_s = \pm \alpha_m$ according when $j_s + 1$ equals i_m or j_m,

(b) if there is no such m, choose α_s to be any integer.

By this choice of $\{\alpha_s\}$, we ensure that in $v_{2k} + \sum_{s=1}^{k} \alpha_s(t_{i_s} - t_{j_s} - r_s)$, the coefficient of each $v_i, i \notin S$ cancels out. Hence, we get

$$v_{2k} = v_{2k} + \sum_{s=1}^{k} \alpha_s(t_{i_s} - t_{j_s} - r_s) = L(v_S) + a, \quad \text{some linear combination.}$$

However, from (2.32) $v_0 = L_{2k}^H(v_S) + a_{2k}^{(r)}$. Hence, because only generating vertices are left in both the linear combinations,

$$v_{2k} + \sum_{s=1}^{k} \alpha_s(t_{i_s} - t_{j_s} - r_s) - v_0 = 0 \tag{2.34}$$

and thus the coefficient of each v_i in the left side has to be zero including the constant term.

Now consider the coefficients of t_i in (2.34). First, since $\alpha_k = 1$, the coefficient of t_{2k} is -1. On the other hand, the coefficient of v_{2k-1} is 0. Hence the coefficient of t_{2k-1} has to be $+1$.

Proceeding to the next step, we know that the coefficient of v_{2k-2} is 0. However, we have just observed that the coefficient of t_{2k-1} is $+1$. Hence the coefficient of t_{2k-2} must be -1. If we continue in this manner, in the expression (2.34) for all odd i, t_i must have coefficient $+1$ and for all even i, t_i must have coefficient -1.

Now suppose for some s, i_s and j_s both are odd or both are even. Then for any choice of α_s, t_{i_s} and t_{j_s} will have opposite signs in the expression (2.34). This contradicts the fact stated in the last paragraph. Hence either i_s is odd and j_s is even or i_s is even and j_s is odd. Since this happens for all $s, 1 \leq s \leq k$, w must be a Symmetric word, proving (i).

(ii) Let $w \in S_{2k}$. First fix the generating vertices. Then we determine the non-generating vertices from left to right. Consider $L(\pi(i-1), \pi(i)) = L(\pi(j-1), \pi(j))$ where $i < j$ and $\pi(i-1), \pi(i)$ and $\pi(j-1)$ have been determined. We rewrite it as

$$\pi(j) = Z + dn \text{ for some integer } d \text{ and } Z = \pi(i-1) + \pi(i) - \pi(j-1).$$

Clearly $\pi(j)$ can be determined uniquely from the above equation since $1 \leq \pi(j) \leq n$. Continuing, we obtain the whole circuit uniquely. Hence the first part of (ii) is proved.

As a consequence, for $w \in S_{2k}$, only one term in the sum (2.31) will be non-zero and that will be equal to 1. Since there are exactly $k!$ Symmetric words, (ii) is proved completely. This completes the proof of the theorem. $\square$

2.4 Symmetric Circulant and related matrices

If we consider the usual Circulant matrix and impose the restriction of symmetry on the matrix, we obtain the *Symmetric Circulant* matrix $C_n^{(s)}$

$$C_n^{(s)} = \begin{bmatrix} x_0 & x_1 & x_2 & \dots & x_2 & x_1 \\ x_1 & x_0 & x_1 & \dots & x_3 & x_2 \\ x_2 & x_1 & x_0 & \dots & x_2 & x_3 \\ & & & \vdots & & \\ x_1 & x_2 & x_3 & \dots & x_1 & x_0 \end{bmatrix}.$$

This may also be considered as a *Doubly Symmetric Toeplitz matrix*. Note that the link function of the Symmetric Circulant matrix is

$$L(i,j) = n/2 - |n/2 - |i-j||.$$

Figure 2.6 gives the simulated ESD of $n^{-1/2} C_n^{(s)}$. In a similar manner the *Doubly Symmetric Hankel matrix* DH_n is defined as

$$DH_n = \begin{bmatrix} x_0 & x_1 & x_2 & \dots & x_3 & x_2 & x_1 \\ x_1 & x_2 & x_3 & \dots & x_2 & x_1 & x_0 \\ x_2 & x_3 & x_4 & \dots & x_1 & x_0 & x_1 \\ & & & \vdots & & & \\ x_2 & x_1 & x_0 & \dots & x_5 & x_4 & x_3 \\ x_1 & x_0 & x_1 & \dots & x_4 & x_3 & x_2 \end{bmatrix}.$$

Its link function is

$$L(i,j) = n/2 - |n/2 - ((i+j-2) \mod n)|, \quad 1 \leq i, j \leq n. \tag{2.35}$$

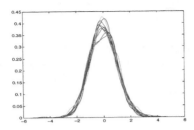

FIGURE 2.6
Smoothed ESD of $C_n^{(s)}$ for $n = 400$, 15 replications. Input is $N(0, 1)$. (Color figure appears following page 120)

Massey et al. (2007)[74] defined a (symmetric) matrix to be *palindromic* if its first row is a palindrome. In particular, the palindromic Toeplitz matrix PT_n is defined below. PH_n is defined similarly.

$$
PT_n = \begin{bmatrix}
x_0 & x_1 & x_2 & \cdots & x_2 & x_1 & x_0 \\
x_1 & x_0 & x_1 & \cdots & x_3 & x_2 & x_1 \\
x_2 & x_1 & x_0 & \cdots & x_4 & x_3 & x_2 \\
& & & \vdots & & & \\
x_1 & x_2 & x_3 & \cdots & x_1 & x_0 & x_1 \\
x_0 & x_1 & x_2 & \cdots & x_2 & x_1 & x_0
\end{bmatrix}.
$$

They established the Gaussian limit for $F^{n^{-1/2}PT_n}$ and $F^{n^{-1/2}PH_n}$ with the following approach: it is known from Hammond and Miller (2005)[63] that the LSD for $n^{-1/2}T_n^{(s)}$ is linked to solutions of some Diophantine equations with certain obstructions. Massey et al. [74] showed that for $n^{-1/2}PT_n$, these restrictions are absent, yielding the Gaussian limit. They also observed a direct relation between PH_n and PT_n to obtain the same conclusion for $n^{-1/2}PH_n$. We now state this result and present the moment method proof from Bose and Sen (2008)[29]. Note that for both the above matrices, $d = 1$, $\alpha_n = \mathrm{O}(n)$ and $k_n = \mathrm{O}(n)$. Hence, the conditions of Lemma 1.4.1 are satsified.

We need the well-known *interlacing inequality* from matrix algebra. See Bhatia (1997) [21].

Lemma 2.4.1. (Interlacing inequality) Suppose A is an $n \times n$ symmetric real matrix with eigenvalues $\lambda_n \leq \lambda_{n-1} \leq \cdots \leq \lambda_1$. Let B be the $(n-1) \times (n-1)$ principal submatrix of A with eigenvalues $\mu_{n-1} \leq \cdots \leq \mu_1$. Then

$$
\lambda_n \leq \mu_{n-1} \leq \lambda_{n-1} \leq \mu_{n-2} \leq \cdots \leq \lambda_2 \leq \mu_1 \leq \lambda_1.
$$

As a consequence

$$
\|F^A - F^B\|_\infty \leq \frac{1}{n}
$$

where $\|f\|_\infty = \sup_x |f(x)|$.

$\diamond$

Theorem 2.4.2. (a) For any input sequence, if the LSD of any one of $n^{-1/2}DH_n$ and $n^{-1/2}PH_n$ exists almost surely, then the other also exists almost surely and they are equal.

(b) For any input sequence, (i) if the LSD of any one of $n^{-1/2}PT_n$ and $n^{-1/2}SC_n$ exist almost surely, then the other also exists almost surely and they are equal and (ii) $(PT_n)^{2k} = (PH_n)^{2k}$ for every k.

(c) If the input sequence $\{x_i\}$ satisfies Assumption I, II or III, then the almost sure LSD of all four matrices $n^{-1/2}PT_n$, $n^{-1/2}C_n^{(s)}$, $n^{-1/2}PH_n$ and $n^{-1/2}DH_n$ are standard Gaussian N with $\beta_{2k}(N) = \frac{(2k)!}{2^k k!}$. ◇

Proof. Note that the above four matrices are closely related to each other. In particular,

- the $n \times n$ principal minor of DH_{n+3} is PH_n and

- the $n \times n$ principal minor of SC_{n+1} is PT_n.

Hence by Lemma 2.4.1, (a) and b(i) are immediate.

Since for all the matrices $k_n \to \infty$ and $k_n \alpha_n = O(n^{-2})$, it is enough to work under Assumption I.

Let J_n be the matrix with entries 1 in the main anti-diagonal and zero elsewhere. Then it is easy to see that

$$(PH_n)J_n = J_n(PH_n) = PT_n$$

and since $J_n^2 = I_n$,

$$(PT_n)^{2k} = (PH_n)^{2k}.$$

This proves (b)(ii). Moreover, it also shows that all four matrices will have the same LSD.

It is thus enough to show that the even moments of the Symmetric Circulant converges to the Gaussian moments. Since $\#\mathcal{W}_{2k} = \frac{(2k)!}{2^k k!}$ it is enough to show that

$$\lim_{n \to \infty} \frac{1}{n^{k+1}} \#\Pi^*(w) = 1, \text{ for each } w \in \mathcal{W}_{2k}(2). \tag{2.36}$$

Let the slopes be defined as before by

$$s(l) = \pi(l) - \pi(l-1). \tag{2.37}$$

Vertices i, j, $i < j$, match if and only if

$$\begin{aligned} |n/2 - |s_i|| &= |n/2 - |s_j|| \\ \Leftrightarrow s(i) - s(j) &= 0, \pm n \quad \text{OR} \quad s(i) + s(j) = 0, \pm n. \end{aligned}$$

That is six possibilities in all. We first show that the first three possibilities do not contribute asymptotically.

Lemma 2.4.3. Fix a pair-matched word w with $|w| = k$. Let N be the number of pair-matched circuits of w which have at least one pair $i < j$, $s(i) - s(j) = 0, \pm n$. Then, as $n \to \infty$, $N = O(n^k)$ and hence $n^{-(k+1)}N \to 0$. ◇

Proof. Let $(i_1, j_1), (i_2, j_2), \ldots, (i_k, j_k)$ denote the pair-partition corresponding to the word w, i.e., $w[i_l] = w[j_l], 1 \leq l \leq k$. Suppose, without loss of generality, $s(i_k) - s(j_k) = 0, \pm n$. Clearly a circuit π becomes completely specified if we know $\pi(0)$ and all the $\{s(i)\}$.

As already observed, if we fix some value for $s(i_l)$, there are at most six options for $s(j_l)$. We may choose the values of $\pi(0), s(i_1), s(i_2), \ldots, s(i_{k-1})$ in $O(n^k)$ ways and then we may choose values of $s(j_1), s(j_2), \ldots, s(j_{k-1})$ in $O(6^k)$ ways. For any such choice, from the sum restriction $\sum_{i=1}^{2k} s(i) = \pi(2k) - \pi(0) = 0$, we know $s(i_k) + s(j_k)$. On the other hand, by hypothesis, $s(i_k) - s(j_k) = 0, +n, -n$. Thus, the pair $(s(i_k), s(j_k))$ has 6 possibilities. Thus, there are at most $O(n^k)$ circuits with the given restrictions and the proof of the lemma is complete. □

Now we continue with the proof of Theorem 2.4.2. Due to the above lemma, it now remains to show that with

$$\Pi'(w) = \{\pi : \pi \text{ is a circuit }, w[i] = w[j] \Rightarrow s_i + s_j = 0, \pm n\}, \quad (2.38)$$

$$\lim_{n \to \infty} \frac{1}{n^{k+1}} \#\Pi'(w) = 1. \quad (2.39)$$

Suppose for some $i < j, s_i + s_j = 0, \pm n$. If we know the circuit up to position $(j-1)$ then $\pi(j)$ has to take one of the values $A - n, A, A + n$ where $A = \pi(j-1) - \pi(i) + \pi(i-1)$. Noting that $-(n-2) \leq A \leq (2n-1)$, exactly one of the three values will fall between 1 and n and be a valid choice for $\pi(j)$. Thus, we first choose the generating vertices arbitrarily, then the non-generating vertices are determined, from left to right uniquely, so that $s(i) + s(j) = 0, \pm n$. This automatically yields $\pi(0) = \pi(2k)$ as follows:

$$\pi(2k) - \pi(0) = \sum_{i=1}^{2k} s_i = dn \text{ for some } d \in \mathbb{Z}. \quad (2.40)$$

But since $|\pi(2k) - \pi(0)| \leq n - 1$, we must have $d = 0$. Thus, $\#\Pi'(w) = n^{k+1}$ and hence $\lim_{n\to\infty} \frac{1}{n^{k+1}} \#\Pi'(w) = 1$, proving the theorem. □

2.5 Additional properties of the LSDs

2.5.1 Moments of the Toeplitz and Hankel LSDs

Tables 2.1 and 2.2 provide a full list of words w up to order six with their type, corresponding range of integration, and the value of $p(w)$ for Toeplitz and Hankel matrices. The second column shows the role of Catalan and Symmetric words. Here I denotes the unit interval $[0, 1]$. The reader is invited to identify the linear functions in the formulae (2.21) and (2.24) and then perform the necessary integration to verify these values.

TABLE 2.1

$\beta_4(T) = 8/3$ and $\beta_4(H) = 2$.

Word	Type	T Range	T Vol	H Range (HR)	H Vol
abba	Cat	I^3	1	$HR = I^3$	1
abab	asym	$v_0 - v_1 + v_2 \in I$	$\frac{2}{3}$	HR $\subset$ hyperplane in $\mathbb{R}^3$	0
aabb	Cat	I^3	1	$HR = I^3$	1

2.5.2   Contribution of words and comparison of LSDs

Information on the contribution of each word to the moments of W, T, H and R and their interrelation is presented in the next theorem.

Theorem 2.5.1. (a) For any word w,

$$0 \le p_W(w) \le p_T(w),\ p_H(w),\ p_R(w) \le 1.$$

(b) For $w \in \mathcal{C}_{2k}$,

$$p_W(w) = p_T(w) = p_H(w) = p_R(w) = p_N(w) = 1.$$

(c) If $w \notin \mathcal{S}_{2k}$, then

$$p_H(w) = p_R(w) = 0.$$

(d)

$$\beta_{2k}(W) = \frac{(2k)!}{k!(k+1)!} \le \beta_{2k}(T) \le \beta_{2k}(N) = \frac{(2k)!}{k!2^k},$$

$$\beta_{2k}(W) = \frac{(2k)!}{k!(k+1)!} \le \beta_{2k}(H) \le \beta_{2k}(R) = k! \le \beta_{2k}(N).$$

(e) If $w \in \mathcal{S}_{2k}$ then

$$p_T(w) = p_H(w) \ \text{ and } \ p_R(w) = 1.$$

(f) For each k,

$$\beta_{2k}(T) \ge \beta_{2k}(H).$$

$\diamond$

Proof. We prove only the parts that have not been established yet. By the integral representations (2.21) and (2.24) of p_T and p_H, (a) is clear.

(b) Suppose $w \in \mathcal{C}_{2k}$ and let S be its set of generating vertices. Note that it is enough to show, for both Toeplitz and Hankel, that for each $j \notin S, \exists\, i \in S, i < j$ such that $v_j = v_i$ with $v_{2k} = v_0$.

To prove the above assertion, we apply induction on $|w|$. So assume that

TABLE 2.2

$\beta_6(T) = 11$ and $\beta_6(H) = 11/2$.

Word	Type	T Range	T Vol	H Range	H Vol
abccba	Cat	I^4	1	I^4	1
abccab	asym	$v_0 - v_1 + v_2 \in I$	2/3	contained in a hyperplane in $\mathbb{R}^4$	0
abcbca	asym	$v_1 - v_2 + v_3 \in I$	2/3	as above	0
abcbac	asym	$v_1 - v_2 + v_3 \in I$ $v_0 - v_2 + v_3 \in I$	1/2	as above	0
abcacb	asym	$v_0 - v_1 + v_3 \in I$ $v_0 - v_1 + v_2 \in I$	1/2	as above	0
abcabc	sym, not Cat	$v_0 - v_1 + v_3 \in I$ $v_0 - v_2 + v_3 \in I$	1/2	$v_0 + v_1 - v_3 \in I$ $-v_0 + v_2 + v_3 \in I$	1/2
abbcca	Cat	I^4	1	I^4	1
abbcac	asym	$v_0 - v_1 + v_3 \in I$	2/3	contained in a hyperplane in $\mathbb{R}^4$	0
abbacc	Cat	I^4	1	I^4	1
abaccb	asym	$v_0 - v_1 + v_2 \in I$	2/3	contained in a hyperplane in $\mathbb{R}^4$	0
abacbc	asym	$v_0 - v_1 + v_2 \in I$ $v_1 - v_2 + v_3 \in I$	1/2	as above	0
ababcc	asym	$v_0 - v_1 + v_2 \in I$	2/3	as above	0
aabccb	Cat	I^4	1	I^4	1
aabcbc	asym	$v_0 - v_2 + v_3 \in I$	2/3	contained in a hyperplane in $\mathbb{R}^4$	0
aabbcc	Cat	I^4	1	I^4	1

the above claim holds for all words w' such that $|w'| < |w|$. Let i and $i + 1$ be the positions of the first occurrence of a double letter xx in w. It implies that $i - 1, i \in S$ and $i + 1 \notin S$. Recall the notation $v_i = \pi(i)/n$. Note that, for both Toeplitz and Hankel, the restriction due to the presence of this double letter xx translates into the following: $v_{i-1} = v_{i+1}$ and at the same time v_i is arbitrary (between 0 and 1).

Clearly w can be written as $w = w_1 xx w_2$ where the word $w' := w_1 w_2 \in \mathcal{C}_{2(k-1)}$. Further, note that $\mathbf{v} = (v_1, v_2, \ldots, v_{i-1}, v_{i+2}, \ldots, v_{2k})$ satisfies the $(k - 1)$ equations imposed by w'. Moreover, if v_s in $\mathbf{v}$ is a generating vertex in w', it is also a generating vertex of w (of course w has an extra generating vertex v_i). Since $|w'| < |w|$, by assumption the assertion holds for w'. It is now easy to see that it also holds for w.

(c) We will check that $v_0 = L_{2k}^H(v_S)$ iff $w \in \mathcal{S}_{2k}$. Thus, if $w \notin \mathcal{S}_{2k}$, $p_H(w)$ will be the integral of a k-dimensional object in a $(k + 1)$-dimensional space, and hence will equal zero, proving (c).

Let $t_i = v_{i-1} + v_i$. Since $w \in \mathcal{W}_{2k}(2)$, let $\{(i_s, j_s), s \geq 1\}$ be such that

$w[i_l] = w[j_l]$ and where $j_s, 1 \leq s \leq k$ is written in ascending order. Then $j_k = 2k$ and

$$t_{i_l} - t_{j_l} = 0, \ 1 \leq l \leq k. \tag{2.41}$$

If w is a Symmetric word,

$$v_0 - v_{2k} = (t_1 + t_3 + \ldots + t_{2k-1}) - (t_2 + t_4 + \ldots + t_{2k}) = \sum_{s=1}^{k} \alpha_s(v_{i_s} - v_{j_s}) = 0$$

for suitable $\alpha_s \in \{-1, 1\}$ and hence $v_0 = L_{2k}^{H}(v_S)$.

To show the converse, let $v_0 = L_{2k}^{H}(v_S)$. Now follow the arguments given in the proof of Theorem 2.3.2 to conclude that w must be a Symmetric word.

(d) The lower bounds follow from (a). As we have seen already, for each word w, $0 \leq p_T(w)$, $p_T(w) \leq 1$ and hence $\beta_{2k}(T) \leq \frac{(2k)!}{k!2^k}$.

Since $\#S_{2k} = k!$, $\beta_{2k}(H) \leq k!$, and (d) is proved completely.

(e) First consider $p_T(w)$. As pointed out earlier (see proof of Theorem 2.2.1), $p_T(w) = \lim_{n \to \infty} \frac{1}{n^{1+k}} \#\Pi^{**}(w)$. Transforming $v_i \mapsto y_i = v_i - 1/2$, we see that $\#\Pi^{**}(w)$ is equal to the size of set

$$\{(y_0, \ldots, y_{2k}) : y_0 = y_{2k}, \ y_i \in \{-1/2, -1/2 + 1/n, \ldots, -1/2 + (n-1)/n\}$$
$$\text{and } y_{i-1} - y_i + y_{j-1} - y_j = 0 \text{ if } w[i] = w[j]\}$$

and for any fixed Symmetric word w,

$$p_T(w) = \int_{-1/2}^{1/2} \int_{-1/2}^{1/2} \cdots \int_{-1/2}^{1/2} \mathbb{I}(-1/2 \leq L_i^{T}(y_S) \leq 1/2, \ \forall \ i \notin S) dy_S. \tag{2.42}$$

Similarly,

$$p_H(w) = \int_{-1/2}^{1/2} \int_{-1/2}^{1/2} \cdots \int_{-1/2}^{1/2} \mathbb{I}(-1/2 \leq L_i^{H}(y_S) \leq 1/2, \ \forall \ i \notin S) dy_S. \tag{2.43}$$

Note that the functions L_i^{T} and L_i^{H} are determined by the k equations as imposed by the word restriction. Let

$$p_i = y_{i-1} + y_i \ \text{ and } \ q_i = y_{i-1} - y_i.$$

Suppose $(y_0, y_1, \ldots, y_{2k})$ satisfies a Hankel restriction, i.e., $p_i - p_j = 0$ if $w[i] = w[j]$. Now we make a transformation

$$(y_0, y_1, y_2, y_3, \ldots, y_{2k}) \mapsto (y_0, -y_1, y_2, -y_3, \ldots, y_{2k}).$$

This transformation does not change the value of the integral in $p_H(w)$. Under this volume-preserving transformation,

$$(p_1, p_2, p_3, p_4, \ldots, p_{2k}) \mapsto (q_1, -q_2, q_3, -q_4, \ldots, -q_{2k}).$$

Now a typical equation in the Hankel case looks like $p_r = p_s$ if $w[r] = w[s]$ before transformation. Since w is a Symmetric word, one of r and s will be odd and the other will be even. So, under the transformation the equation becomes

$$(-1)^{r-1}q_r = (-1)^{s-1}q_s \quad \text{or,} \quad q_r + q_s = 0.$$

But this is nothing but the corresponding Toeplitz equation for the word w. Hence, $p_H(w) = p_T(w)$. It is easy to see that the total number of eligible circuits is of exact order n^{k+1}. This proves (e).

(f) This now follows trivially from (b), (c) and (e). □

Table 2.3 provides a summary of the moments for the different LSDs derived so far.

TABLE 2.3
Words and moments for symmetric X.

MATRIX	w Cat	w Symmetric, not Cat	Other w	β_{2k} or LSD
$C_n^{(s)}$	1	1	1	$\frac{(2k)!}{2^k k!}$, $N(0,1)$
PT_n	1	1	1	as above
PH_n	1	1	1	as above
DH_n	1	1	1	as above
$R_n^{(s)}$	1	1	0	$k!$, $\mathcal{L}_R$
$T_n^{(s)}$	1	$0 < p_T(w) < 1$	$0 < p_T(w) < 1$	$\frac{(2k)!}{k!(k+1)!} \le \beta_{2k} \le \frac{(2k)!}{k!2^k}$
$H_n^{(s)}$	1	$0 < p_H(w) = p_T(w) < 1$	0	$\frac{(2k)!}{k!(k+1)!} \le \beta_{2k} \le k!$
$W_n^{(s)}$	1	0	0	$\frac{(2k)!}{k!(k+1)!}$, $\mathcal{L}_W$

2.5.3 Unbounded support of the Toeplitz and Hankel LSDs

Not many detailed properties of the limit distributions T and H are known. In Bryc, Dembo and Jiang (2006)[34], the unboundedness of LSD has been proved, separately for T and H. Now with Theorem 2.5.1 in our hand we can give a short combined proof of this fact as follows.

Theorem 2.5.2.

$$\liminf_{k \to \infty} \beta_{2k}^{1/k}(T) = \infty \tag{2.44}$$

and

$$\liminf_{k \to \infty} \beta_{2k}^{1/k}(H) = \infty. \tag{2.45}$$

Hence both T and H have unbounded support. ◇

Proof. The basic approach of the proof is borrowed from Bryc, Dembo and Jiang (2006)[34]. Recall relation (2.21). Let $U_0, U_1, \ldots, U_k$ be i.i.d. uniform random variables on the interval $[0, 1]$. Then for any pair-matched word w of length $2k$, not necessarily Symmetric, we can write

$$p_T(w) = \mathrm{P}\Big(\bigcap_{i=1}^{k} \{\sum_{j=0}^{k} \eta_{ij} U_j \in [0, 1]\}\Big) \tag{2.46}$$

for some suitable $\eta_{ij} \in \{-1, 0, 1\}$ with $\sum_j \eta_{ij} = 1$.

Let

$$S_i = \sum_{j=0}^{k} \eta_{ij} Y_j, \quad \text{where} \quad Y_j := U_j - 1/2 \text{ are i.i.d. } U(-1/2, 1/2).$$

Fix $\epsilon > 0$ small. As in Bryc, Dembo and Jiang (2006)[34], define the event

$$A = \bigcap_j \Big(|Y_j| \le \frac{1}{2\epsilon(k+1)}\Big). \tag{2.47}$$

Note that conditional on A, $Z_j := \epsilon(k+1)Y_j$ are again i.i.d. $U(-1/2, 1/2)$. Hence,

$$
\begin{aligned}
\mathrm{P}\big(|S_i| > 1/2 | A\big) &= \mathrm{P}\big(|\sum_{j=0}^{k} \eta_{ij} Y_j| > \epsilon(k+1)/2\big) \\
&= 2\mathrm{P}\big(\sum_{j=0}^{k} Y_j > \epsilon(k+1)/2\big).
\end{aligned}
$$

Let T_i be the number of $\{\eta_{ij} \ne 0, 1 \le j \le k+1\}$. Note that $T_i \le k+1$. Then by Markov's inequality, the above expression is bounded above by

$$2e^{-\epsilon^2(k+1)/2}[\mathrm{E}(e^{\epsilon Y_1})]^{T_i} = 2e^{-\epsilon^2(k+1)/2}\Big(\frac{e^{\epsilon/2} - e^{-\epsilon/2}}{\epsilon}\Big)^{T_i}.$$

Noting that for every $x > 0$, $\frac{e^{x/2} - e^{-x/2}}{2x} \le e^{x^2/2}$, and hence for every i,

$$\mathrm{P}\big(|S_i| > 1/2 | A\big) \le 2\exp\big(-\epsilon^2(k+1)/2 + \epsilon^2 T_i/4\big) \le 2e^{-\epsilon^2(k+1)/4} \le 1/2$$

for all large k.

Hence it follows that for all large k,

$$p_T(w) \leq \frac{1}{2}\mathrm{P}(A) = \frac{1}{2}(\epsilon(k+1))^{-(k+1)}.$$

Also recall from Theorem 2.5.1 that for symmetric words $p_T(w) = p_H(w)$. Hence using the fact that there are $k!$ symmetric words,

$$\beta_{2k}(T), \beta_{2k}(H) \geq k! \frac{1}{2}(\epsilon(k+1))^{-(k+1)} \geq (3\epsilon)^{-k}. \tag{2.48}$$

As a consequence, $\beta_{2k}(T)^{1/(2k)}, \beta_{2k}(T)^{1/(2k)} \to \infty$ as $k \to \infty$ and this proves the result. □

Hammond and Miller (2005)[63]) have shown that the *ratio* of the (even) moments of the LSD to the standard Gaussian moments tend to 0 as the order increases.

2.5.4 Non-unimodality of the Hankel LSD

It is clear from the simulation results displayed in Figure 2.4 that the distribution of H is *bimodal*. However, there is no known proof for this. Bryc, Dembo and Jiang (2006)[34] proved that the distribution is not unimodal.

Theorem 2.5.3. *The distribution of H is not unimodal.* ◇

Proof. Suppose, if possible, the distribution of H is unimodal. Note that the distribution of H is symmetric and also has all moments finite. Hence by Khinchin's theorem (Khinchin (1938)[111]) it follows that if $\phi(t) = \int e^{itx} H(dx)$ is the characteristic function of H, then

$$g(t) = \phi(t) + t\phi'(t) \tag{2.49}$$

must also be a characteristic function. As a consequence the even moments corresponding to $g(t)$ must be a non-negative definite sequence. On the other hand, it is easy to check that the $(2k)$-th derivative of g at 0 satisfies

$$g^{(2k)}(0) = (2k+1)\phi^{(2k)}(0). \tag{2.50}$$

Hence the moment sequence, $(2k+1)\beta_{2k}(H)$, must be a positive definite sequence. In other words the Hankel matrices

$$(((2(i+j) - 3)\beta_{2(i+j-2)}(H)))_{1 \leq i,j \leq n} \tag{2.51}$$

should be non-negative definite for all n. For $n = 3$, observe that this matrix equals

$$M = \begin{bmatrix} 1 & 3\beta_2 & 5\beta_4 \\ 3\beta_2 & 5\beta_4 & 7\beta_6 \\ 5\beta_4 & 7\beta_6 & 9\beta_8 \end{bmatrix}.$$

Now using the facts that $b_2 = \beta_2(H) = 1, b_4 = \beta_4(H) = 2, b_6 = \beta_6(H) = 11/2$ and $b_8 = \beta_8 = 281/15$, it follows that $det(M) = -73/20$. This completes the proof. □

2.5.5   Density of the Toeplitz LSD

The simulations for T and H also provide evidence of the existence of their density. However, the moment method is not sophisticated enough to yield results on the density.

Sen and Virag (2011)[93] have shown that the distribution of T is *absolutely continuous* with respect to the Lebesgue measure and the density is bounded. Their proof involves working with the Stieltjes transform, which unfortunately we shall not discuss in this book. The existence of the density has *not* yet been proved for the Hankel limit H.

2.5.6   Pyramidal multiplicativity

As we have mentioned, not much is known about the LSDs $\mathcal{L}_T$ and $\mathcal{L}_H$. Some asymptotic bounds on the moments have been derived in Hammond and Miller (2005)[63]. Attempts by different researchers to discover patterns in the sequence of moments and to compute them for Toeplitz and Hankel limits, have largely failed so far.

However, the following is worth mentioning. We need a definition from Bożejko and Speicher (1996)[32] which was later used by Bryc, Dembo and Jiang (2006)[34] to compute the LSD of Markov matrices. A function $f : \mathcal{D} \subset \cup_k W_k(2) \to \mathbb{R}$ is called *pyramidally multiplicative*, if for every $w \in \mathcal{D}$ of the form $w = x w_1 y$ where w_1 is pair-matched, we have

$$f(w) = f(w_1) f(xy).$$

An easy example of this is the Wigner limit $p(w)$ which is pyramidally multiplicative on the entire class of pair-matched words. This is because $p(w) = 1$ if w is Catalan and $p(w)$ is zero if it is non-Catalan. Similarly $p(w)$ is trivially pyramidally multiplicative for the Symmetric Circulant limit.

On the other hand $p(w)$ is pyramidally multiplicative on the set of all Catalan words for Toeplitz, Hankel, Reverse Circulant. But $p(w)$ is not pyramidally multiplicative for the Toeplitz or the Hankel limit on the full set. For example, consider the Toeplitz matrix and the word $w = abab cdcd = w_1 w_2$ where $abab = w_1$ and $cdcd = w_2$. It is known that $p(w_1) = 2/3 = p(w_2)$ but $p(w) = 9/20 \neq 4/9$ and hence pyramidal multiplicativity is absent.

However, this multiplicativity appears in a restricted way. We have already observed that $p(w)$ is pyramidally multiplicative on the sub-class of Catalan words. Now we claim that the multiplicativity holds whenever w_1 is Catalan. To see this, first consider the Toeplitz limit and consider a word of length $2k$ of the form $x w_1 y$ where w_1 is Catalan of reduced length. Consider a typical double letter $w[i] = w[i + 1]$ in w_1. First note that $p(w) = 1$. Now we know that the value $p(w)$ is identified by the system of k equations in $2k + 1$ unknowns $x_i, 0 \leq i \leq 2k$, one for each pair-match. In particular, the pair-match from the double letter gives rise to the equation $x_{i-1} - x_i + x_{i+1} - x_i = 0$, implying $x_{i-1} = x_{i+1}$. Now consider the reduced word $w' = a \cdots cd \cdots z$ after dropping w_1 and relabel the unknowns as $y_0 = x_0, \ldots, y_{i-1} = x_{i-1}, y_i = x_{i+2}, \ldots, y_{2k-1} = x_{2k+1}$ and consider the $(k-1)$ equations that still exist (one equation $x_{i-1} = x_{i+1}$ is dropped). But this is the same as the equations for the reduced pair-matched word of length $2(k-1)$. Inductively, one can again drop a double letter and eventually the entire w_1 is dropped. So we have $p(w) = p(w') = p(w_1)p(w')$ since $p(w_1) = 1$. The same argument holds for Hankel matrices too.

2.6   Exercises

1. Derive the formula (2.1).

2. Simulate the ESD of matrices studied in this chapter with different input sequences. Also see what happens if you use the input sequence as the i.i.d. standard Cauchy variables. Try different normalizations.

3. Show by integration that for every positive integer k,
$$\int_2^2 s^{2k} p_W(s) = \int_{-2}^2 s^{2k} \frac{1}{2\pi}\sqrt{4-s^2}ds = \frac{(2k)!}{k!(k+1)!}.$$

4. Show directly that $\beta_{2k} = \frac{(2k)!}{k!(k+1)!}$ satisfies Carleman's/Riesz's condition.

5. Let D_k be the number of Catalan words of length $2k$ and define $D_0 = 1$. Argue combinatorially to show that for all positive integers k,
$$D_k = D_1 D_{k-2} + \cdots + D_{k-1} D_0.$$
Use this to derive that for all positive integers k, $D_k = \frac{(2k)!}{k!(k+1)!}$.

6. Suppose U_1 and U_2 are i.i.d. random variables uniformly distributed on the interval $[0, 1]$. Show that $W = 2U_1 \cos(2\pi U_2)$ is distributed as the standard semi-circle law.

7. Is $p_T(w) > 0$ for every pair-matched word w?

8. Show that the Symmetric Reverse Circulant matrix $RC_n^{(s)}$ is the only k-Circulant matrix which is symmetric for arbitrary input sequences.

9. Show that for any three arbitrary Reverse Circulant matrices A, B, C of the same order, $ABC = CBA$. Further, show by example that they need not fully commute.

10. Show that arbitrary Symmetric Circulant matrices of the same order commute.

11. Show that the distribution of the symmetrized square root of $\sqrt{\chi_2^2/2}$, where χ_2^2 is a chi-squared random variable with two degrees of freedom, is $\mathcal{L}_R$.

12. Prove Lemma 2.3.1.

13. Suppose the input sequence $\{x_i\}$ of $RC_n^{(s)}$ is i.i.d. with $\mathrm{E}(x_i) = 0$ and $V(x_i) = 1$ and $\mathrm{E}|x_i|^3 < \infty$. Show by using normal approximation results (Berry-Esseen theorem) and the formula for eigenvalues that $F^{n^{-1/2}RC_n^{(s)}} \to R$ in probability.

14. Verify the information given in Tables 2.1 and 2.2.

15. Show that the 8th moment of $\mathcal{L}_H$ equals $281/15$.

16. Consider the tri-diagonal symmetric non-random Toeplitz matrix $T_{3,n}$ with t_0 on the main diagonal and t_1 on the next upper and lower diagonals. Show directly by using the moment method that the LSD of $T_{3,n}$ exists. Also find the limit.

3

Patterned XX' matrices

Let X_p be a sequence of $p \times n$ patterned random matrices whose link function does not necessarily satisfy the symmetry condition $L(i,j) = L(j,i)$. The goal of this chapter is to enlarge the scope of our approach and study the LSD of matrices of the form

$$A_p(X) = \frac{1}{n}X_pX_p'. \tag{3.1}$$

Note that the eigenvalues of A_p are the squared singular values of $X_p/\sqrt{n}$. The best example of such matrices is of course the (unadjusted) Sample variance-covariance matrix. This matrix is useful in multivariate statistical analysis. In particular if the rows of X_p are i.i.d. Gaussian, then this is called a Wishart matrix and the joint distribution of the eigenvalues is known (see Anderson (1984)[2]).

When all the entries of X_p are i.i.d. with mean zero and variance one, the LSD of A_p is the well-known Marčenko-Pastur law when $p/n \to y$, $y > 0$. This law is defined later.

When $y = 0$, then a different situation arises. Let

$$N_p(X) = \left(\frac{n}{p}\right)^{1/2}\left(\frac{1}{n}X_pX_p' - I_p\right) = \left(\frac{n}{p}\right)^{1/2}\left(A_p(X) - I_p\right) \tag{3.2}$$

where the entries of X_p are as above, but, in addition, let the entries have a finite fourth moment. Then the LSD of $N_p(X)$ is the semi-circle law.

One may then be curious if the LSD of A_p and N_p exist for other patterned matrices X_p. One motivation to investigate this, comes from wireless communication theory. Many results on the information-theoretic limits of various wireless communication channels make substantial use of asymptotic random matrix theory, as the size of the matrix increases. Tulino and Verdu (2004)[101] provide an extended survey of results and works in this area. Silverstein and Tulino (2006)[96] review some of the existing mathematical results that are relevant to the analysis of the properties of random matrices arising in wireless communication. Couillet and Debbah (2011)[45] is an excellent source for detailed information and results on the use of random matrix methods in wireless communication.

A typical wireless communication channel may be described by the linear vector memory-less channel

$$y = X_p\theta + \epsilon$$

where θ is the n-dimensional vector of the signal input, y is the p-dimensional vector of the signal output, and the p-dimensional vector ϵ is the additive noise. The matrix X_p, is a $p \times n$ random matrix, generally with complex entries.

Silverstein and Tulino (2006)[96] emphasize the use of the asymptotic distribution of the squared singular values of X_p under various assumptions where n and p tend to infinity while the *aspect ratio*, $p/n \to y$, $0 < y < \infty$. In their model, generally speaking, the *channel matrix* X_p may be viewed as $X_p = f(A_1, A_2, ..., A_k)$ where $\{A_i\}$ are some independent random (rectangular) matrices with complex entries, each having its own meaning in terms of the channel. In most of the cases they studied, some A_i have all i.i.d. entries, while the LSD of the other matrices $A_j A_j^*, j \neq i$ are assumed to exist. Then the LSD of $X_p X_p^*$ is computed in terms of the LSD of $A_j A_j^*, j \neq i$. Recall that for any matrix $M = ((m_{ij}))$, M^* is the matrix $M^* = ((\bar{m}_{ji}))$ where $\bar{x}$ denotes the complex conjugate of x.

Moreover, certain CDMA channels can be modeled as $X_p = CSA$ where C is a $p \times p$ random symmetric Toeplitz matrix, S is a matrix with i.i.d. complex entries independent of C, and A is an $n \times n$ deterministic diagonal matrix. See Tulino and Verdu (2004)[101, Chapter 3] for more information. One of the main theorems in Silverstein and Tulino (2006)[96] establishes the LSD of X_p when C is not necessarily Toeplitz, under the added assumption that the LSD of CC' exists. When C is random *symmetric* Topelitz, the existence of the LSD of CC' is immediate from Theorem 2.2.1 of Chapter 2. This also is a motivation to study the LSD of A_p and N_p in some generality where X_p is not necessarily square or symmetric.

In this chapter, we establish the existence of the LSD of matrices A_p and N_p, respectively, with $p \to \infty$ and $n = n(p) \to \infty$ and $p/n \to y$ when (i) $0 < y < \infty$ and (ii) $y = 0$.

As examples we show the existence of the LSD when X_p is taken to be the appropriate non-symmetric or symmetric Hankel, Toeplitz, Circulant and Reverse Circulant matrices. In particular, when $y = 0$, the limit of N_p for all these matrices coincide and is the same as the limit for the symmetric Toeplitz derived in the previous chapter. In other cases, the LSDs are different and no closed-form expressions are known. We demonstrate the nature of these limits through some simulation results.

3.1   A unified setup

Recall the main assumptions introduced in Chapter 1.

Assumption I $\{x_i, x_{ij}\}$ are i.i.d. and uniformly bounded with mean 0 and variance 1.

Assumption II $\{x_i, x_{ij}\}$ are i.i.d. with mean zero and variance 1.

Assumption III $\{x_i, x_{ij}\}$ are independent with mean zero and variance 1 and with uniformly bounded moments of all order.

Let $\{x_\alpha\}_{\alpha \in \mathcal{Z}}$ be an *input sequence* where $\mathcal{Z} = \mathbb{Z}$ or $\mathbb{Z}^2$. They will be required to satisfy one or more of the above assumptions.

Now the domain of the link function is different due to the presence of two parameters n and p as opposed to only n in the previous two chapters. Thus, a *link function* now takes the form:

$$L_p : \{1, 2, \ldots, p\} \times \{1, 2, \ldots, n = n(p)\} \to \mathcal{Z}.$$

As before, we will write L for L_p. Define the matrices

$$
\begin{aligned}
X_p &= ((x_{L_p(i,j)}))_{1 \leq i \leq p,\ 1 \leq j \leq n} \quad \text{and} \\
A_p &= A_p(X) = n^{-1} X_p X_p'.
\end{aligned}
\tag{3.3}
$$

We wish to investigate the LSD of A_p for different link functions as $p, n \to \infty$ in some fashion to be spelled out soon. Some examples of link functions that correspond to matrices of interest are as follows:

(i) Sample variance-covariance matrix: $L_p(i, j) = (i, j)$.

(ii) Non-symmetric Toeplitz matrix: $L_p(i, j) = i - j$.

(iii) Non-symmetric Hankel matrix: $L_p(i, j) = \mathrm{sgn}(i - j)(i + j)$ where

$$
\mathrm{sgn}(l) = \begin{cases} 1 & \text{if } l \geq 0, \\ -1 & \text{if } l < 0. \end{cases}
\tag{3.4}
$$

(iv) Non-symmetric Reverse Circulant: $L_p(i, j) = \mathrm{sgn}(i - j)\,[(i + j - 2) \mod n]$.

(v) Non-symmetric Circulant: $L_p(i, j) = (j - i) \mod n$.

We have natural extension of the concepts that were introduced in Chapter 1. The trace-moment formula is now

$$
\begin{aligned}
\beta_h(A_p) &= p^{-1} \operatorname{Tr} A_p^h \\
&= p^{-1} n^{-h} \sum_{1 \leq i_1, \ldots, i_h \leq n} x_{L_p(i_1, i_2)} x_{L_p(i_3, i_2)} \cdots x_{L_p(i_{2h-1}, i_{2h})} x_{L_p(i_1, i_{2h})}.
\end{aligned}
\tag{3.5}
$$

Any circuit now is always of even length, say $2h$, and vertices do not necessarily have the same range of values. Clearly

(i) $\pi(0) = \pi(2h)$,
(ii) $1 \leq \pi(2i) \leq p$, $\forall\, 0 \leq i \leq h$ and
(iii) $1 \leq \pi(2i - 1) \leq n$, $\forall\, 1 \leq i \leq h$.

Let

$$\xi_\pi(2i-1) = L(\pi(2i-2),\pi(2i-1)), \ 1 \le i \le h \ \text{ and}$$
$$\xi_\pi(2i) = L(\pi(2i),\pi(2i-1)), \ 1 \le i \le h.$$

Matched circuits, words, and generating vertices are defined in a natural way. For example, π is matched if given any i, there is at least one $j \ne i$ such that $\xi_\pi(i) = \xi_\pi(j)$. To distinguish between the two types of vertices, we make the following definition.

Definition 3.1.1. Vertices $\pi(2i), 0 \le i \le h$ will be called *even vertices* or *p-vertices*. Vertices $\pi(2i-1), 1 \le i \le h$ will be called *odd vertices* or *n-vertices*. An odd generating vertex $\pi(2i-1)$ is of *Type I* if $\pi(2i)$ is also generating. Otherwise it is of *Type II*. ◇

Define for any circuit π,

$$X_\pi = \prod_{i=1}^{h} x_{\xi_\pi(2i-1)} x_{\xi_\pi(2i)}. \tag{3.6}$$

Since the input sequence has mean zero and is independent, if π is non-matched, then $E(X_\pi) = 0$.

For any w, the two classes $\Pi(w)$ and $\Pi^*(w)$ are now defined as

$$\Pi(w) = \{\pi : w[i] = w[j] \Leftrightarrow \xi_\pi(i) = \xi_\pi(j), \ \forall i,j\},$$
$$\Pi^*(w) = \{\pi : w[i] = w[j] \Rightarrow \xi_\pi(i) = \xi_\pi(j) \ \forall i,j\}.$$

Note that

$$\#\{\xi_\pi(i) : 1 \le i \le 2h\} = |w|.$$

For any w, let $\widetilde{\Pi}(w)$ be the possibly larger class of circuits with the range $1 \le \pi(i) \le \max(p,n), 1 \le i \le 2k$. Likewise define $\widetilde{\Pi}^*(w)$.

The trace formula (3.5) now can be rewritten as

$$p^{-1}\operatorname{Tr} A_p^h = p^{-1}n^{-h} \sum_{\pi:\ \pi \text{ circuit}} X_\pi = p^{-1}n^{-h} \sum_w \sum_{\pi \in \Pi(w)} X_\pi,$$

where the first sum is taken over all words of length $(2h)$. When we take expectation, only the matched words survive, and hence

$$E\left[p^{-1}\operatorname{Tr} A_p^h\right] = p^{-1}n^{-h} \sum_{w \text{ matched}} \sum_{\pi \in \Pi(w)} E X_\pi.$$

We shall consider two different regimes, and the treatment and behavior of the limits of $A_p(X)$ and $N_p(X)$ respectively in the following two regimes are quite different:

Regime I. $p \to \infty, \ p/n \to y \in (0, \infty)$.

Regime II. $p \to \infty, n = n(p) \to \infty, \ p/n \to y = 0$.

The quantity y will be called the *aspect ratio*.

3.2   Aspect ratio $y \neq 0$

3.2.1   Preliminaries

In Regime I, we assume that $\{x_\alpha\}$ satisfy Assumption I, II or III stated in Chapter 1 and consider the spectral distribution of $A_p(X)$. As in Chapter 1, we will first prove a truncation result which shows that under appropriate assumption on the link function, without loss of generality, we may assume the input sequence to be uniformly bounded and work under Assumption I.

Here also under reasonable assumptions on the allowed pattern, *if* the limit of empirical moments exist, they automatically satisfy Carleman's/Riesz's condition and thus ensure the existence of the LSD. Specific patterned matrices are covered in the later sections.

Recall the definition of Mallow's metric given in (1.7). The following result is an extension of Lemma 1.3.2 to $X_p X_p'$ matrices. As before, a proof may be found in Bai and Silverstein (2006)[8] or Bai (1999)[5].

Lemma 3.2.1. Suppose A and B are $p \times n$ real matrices. Let $X = AA'$ and $Y = BB'$. Then

$$
\begin{aligned}
W_2^2(F^X, F^Y) &\leq \left(\frac{1}{p}\sum_{i=1}^{p} |\lambda_i(X) - \lambda_i(Y)|\right)^2 \\
&\leq \frac{2}{p^2}\operatorname{Tr}(X+Y)\operatorname{Tr}\left[(A-B)(A-B)'\right]. \qquad (3.7)
\end{aligned}
$$

$\diamond$

We shall use the following general assumptions on the link function.

Assumption A (Property B). There exists positive integer Δ such that for any $\alpha \in \mathcal{Z}$ and $p \geq 1$,

(i) $\#\{i : L_p(i,j) = \alpha\} \leq \Delta$ for all $j \in \{1, 2, \ldots, n\}$ and

(ii) $\#\{j : L_p(i,j) = \alpha\} \leq \Delta$ for all $i \in \{1, 2, \ldots, p\}$.

Assumption A is really the same Property B of (1.4.1), Chapter 1. As before, it stipulates that the number of times any fixed input variable appears in any given row or column remains bounded. Following is a variant of this assumption to be used in the later sections.

Assumption A' For any $\alpha \in \mathcal{Z}$ and $p \geq 1$, $\#\{i : L_p(i,j) = \alpha\} \leq 1$ for all $1 \leq j \leq n$.

Recall the definitions of k_n and α_n from Chapter 1. We now define the natural extensions of these sequences.

If L_p takes values in $\mathcal{Z} = \mathbb{Z}^2$, we have for *two* increasing sequences of integers a_{1p} and a_{2p},

$$\{\alpha : L_p(i,j) = \alpha, 1 \leq i \leq p, 1 \leq j \leq n\} \subset \{1, \dots a_{1p}\} \times \{1, \dots a_{2p}\}.$$

In that case define $k_p = a_{1p}a_{2p}$.

If L_p takes values in $\mathcal{Z} = \mathbb{Z}$, we have for *one* increasing sequence of integers a_{1p},

$$\{\alpha : L_p(i,j) = \alpha, 1 \leq i \leq p, 1 \leq j \leq n\} \subset \{1, \dots a_{1p}\}.$$

In that case define $k_p = a_{1p}$.

Next define

$$\alpha_p = \max_{\alpha \in \mathcal{Z}} \#(L_p^{-1}(\alpha)).$$

Assumption B $k_p \alpha_p = \mathrm{O}(np)$.

Assumption B is an extension of the condition $k_n^d \alpha_n = \mathrm{O}(n^2)$ that we have used several times earlier, for instance in Lemma 1.4.1. Clearly Assumption A' implies Assumption A(i), but Assumption B is not related to Assumptions A or A'. Consider the link function $L_p(i,j) = (1,1)$ if $i = j$ and $L_p(i,j) = (i,j)$ if $i \neq j$. This link function satisfies Assumption A(ii) and Assumption A' but not Assumption B. On the other hand, if $L_p(i,j) = 1$ for all (i,j), then it satisfies Assumption B but not Assumption A.

It is easy to verify that all the link functions listed earlier in Section 3.1 satisfy Assumptions A and B. It may also be noted that the symmetric Toeplitz link function does *not* satisfy Assumption A'.

Now we prove an extension of Lemma 1.4.1 and show that if the LSD exists under Assumption I, then the same LSD holds under Assumptions II or III. This fact is essentially proved in Bose, Gangopadhyay and Sen (2010)[25]. We provide a detailed proof for clarity and completeness. The proof is similar to the proof of Lemma 1.4.1 but now uses Lemma 3.2.1.

Theorem 3.2.2. Suppose $p \to \infty$, $\frac{p}{n} \to y \in (0,\infty)$ and Assumption B holds. Suppose for any input sequence which satisfies Assumption I, $F^{n^{-1}X_p X_p'}$ converges to some fixed non-random distribution G almost surely. Then the same limit continues to hold if the input sequence satisfies Assumption II or III. In particular, the LSD is universal.   ◇

Proof. For simplicity, we discuss only the case of Assumption II. The proof under Assumption III is left as an exercise. Without loss of generality we shall assume that $\mathcal{Z} = \mathbb{Z}$ and also write X for X_p. For $t > 0$, denote

$$\mu(t) = \mathrm{E}\left[x_0 \mathbb{I}(|x_0| > t)\right] = -\mathrm{E}\left[x_0 \mathbb{I}(|x_0| \leq t)\right]$$

and let

$$\sigma^2(t) = \mathrm{Var}(x_0 \mathbb{I}(|x_0| \leq t)) = \mathrm{E}\left[x_0^2 \mathbb{I}(|x_0| \leq t)\right] - \mu(t)^2.$$

Since $\mathrm{E}(x_0) = 0$ and $\mathrm{E}(x_0^2) = 1$, we have $\mu(t) \to 0$ and $\sigma(t) \to 1$ as $t \to \infty$ and $\sigma^2(t) \leq 1$. Define random variables

$$
\begin{aligned}
x_i^* &= \frac{x_i \mathbb{I}(|x_i| \leq t) + \mu(t)}{\sigma(t)} = \frac{x_i - \bar{x}_i}{\sigma(t)}, \\
\bar{x}_i &= x_i \mathbb{I}(|x_i| > t) - \mu(t) = x_i - \sigma(t) x_i^*.
\end{aligned}
$$

It is easy to see that $\mathrm{E}(\bar{x}_0^2) = 1 - \sigma^2(t) - \mu(t)^2 \to 0$ as t tends to ∞. Further, $\{x_i^*\}$ are i.i.d. bounded random variables with mean zero and variance one. Let us replace the entries $x_{L_p(i,j)}$ of the matrix X_p by the truncated version $x_{L_p(i,j)}^*$ (respectively $\bar{x}_{L_p(i,j)}$) and denote this matrix by Y (respectively $\bar{X}_p$). By triangle inequality and Lemma 3.2.1,

$$
\begin{aligned}
&W_2^2\left(F^{n^{-1}X_p X_p'}, F^{n^{-1}YY'}\right) \\
\leq\ & 2W_2^2\left(F^{n^{-1}X_p X_p'}, F^{n^{-1}\sigma(t)^2 YY'}\right) + 2W_2^2\left(F^{n^{-1}YY'}, F^{n^{-1}\sigma(t)^2 YY'}\right) \\
\leq\ & \frac{2}{p^2 n^2}\, \mathrm{Tr}(X_p X_p' + \sigma(t)^2 YY')\, \mathrm{Tr}(X_p - \sigma(t)Y)(X_p - \sigma(t)Y') \\
& + \frac{2}{p^2 n^2}\, \mathrm{Tr}(YY' + \sigma(t)^2 YY')\, \mathrm{Tr}(\sigma(t)Y - Y)(\sigma(t)Y - Y)'.
\end{aligned}
$$

To tackle the first term on the right side above,

$$
\begin{aligned}
\mathrm{Tr}(X_p X_p' + \sigma(t)^2 YY') &= \sum_{i=1}^{p}\sum_{k=1}^{n} x_{L_p(i,k)}^2 + \sigma(t)^2 \sum_{i=1}^{p}\sum_{k=1}^{n} x_{L_p(i,k)}^{*2} \\
&\leq \alpha_p k_p \left(\frac{\sum_{i=1}^{k_p} x_i^2}{k_p}\right) + \sum_{i=1}^{p}\sum_{k=1}^{n}\left(x_{L_p(i,k)} - \bar{x}_{L_p(i,k)}\right)^2 \\
&\leq \alpha_p k_p \left(\frac{\sum_{i=1}^{k_p} x_i^2}{k_p}\right) + \alpha_p k_p \left(\frac{\sum_{i=1}^{k_p} x_i^2 I\{|x_i| > t\}}{k_p}\right) \\
&\quad + |\mu(t)| \alpha_p k_p \left(\frac{\sum_{i=1}^{k_p} |x_i|}{k_p}\right) + \alpha_p k_p \left(\frac{\sum_{i=1}^{k_p} \bar{x}_i^2}{k_p}\right).
\end{aligned}
$$

Therefore using $\alpha_p k_p = \mathrm{O}(np)$ and the SLLN, we can see that

$(np)^{-1}(\text{Tr}(X_p X_p' + \sigma(t)^2 YY'))$ is bounded. Now,

$$\frac{1}{np}\left|\text{Tr}[(X_p - \sigma(t)Y)(X_p - \sigma(t)Y)']\right| = \frac{1}{np}\left|\text{Tr}(\bar{X}_p \bar{X}_p')\right|$$

$$\leq \frac{1}{np}\alpha_p k_p\Big(\frac{1}{k_p}\sum_{i=1}^{k_p}\bar{x}_i^2\Big)$$

which is bounded by $C\,\text{E}(\bar{x}_i^2)$ almost surely, for some constant C. Here we use the condition $\alpha_p k_p = O(np)$ and the SLLN for $\{\bar{x}_i^2\}$. Since $\text{E}(\bar{x}_i^2) \to 0$ as $t \to \infty$ we can make the right side tend to 0 almost surely, by first letting p tend to ∞, and then letting t tend to ∞. This takes care of the first term.

To tackle the second term, noting that for large t, $\sigma(t)$ is bounded above and also stays away from 0, for some chosen constant C,

$$T_2 = \frac{2}{n^2 p^2}\left[\text{Tr}\big((\sigma(t)^2)YY' + YY')\big)\,\text{Tr}\big((\sigma(t)Y - Y)(\sigma(t)Y - Y)'\big)\right]$$

$$= \frac{2}{n^2 p^2}(\sigma(t)^2 + 1)(\sigma(t) - 1)^2(\text{Tr}(YY'))^2$$

$$\leq \frac{2C}{n^2 p^2}\Big(\sum_{i=1}^{p}\sum_{k=1}^{n}x_{L_p(i,k)}^{*2}\Big)^2$$

$$\leq \frac{2C}{n^2 p^2}\Big[\alpha_p k_p\Big(\frac{\sum_{i=1}^{k_p}x_i^2}{k_p}\Big) + \alpha_p k_p\Big(\frac{\sum_{i=1}^{k_p}x_i^2 I\{|x_i| > t\}}{k_p}\Big)$$

$$+ |\mu(t)|\alpha_p k_p\Big(\frac{\sum_{i=1}^{k_p}|x_i|}{k_p}\Big) + \alpha_p k_p\Big(\frac{\sum_{i=1}^{k_p}\bar{x}_i^2}{k_p}\Big)\Big]^2 .$$

Again using the condition $\alpha_p k_p = O(np)$ and $\sigma(t) \to 1$ as $t \to \infty$, we get $T_2 \to 0$ almost surely. This completes the proof of the theorem. $\qquad\square$

The result below is inspired by Lemma 1.4.2 and Theorem 1.4.4. It is essentially taken from Bose, Gangopadhyay and Sen (2010)[25]. Recall that $\mathcal{W}_{2h}(2)$ is the set of all pair-matched words of length $2h$ (see (1.22)).

Theorem 3.2.3. Let $A_p = (1/n)X_p X_p'$ where the entries of X_p satisfy Assumption I and L_p satisfies Assumptions A and B. Let $p/n \to y$, $0 < y < \infty$. Then the following hold:

(a) If w is of length $2h$ with at least one edge of order ≥ 3, then
$$p^{-1}n^{-h}\sum_{\pi \in \Pi(w)}\text{E}\,X_\pi \to 0.$$

(b) For each $h \geq 1$, $\text{E}\left[p^{-1}\text{Tr}\,A_p^h\right] - \sum_{w \in \mathcal{W}_{2h}(2)}p^{-1}n^{-h}\#\Pi^*(w) \to 0.$

(c) For each $h \geq 1$, $p^{-1}\text{Tr}\,A_p^h - \text{E}\left[p^{-1}\text{Tr}\,A_p^h\right] \to 0$ almost surely.

(d) The sequence $\beta_h = \limsup_p \sum_{w \in \mathcal{W}_{2h}(2)} p^{-1} n^{-h} \#\Pi^*(w)$, satisfies Carleman's/Riesz's condition.

(e) The sequence $\{F^{A_p(X)}\}$ is almost surely tight. The moments of any subsequential limit is dominated by some Gaussian moments. The LSD exists *iff* $\lim_p \sum_{w \in \mathcal{W}_{2h}(2)} p^{-1} n^{-h} \#\Pi^*(w)$ exists for each h. The same LSD continues to hold if the input sequence satisfies Assumption II or III.   ◇

Without assuming anything further on the link function L_p, the limit in (e) above need not exist. However, as in the previous chapter, we shall provide several examples where the limit does indeed exist for every h.

Proof of Theorem 3.2.3. For the special case of the Sample covariance matrix, a proof can be found in Bose and Sen (2008)[29]. The same arguments hold for general link functions. Here are the details. The proof of the last part will be given only under Assumption II.

(a) and (b). Recall that circuits which have at least one edge of order one contribute zero to the expectation. Thus, consider all circuits which have at least one edge of order at least three and all other edges of order at least two. Let $N_{h,3+}$ be the number of such circuits of length $(2h)$.

Now recall that $1 \le \pi(2i+1) \le n$ and $1 \le \pi(2i) \le p$. However, since the aspect ratio is bounded, to prove negligibility for $N_{h,3+}$, we may argue as follows: Let $\widetilde{\Pi}(w)$ be the possibly larger class of the same circuits but with range $1 \le \pi(i) \le \max(p,n)$, $1 \le i \le 2h$. Then, there is a constant C, such that

$$p^{-1} n^{-h} \sum_{\substack{w \text{ has one edge} \\ \text{of order at least 3}}} \#\Pi(w) \le C \left[\max(p,n)\right]^{-(h+1)} \sum_{\substack{w \text{ has one edge} \\ \text{of order at least 3}}} \#\widetilde{\Pi}(w) \to 0$$

from Lemma 1.4.2 of Chapter 1. This proves (a), and (b) is then a consequence.

(c) It is enough to show that

$$\mathrm{E}\left[p^{-1} \operatorname{Tr} A_p^h - \mathrm{E}\, p^{-1} \operatorname{Tr} A_p^h\right]^4 = \mathrm{O}(p^{-2}).$$

We follow the arguments in the proof of Lemma 1.4.3. Clearly

$$\frac{1}{p^4} \mathrm{E}\left[\operatorname{Tr} A_p^h - \mathrm{E}\left(\operatorname{Tr} A_p^h\right)\right]^4 = \frac{1}{p^4 n^{4h}} \sum_{\pi_1, \pi_2, \pi_3, \pi_4} \mathrm{E}\left[\prod_{i=1}^4 (\mathrm{X}_{\pi_i} - \mathrm{E}\,\mathrm{X}_{\pi_i})\right].$$

If $(\pi_1, \pi_2, \pi_3, \pi_4)$ are not jointly-matched, then one of the circuits, say π_j, has an L-value which does not occur anywhere else. Also note that $\mathrm{E}\,\mathrm{X}_{\pi_j} = 0$. Hence, using independence,

$$\mathrm{E}\left[\prod_{i=1}^4 (\mathrm{X}_{\pi_i} - \mathrm{E}\,\mathrm{X}_{\pi_i})\right] = \mathrm{E}\left[\mathrm{X}_{\pi_j} \prod_{i=1, i \ne j}^4 (\mathrm{X}_{\pi_i} - \mathrm{E}\,\mathrm{X}_{\pi_i})\right] = 0.$$

Further, if $(\pi_1, \pi_2, \pi_3, \pi_4)$ is jointly-matched but is not cross-matched, then one of the circuits, say π_j, is only self-matched, that is, none of its L-values is shared with those of the other circuits. Then again, by independence,

$$\mathrm{E}\left[\prod_{i=1}^{4}(X_{\pi_i} - \mathrm{E}\,X_{\pi_i})\right] = \mathrm{E}\left[(X_{\pi_j} - \mathrm{E}\,X_{\pi_j})\right]\mathrm{E}\left[\prod_{i=1, i\neq j}^{4}(X_{\pi_i} - \mathrm{E}\,X_{\pi_i})\right] = 0.$$

So it is clear that for non-zero contribution, the quadruple of circuits must be jointly-matched and cross-matched. Observe that the total number of edges in each circuit of the quadruple is $(2h)$. So the total number of edges over 4 circuits is $(8h)$. Since they are at least pair-matched, there can be at most $(4h)$ distinct edges, i.e., distinct L-values.

Since $\frac{p}{n} \to y$, $0 < y < \infty$, enlarging the range of the odd and even vertices as earlier, it is easy to see, using Lemma 1.4.3, that there is a constant K which depends on y and h, such that

$$\frac{1}{p^4}\,\mathrm{E}\left[\,\mathrm{Tr}\,A_p^h - \mathrm{E}\left(\mathrm{Tr}\,A_p^h\right)\right]^4 \leq K\frac{p^{4h+2}}{p^{4h+4}} = \mathrm{O}(p^{-2}). \tag{3.8}$$

This guarantees that if there is convergence, it is almost sure. So (c) is proved.

By Assumption A, for any matched word w of length $(2h)$ with $|w| = h$, we have $\#(\Pi^*(w) \leq n^{h+1}\Delta^h$ which easily yields the validity of Carleman's/Riesz's condition, proving (d).

(e) These claims are now easy consequences of the discussion so far. □

In the next few sections, we shall focus on specific matrices. In particular we let X_p be the I.I.D., the non-symmetric Toeplitz, Hankel, Reverse Circulant and Circulant matrix. We shall show that in each case, the LSD exists in Regime I. Closed-form expressions for the LSD or for its moments do not seem to be obtainable except for the Sample variance-covariance matrix. Not much is known for most of the other LSDs.

3.2.2   Sample variance-covariance matrix

The Sample variance-covariance matrix is defined as

$$A_p(W) = n^{-1}W_p W_p' \quad \text{where} \quad W_p = ((x_{ij}))_{1\leq i\leq p, 1\leq j\leq n}. \tag{3.9}$$

In the random matrix literature, this matrix is denoted by S. In the statistics literature, there is an adjustment by the sample mean. This does not make any difference to the LSD under the assumptions that we will work with.

It will help us to think of the link function also as a pair:

$$L1, L2 : \mathbb{N}^2 \to \mathbb{Z}_+^2, \quad L1(i,j) = (i,j), \quad L2(i,j) = (j,i).$$

3.2.2.1   Catalan words and the Marčenko-Pastur law

Recall the set of all Catalan words of length $2k$, $\mathcal{C}_{2k}$, from Section 2.1.1 of Chapter 2 and its crucial role in the LSD of Wigner matrices. They continue to play an important role here but due to the possible unequal values of p and n, we now need to count suitable subclasses of these words. Let

$$M_{t,k} = \{w \in \mathcal{C}_{2k} : w \text{ has exactly } (t+1) \text{ even generating vertices}\}. \quad (3.10)$$

Hence, any $w \in M_{t,k}$ has exactly $(k-t)$ odd generating vertices. The following lemma gives the count of $M_{t,k}$.

Lemma 3.2.4.

$$\#M_{t,k} = \binom{k-1}{t}^2 - \binom{k-1}{t+1}\binom{k-1}{t-1} = \frac{1}{t+1}\binom{k}{t}\binom{k-1}{t}.$$

$\diamond$

Proof. We know from the proof of Lemma 2.1.2 that $\mathcal{C}_{2k}$ is in bijection with the following set of paths:

$$\{\{u_l\}_{1 \leq l \leq 2k} : u_l = \pm 1, \ S_l \geq 0, \forall l, \ S_{2k} = 0\}, \quad \text{where} \quad S_l = \sum_{j=1}^{l} u_j, l \geq 1.$$

Note that we always have $u_1 = 1$ and $S_{2k-1} = 1$. So, consider only those paths starting at $(1,1)$ and ending at $(2k-1, 1)$. We relate $M_{t,k}$ to a certain subset of these paths. Since the words in $M_{t,k}$ have exactly $(t+1)$ even generating vertices, the corresponding paths have $u_l = 1$ at exactly t of the $k-1$ even steps.

First we count the number of paths without the constraint $S_l \geq 0$ for all l. Choose these t even positions from the $(k-1)$ even positions in $\binom{k-1}{t}$ ways. Then we are left to choose $k-1-t$ odd positions for $u_l = 1$ from a total of $k-1$ odd positions (so that we have $k-1$ many $u_l = 1$) and this can be done in $\binom{k-1}{k-1-t} = \binom{k-1}{t}$ ways. So, the total number of such choices is $\binom{k-1}{t}^2$.

Now we need to eliminate the paths which have $S_l \leq 0$ for at least one l. We count the total number of such paths by a reflection principle. Any such path touches $y = -1$ line at least once and hence has two consecutive upward movements. Consider the *last* time this occurs so that $u_l = +1$ and $u_{l+1} = +1$. We consider a transformation

$$(u_2, \ldots, u_l = 1, u_{l+1} = 1, \ldots, u_{2k-1}) \mapsto (u_2, \ldots, u_{l-1}, -1, -1, u_{l+2}, \ldots, u_{2k-1}).$$

The resulting sequence is a path from $(1,1)$ to $(2k-1, -3)$ and defines a bijection from the set of all required paths from $(1,1)$ to $(2k-1, 1)$ with exactly t $+1$'s to the set of all paths from $(1,1)$ to $(2k-1, -3)$ having $u_l = 1$

at $(t-1)$ many of the $k-1$ even steps and $u_l = 1$ at $(k-2-t)$ many of the $k-1$ odd steps. The number of such paths is $\binom{k-1}{t-1}\binom{k-1}{t+1}$. Hence,

$$\#M_{t,k} = \binom{k-1}{t}^2 - \binom{k-1}{t+1}\binom{k-1}{t-1} = \frac{1}{t+1}\binom{k}{t}\binom{k-1}{t}.$$

$\square$

Definition 3.2.1. (*Marčenko-Pastur law*) This law, denoted by $\mathcal{L}_{MPy}$, has a positive mass $1 - \frac{1}{y}$ at the origin if $y > 1$. Elsewhere it has a density:

$$f_{MPy}(x) = \begin{cases} \frac{1}{2\pi xy}\sqrt{(b-x)(x-a)} & \text{if } a \le x \le b, \\ 0 & \text{otherwise} \end{cases} \tag{3.11}$$

where $a = a(y) = (1 - \sqrt{y})^2$ and $b = b(y) = (1 + \sqrt{y})^2$. $\diamond$

We shall denote by MP_y a random variable having the law $\mathcal{L}_{MPy}$. Recall that W denotes a random variable with the semi-circle law. The following lemma gives information on the moments of MP_y.

Lemma 3.2.5. (a) For every integer $k \ge 1$,

$$\beta_k(MP_y) = \sum_{t=0}^{k-1}\frac{1}{t+1}\binom{k}{t}\binom{k-1}{t}y^t = \sum_{t=0}^{k-1}\#M_{t,k}y^t, \quad 0 < y < \infty.$$

(b) For every integer $k \ge 1$,

$$\beta_k(MP_1) = \sum_{t=0}^{k-1}\frac{1}{t+1}\binom{k}{t}\binom{k-1}{t} = \frac{(2k)!}{(k+1)!k!} = \beta_{2k}(W). \tag{3.12}$$

If $y \le 1$ then $\beta_k(MP_y) \le \beta_{2k}(W)$. $\diamond$

Proof. (a)

$$\beta_k = \frac{1}{2\pi y}\int_a^b x^{k-1}\sqrt{(b-x)(x-a)}\,dx$$

$$= \frac{1}{2\pi y}\int_{-2\sqrt{y}}^{2\sqrt{y}}(1+y+z)^{k-1}\sqrt{4y-z^2}\,dz, \quad (x = 1+y+z)$$

$$= \frac{1}{2\pi y}\sum_{l=0}^{k-1}\binom{k-1}{\ell}(1+y)^{k-1-l}\int_{-2\sqrt{y}}^{2\sqrt{y}}z^l\sqrt{4y-z^2}\,dz$$

$$= \frac{1}{2\pi y}\sum_{2l=0}^{k-1}\binom{k-1}{2l}(1+y)^{k-1-2l}(4y)^{l+1}\int_{-1}^{1}u^{2l}\sqrt{1-u^2}\,du, \quad (z = 2\sqrt{y}u)$$

$$= \frac{1}{2\pi y}\sum_{2l=0}^{k-1}\binom{k-1}{2l}(1+y)^{k-1-2\ell}(4y)^{l+1}\int_{0}^{1}w^{l-1/2}\sqrt{1-w}\,dw, \quad (u = \sqrt{w})$$

$$= \sum_{2l=0}^{k-1} \frac{(k-1)!}{l!(l+1)!(k-1-2\ell)!} y^l (1+y)^{k-1-2\ell}$$

$$= \sum_{2l=0}^{k-1} \sum_{s=0}^{k-1-2\ell} \frac{(k-1)!}{l!(l+1)!s!(k-1-2\ell-s)!} y^{s+l}$$

$$= \sum_{2l=0}^{k-1} \sum_{r=\ell}^{k-1-\ell} \frac{(k-1)!}{l!(l+1)!(r-l)!(k-1-r-l)!} y^r, \quad (r=s+l)$$

$$= \frac{1}{k} \sum_{r=0}^{k-1} \binom{k}{r} y^r \sum_{\ell=0}^{\min(r,k-1-r)} \binom{s}{l} \binom{k-r}{k-r-\ell-1}$$

$$= \frac{1}{k} \sum_{r=0}^{k-1} \binom{k}{r} \binom{k}{r+1} y^r = \sum_{r=0}^{k-1} \frac{1}{r+1} \binom{k}{r} \binom{k-1}{r} y^r.$$

(b) Observe that the summation term in (3.12) equals the coefficient of x^k in $(1+x)^{k-1} \int (1+x)^k dx$. Which in turn is the coefficient of x^k in $\frac{1}{k+1}(1+x)^{k-1}(1+x)^{k+1} =$ coefficient of x^k in $\frac{1}{k+1}(1+x)^{2k}$. The latter is indeed the right side of (3.12). The rest of the claim is easily established. □

3.2.2.2 LSD

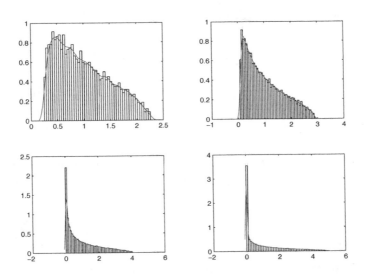

FIGURE 3.1
Histogram and smoothed ESD of the Sample variance-covariance matrix for $n = 400$, 15 replications. Input is i.i.d. N(0, 1). Top left, $y = 0.25$; top right, $y = 0.5$; bottom left, $y = 1.0$ and bottom right, $y = 1.5$.

Figure 3.1 provides some simulations for the S matrix. The first success in finding the LSD of S was due to Marčenko and Pastur (1967)[73]. Several authors have worked on this matrix over the years, See for example, Grenander and Silverstein (1977)[62], Wachter (1978)[104], Jonsson (1982)[67], Yin (1986)[112], Bai and Yin (1988)[6] and Bai (1999)[5]. When the entries of X are not independent, Yin and Krishnaiah (1985)[113] investigated the LSD of the S matrix when the underlying distribution is isotropic. For further developments on the S matrix see Bai and Zhou (2008)[7].

The following theorem gives the LSD of the S matrix in Regime I. We shall give a moment method proof following Bose and Sen (2008)[29]. See Bai (1999)[5] for a proof based on the method of Stieltjes transform.

Theorem 3.2.6. Suppose $\{x_{ij}\}$ satisfy Assumption I, II or III and $p \to \infty$. If $p/n \to y \in (0, \infty)$, then $\{F^{A_p(W_p)}\}$ converges to $\mathcal{L}_{MPy}$ a.s.   ◇

Remark 3.2.1. Suppose $p = n$. Then W_p is a square matrix with i.i.d. elements. By virtue of the above result and Lemma 3.2.5(b), the LSD W of $n^{-1/2}W_n^{(s)}$ and the LSD MP_1 of $A_p(W_p)$ $(p = n)$ bear the relation $MP_1 \overset{\mathcal{D}}{=} W^2$. We shall see later that this squaring relation does not necessarily hold for the corresponding symmetric and non-symmetric versions of other patterned matrices. Moreover, we shall also see that when $p/n \to 0$, the LSD of $N_P(W_p)$ is our familiar semi-circle law.   ◇

Proof of Theorem 3.2.6. We apply mutatis mutandis the proof given for the Wigner matrix. It is enough to show that for every $k \geq 1$,

$$\lim_{p \to \infty} \mathrm{E}[\beta_k(S)] = \sum_{t=0}^{k-1} \#M_{t,k} y^t \qquad (3.13)$$

where

$$\beta_k(S) = \frac{1}{n^k p} \sum_\pi x_{L1(\pi(0),\pi(1))} x_{L2(\pi(1),\pi(2))} x_{L1(\pi(2),\pi(3))} \cdots x_{L2(\pi(2k-1),\pi(2k))}.$$

As seen in the proof of Theorem 2.1.3, Lemma 1.4.2 remains valid in the present case. Hence only the pair-matched circuits potentially contribute and we need to calculate

$$\lim_n \sum_{w \in \mathcal{W}_{2k}(2)} \sum_{\pi \in \Pi(w)} \frac{1}{n^k p} \mathrm{E}[x_{L1(\pi(0),\pi(1))} \cdots x_{L2(\pi(2k-1),\pi(2k))}] = \lim_n \sum_{w \in \mathcal{W}_{2k}(2)} \frac{\#\Pi^*(w)}{n^k p}.$$

We need *exactly* $(k+1)$ generating vertices (hence k non-generating vertices) for a contribution. There is an obvious similarity between the matching restrictions here and the ones we encountered for the Wigner link function.

Note that for $i \neq j$,

$$L1(\pi(i-1), \pi(i)) = L2(\pi(j-1), \pi(j)) \qquad (3.14)$$

implies a (C2) constraint, as defined in (2.5).

On the other hand,

$$Lt(\pi(i-1), \pi(i)) = Lt(\pi(j-1), \pi(j)), \ t = 1 \text{ or } 2, \tag{3.15}$$

yields a (C1) constraint as defined in (2.5).

However, unlike the Wigner matrix here, $w[i] = w[j]$ implies exactly one of the constraints is satisfied; (C1) if i and j have the same parity and (C2) otherwise. Hence there is a *unique* $\bar{\lambda}$ (depending on w) such that $\Pi^*(w) = \Pi^*_{\bar{\lambda}}(w)$.

As before, let λ_0 be the index when all constraints in $\Pi^*_{\lambda_0}(w)$ are (C2). Let $\tilde{\Pi}^*_{\bar{\lambda}}(w)$ denote the class $\Pi^*_{\bar{\lambda}}(w)$ but where $1 \leq \pi(i) \leq \max(p,n)$, $i = 0, 1, \dots 2k$.

If w is not Catalan then it is easy to see that $\bar{\lambda} \neq \lambda_0$. Hence it follows that

$$n^{-k}p^{-1}\#\Pi^*_{\bar{\lambda}}(w) \leq C[\max(p,n)]^{-(k+1)}\# \bigcup_{\lambda \neq \lambda_0} \tilde{\Pi}^*_{\lambda}(w) \to 0.$$

For any $0 \leq t \leq k-1$, if w is Catalan with $(t+1)$ even generating vertices (with range p) and $(k-t)$ odd generating vertices (with range n) then

$$\lim_{n \to \infty} n^{-k}p^{-1}\#\Pi^*_{\lambda_0}(w) = \lim_{n \to \infty} n^{-k}p^{-1}(p^{t+1}n^{k-t}) = y^t.$$

Now Lemma 3.2.4 implies that (3.13) holds and the proof is complete. $\qquad\square$

3.2.3 Other XX' matrices

In Chapter 2 we established the existence of the LSD of the Symmetric Toeplitz, Symmetric Hankel, Reverse Circulant and Symmetric Circulant matrices. In this section we shall consider their suitable non-symmetric versions. So, let

$$T_p = T = ((x_{i-j}))_{p \times n}, \ \ A_p(T) = n^{-1}T_pT_p'$$

be the *non-symmetric Toeplitz* matrix.

Likewise, let $H = H_{p \times n}$ be the *non-symmetric Hankel* matrix where the (i,j)th entry is x_{i+j} if $i \geq j$ and $x_{-(i+j)}$ if $i < j$.

Also let $H_p^{(s)} = ((x_{i+j}))_{p,n}$ be the rectangular Hankel matrix with symmetric link function. Let

$$A_p(H) = n^{-1}H_pH_p' \ \ \text{and} \ \ A_p(H^{(s)}) = n^{-1}H_p^{(s)}H_p^{(s)'}.$$

Let $R_p^{(s)}$ be the $p \times n$ matrix with the Reverse Circulant link function $L(i,j) = (i+j-2) \bmod n$ and $R_p = R_{p \times n}$ be the *non-symmetric* version of $R_p^{(s)}$ with the link function

$$\begin{aligned} L(i,j) &= (i+j-2) \bmod n \text{ for } i \leq j \\ &= -[(i+j-2) \bmod n] \text{ for } i > j. \end{aligned} \tag{3.16}$$

So, in effect, the rows of R_p are the first p rows of a (non-symmetric) Reverse Circulant matrix. Let

$$A_p(R) = n^{-1} R_p R_p' \quad \text{and} \quad A_p(R^{(s)}) = n^{-1} R_p^{(s)} R_p^{(s)'}.$$

Finally, define:

$$C_p = C_{p \times n} = ((x_{L(i,j)})), \quad L(i,j) = (j-i) \bmod n, \quad \text{and} \quad A_p(C) = p^{-1} C_p C_p'.$$

Recall that for the S matrix, the set of Catalan words with a fixed number of even generating vertices played a crucial role. When we move to other matrices, other types of words come into the picture. For convenience of counting, the pair-matched words are classified as follows. Let

$$\mathcal{W}_{t,h} = \{w \in \mathcal{W}_{2h}(2) : w \text{ has exactly } (t+1) \text{ } even \text{ generating vertices}\}. \tag{3.17}$$

Note that there is always at least one even generating vertex $\pi(0)$, and the number of even generating vertices is bounded by h. The total number of odd generating vertices for a word in $\mathcal{W}_{t,h}$ is $(h-t)$. Also let

$$\mathcal{W}_{t,h}^0 = \{w : w \in \mathcal{W}_{t,h} \text{ and } w \text{ is symmetric}\}. \tag{3.18}$$

Theorem 3.2.7. If Assumption I, II or III holds and $p \to \infty$, $\frac{p}{n} \to y \in (0, \infty)$, then for X equal to T, H, R or C, $\{F^{A_p(X)}\}$ converges almost surely to a non-random distribution which is universal.   ◇

Proof. We break the proof into different parts for each matrix.

(i) *Toeplitz case* In view of Theorem 3.2.3 we will only have to show that, for each $h \geq 1$ and each word $w \in \mathcal{W}_{t,h}$, $\lim_p p^{-1} n^{-h} \#\Pi^*(w)$ exists.

We first show that the circuit condition implies that words which are *not symmetric* (see Definition 2.3.2) do not contribute in the limit.

Lemma 3.2.8. (Toeplitz case) Suppose $p/n \to y \in (0, \infty)$. Let w be any pair-matched word of length $(2h)$ which is not symmetric. Then in Regime I,

$$p^{-1} n^{-h} \#\Pi^*(w) \to 0 \quad \text{as} \quad p \to \infty. \tag{3.19}$$

                                                                                    ◇

Proof. Let w be a pair-matched non-symmetric word of length $(2h)$. Let S be the set of $(h+1)$ indices corresponding to the generating vertices of w. Now due to the circuit condition $\pi(0) - \pi(2h) = 0$, we must have

$$\xi_\pi(1) - \xi_\pi(2) + \xi_\pi(3) - \ldots + \xi_\pi(2h-1) - \xi_\pi(2h) = 0. \tag{3.20}$$

Let us enumerate S, left to right, as $\{0, i_1, i_2, \ldots, i_h\}$ and for each $i_t \in S \setminus \{0\}$, let j_t be its matching index, so that $\xi_\pi(i_t) = \xi_\pi(j_t)$, $i_t < j_t$.

Since w is not symmetric, there is at least one pair of matched indices of the same parity. Because of (3.20), the number of pairs of matched L-values with odd indices is the same as the number of pairs of matched L-values with even indices. Consider the set $\mathcal{P}$ of all indices of $S \setminus \{0\}$ whose matched counterpart has the same parity. Let $i_{\max} = \max \mathcal{P}$ and let $j_{\max}$ be its matched index.

So for any $i \in S \setminus \{0\}$ with $i > i_{\max}$, $\xi_\pi(i)$ has a matching L-value with the index of opposite parity and hence when they are both substituted in (3.20), they cancel out each other.

Now, according to our convention, we start choosing generating vertices from the left end of the circuit. We stop when we reach $\pi(i_{\max})$. By this process we fully determine the values of $\{\xi_\pi(t)\}$ for all $t < i_{\max}$. On the other hand, if we consider $\xi_\pi(t)$ with $t > i_{\max}$, then we immediately realize that

1. either its value is already determined. This is the case when its matching counterpart appears to the left of $\xi_\pi(i_{\max})$;

2. or, its matching counterpart has an index of opposite parity as we have observed before.

Thus, in (3.20), except $\xi_\pi(i_{\max})$ and $\xi_\pi(j_{\max})$, all other $\{\xi_\pi(t)\}$ are either already determined or get cancelled with their own counterpart. So, (3.20) forces $\xi_\pi(i_{\max}) + \xi_\pi(j_{\max}) = 2\xi_\pi(i_{\max})$ to take some particular value. Therefore, $\pi(i_{\max})$ has no free choice though it is a generating vertex. This is a contradiction and the proof of the lemma is complete. $\qquad\square$

Therefore, going back to the proof of the theorem, for the Toeplitz case,

$$\beta_h = \lim_p \sum_{t=0}^{h-1} \left(\frac{p}{n}\right)^t \sum_{w \in \mathcal{W}_{t,h}^0} \frac{1}{p^{t+1} n^{h-t}} \#\Pi^*(w).$$

Fix a $w \in \mathcal{W}_{t,h}^0$. If $w[2i] = w[2j+1]$ then we have the restriction,

$$\pi(2i+1) - \pi(2i) = \pi(2j) - \pi(2j+1).$$

Fix a word of length h and let S be the set of indices which correspond to the generating vertices. It is easy to see that if we consider the above h linear restrictions on the vertices and do not take into account the circuit condition $\pi(0) = \pi(2h)$, then each dependent vertex can be written in a *unique* manner as an integral linear combination of generating vertices which occur to the left of that particular vertex in the circuit. That is,

$$\pi(i) = \sum_{j: j \leq i,\ j \in S} a_{i,j} \pi(j) \quad \text{for some } a_{i,j} \in \mathbb{Z}.$$

Note that for $i \in S$, $a_{j,i} = \mathbb{I}\{i = j\}$ and since w is *symmetric*, we have $\pi(2h) = \pi(0)$ so that the circuit condition is automatically satisfied.

Now we compute the scaled limit of $\#\Pi^*(w)$. Define

$$t_{2i} = \frac{\pi(2i)}{p}, \quad t_{2i+1} = \frac{\pi(2i+1)}{n} \quad \text{and} \quad y_n = p/n.$$

From the above discussion, if $i \notin G \cup \{2h\}$, t_i can be written in a unique manner as a linear combination of $t_S := \{t_j : j \in S\}$, namely,

$$t_{2i-1} = L_{2i-1,n}^T(t_S) := \sum_{2j-1 \in S,\ 2j-1 \leq 2i-1} a_{2i-1,2j-1} t_{2j-1} + \sum_{2j \in S,\ 2j \leq i} y_n a_{2i-1,2j} t_{2j}$$

$$t_{2i} = L_{2i,n}^T(t_S) := \sum_{2j-1 \in S,\ 2j-1 \leq 2i} (1/y_n) a_{2i,2j-1} t_{2j-1} + \sum_{2j \in S,\ 2j \leq i} a_{2i,2j} t_{2j}.$$

It is obvious that these linear combinations, say $L_{i,n}^T(t_S)$, depend on the word w but we suppress this dependence for notational brevity. Thus, the number of elements in $\Pi^*(w)$ can be expressed alternatively as follows. Let

$$U_j = \{1/j, 2/j, \ldots, j/j\}, \ j \geq 1.$$

Then

$$\begin{aligned}
\#\Pi^*(w) &= \#\Big\{(t_0, \ldots, t_{2h}) : t_{2i} \in U_p, \ t_{2i-1} \in U_n \text{ and} \\
&\qquad \frac{p}{n} t_{2i} - t_{2i+1} = \frac{p}{n} t_{2j} - t_{2j-1} \text{ if } w[2i+1] = w[2j]\Big\} \\
&= \#\{\{t_i, i \in S\} : t_i \in U_p \text{ if } i \text{ is even}, t_i \in U_n \text{ if } i \text{ is odd, and} \\
&\qquad i \in S, 0 < L_{i,n}^T(t_S) \leq 1, \ \forall \ i \notin S \cup \{2h\}\}.
\end{aligned}$$

As in the proof of Theorem 2.2.1 for the symmetric Toeplitz matrix, the above expression is a multi-dimensional Riemann sum. Passing to the limit,

$$\begin{aligned}
\beta_h &= \sum_{t=1}^{h-1} y^t \sum_{w \in \mathcal{W}_{t,h}^0} \lim_{p \to \infty} \frac{1}{p^{t+1} n^{h-t}} \#\Pi^*(w) \\
&= \sum_{t=1}^{h-1} y^t \sum_{w \in \mathcal{W}_{t,h}^0} \underbrace{\int_0^1 \int_0^1 \cdots \int_0^1}_{h+1} \mathbb{I}(0 < L_i^T(t_S) \leq 1, \ \forall \ i \notin S \cup \{2h\}) dt_S,
\end{aligned}$$

where $L_i^T(t_S)$ is the same as $L_{i,n}^T(t_S)$ with all y_n being replaced by y. The above argument shows that when $y_n \to y \in (0, \infty)$, the LSD of $A_p(T)$ exists.

(ii) *The Hankel case.* By Theorem 3.2.3, we know that the h-th moment of the Hankel LSD (provided the limit below exists) would be given by

$$\beta_h = \lim_{p \to \infty} \sum_{t=0}^{h-1} \left(\frac{p}{n}\right)^t \sum_{w \in \mathcal{W}_{t,h}} \frac{1}{p^{t+1} n^{h-t}} \#\Pi^*(w). \tag{3.21}$$

We note that $\Pi^*(w) \subseteq \Pi_{\tilde{H}}^*(w)$ where $\Pi_{\tilde{H}}^*(w)$ is as defined in (2.23) with symmetric Hankel link function $L(i,j) = i + j - 2$ and each vertex has the same range between 1 and $\max(p,n)$ since in the latter case we have more circuits but fewer restrictions. From the arguments given in Chapter 2 for the symmetric Hankel link function, it follows that in this case too for any non-symmetric word w, $n^{-(h+1)} \#\Pi^*(w) \to 0$. Thus, (3.21) reduces to

$$\beta_h = \lim_{p \to \infty} \sum_{t=0}^{h-1} \left(\frac{p}{n}\right)^t \sum_{w \in \mathcal{W}_{t,h}^0} \frac{1}{p^{t+1} n^{h-t}} \#\Pi^*(w). \tag{3.22}$$

Let us first consider the case when the link function is symmetric Hankel, i.e., $L(i,j) = i + j - 2$. In that case for a word $w \in \mathcal{W}_{t,h}^0$, if $w[2i] = w[2j+1]$, we have the restriction,

$$\pi(2i+1) + \pi(2i) = \pi(2j) + \pi(2j+1).$$

As in the Toeplitz case, we can express each vertex in a unique manner as an integral linear combination of generating vertices occurring to its left.

$$\pi(i) = \sum_{j : j \leq i, \ j \in S} b_{i,j} \pi(j) \quad \text{for some} \ \ b_{i,j} \in \mathbb{Z}, \tag{3.23}$$

with $b_{j,i} = \mathbb{I}\{i = j\}$ if $i \in S$ and $\pi(2h) = \pi(0)$ since w is symmetric.

As before let t_i be the scaled vertices. Similar to the Toeplitz case, we define $L_{i,n}^H(t_S)$ as the linear combination which expresses t_i in terms of the "free coordinates" t_S and $L_i^H(t_S)$ as its limiting version. It immediately yields

$$\lim_{p \to \infty} \frac{\#\Pi^*(w)}{p^{t+1} n^{h-t}} = \underbrace{\int_0^1 \int_0^1 \cdots \int_0^1}_{h+1} \mathbb{I}(0 < L_i^H(t_S) \leq 1, \ \forall \, i \notin S \cup \{2h\}) dt_S. \tag{3.24}$$

Now consider the non-symmetric Hankel link. Here instead of calculating the $(h + 1)$-dimensional Euclidean volume of the entire set $\{0 \leq L_i^H(t_S) \leq 1, \ \forall \, i \notin S \cup \{2h\}\}$, we need to take into account the restricting hyperplanes that arise from the non-symmetric nature of the link function since it assumes different signs around the diagonal. So, unlike the symmetric case, this imposes the following extra restrictions in addition to the usual equality between two L-values $\xi_\pi(2i)$ and $\xi_\pi(2j+1)$:

$$\text{either} \quad\quad \pi(2i+1) \leq \pi(2i), \pi(2j+1) \leq \pi(2j)$$
$$\text{or} \quad\quad \pi(2i+1) > \pi(2i), \pi(2j+1) > \pi(2j).$$

Thus, the size of the set $\Pi^*(w)$ is the same as the cardinality of

$$\{t_G : 0 < L_{i,n}^H(t_S) \leq 1, \ \forall \, i \notin S \cup \{2h\}, \text{sgn}(L_{2i+1,n}^H(t_S) - y_n L_{2i,n}^H(t_S))$$
$$= \text{sgn}(L_{2j+1,n}^H(t_S) - y_n L_{2j,n}^H(t_S)) \ \text{if} \ w[2i+1] = w[2j]\},$$

where, each $t_i \in U_n$ or $t_i \in U_p$, according when i is odd or even.

This entire set may be written as a product of indicator functions in terms of $\{L_{i,n}^H(t_S)\}$, albeit in a complicated manner. When summed over t_S, in the limit this equals the corresponding Riemann integral where the indicators are replaced by their limits and y_n is replaced by y. Let us denote the overall indicator function by $f^H(t_S)$. So, we have

$$
\begin{aligned}
\beta_h &= \sum_{t=1}^{h-1} y^t \sum_{w \in \mathcal{W}_{t,h}^0} \lim_{p \to \infty} \frac{1}{pn^h} \#\Pi^*(w) \\
&= \sum_{t=1}^{h-1} y^t \sum_{w \in \mathcal{W}_{t,h}^0} \underbrace{\int_0^1 \int_0^1 \cdots \int_0^1}_{h+1} f^H(t_G) dt_G.
\end{aligned}
\tag{3.25}
$$

This completes the proof for the Hankel case.

(iii) *The Reverse Circulant case.* As in the Hankel case, $\Pi^*(w) \subseteq \Pi_{\tilde{R}}^*(w)$ where $\Pi_{\tilde{R}}^*(w)$ is as in (2.23) with the symmetric link function $L(i,j) = (i+j-2) \bmod n$. In Theorem 2.3.2 we have seen that for this link function, $n^{-(h+1)} \#\Pi^*(w) \to 0$ whenever w is non-symmetric. Thus, again only symmetric words matter, and it is enough to show that each term within $[\ ]$ below has a finite limit.

$$
\beta_h = \lim_{p \to \infty} \sum_{t=0}^{h-1} \left(\frac{p}{n} \right)^t \sum_{w \in \mathcal{W}_{t,h}^0} \left[\frac{1}{p^{t+1} n^{h-t}} \#\Pi^*(w) \right], \ \forall h \geq 1.
\tag{3.26}
$$

Let us first consider the *symmetric* link function $L(i,j) = (i+j) \bmod n$. In this case, if $w[i] = w[j]$, we obtain the restriction of the following type:

$$
(\pi(i) + \pi(i-1) - 2) \bmod n = (\pi(j) + \pi(j-1) - 2) \bmod n,
$$

which is equivalent to

$$
\pi(i) + \pi(i-1) - (\pi(j) + \pi(j-1)) \in n\mathbb{Z}.
$$

We now have exactly the same set of equations as in the Hankel case (3.23), with some added restrictions given as follows. As before, let S denote the set of indices corresponding to the generating vertices. Then

$$
\pi(i) - \sum_{j:j \leq i,\ j \in S} b_{i,j} \pi(j) \in n\mathbb{Z} \quad \text{for all } i \notin S.
$$

If $i \notin S \cup \{2h\}$, let $k_{i,n} = k_{i,n}(\pi(j) : j \in G)$ be the *unique* integer such that

$$
1 \leq \sum_{j:j \leq i,\ j \in S} b_{i,j} \pi(j) + k_{i,n} \leq n.
$$

Thus, once we fix the generating vertices, there is exactly one choice for each of the non-generating odd vertices. For the non-generating even vertices, things are a bit complicated and we argue as follows:

For any real number a, let

$$
\begin{aligned}
L(a) &= \max\{m \in \mathbb{Z} : m < a\}, \\
F(a) &= a - L(a).
\end{aligned}
$$

Then $k_{i,n}$ may be written as

$$
\frac{k_{i,n}}{n} = \frac{1}{n} F\Big(\sum_{j:j\leq i,\ j\in F} b_{i,j}\pi(j) \Big).
$$

Now let us fix $2i \notin S, i \neq h$. We have at least $[y_n]$ choices for the vertex $\pi(2i)$. Moreover, we can have one additional choice for $\pi(2i)$ if

$$
\sum_{j:j\leq 2i,\ j\in G} b_{2i,j}\pi(j) + k_{2i,n} \leq p - [y_n]n.
$$

Dividing by n, the above condition can be rewritten as

$$
F(y_n L^H_{2i,n}(t_S)) \leq y_n - [y_n].
$$

Let

$$
S^- = \{2i : 2i \notin S, i \neq h\}.
$$

From the above discussion for the symmetric link function, when $w \in \mathcal{W}^0_{t,h}$, the cardinality of $\Pi^*(w)$ equals

$$
[y_n]^{h-t-1}p^{t+1}n^{h-t} + \sum_{\emptyset \neq S'\subseteq S^-} \#\{t_S : F(y_n L^H_{2i,n}(t_S)) \leq y_n - [y_n], \mathbb{I}(2i \in S')\}
$$

$$
\times [y_n]^{\#(S^- - S')}.
$$

Note that F has discontinuities only at integer points. Therefore, we still have the following convergence:

$$
\lim_{p\to\infty} \frac{\#\Pi^*(w)}{p^{t+1}n^{h-t}} = [y]^{h-t-1} + y^{\#(S^--S')} \times
$$

$$
\sum_{\emptyset \neq S'\subseteq S^-} \int_0^1 \cdots \int_0^1 \mathbb{I}\left(F(y L^H_{2i}(t_S)) \leq y - [y],\ 2i \in S' \right) dt_S.
$$

Coming back to the non-symmetric case, we now have the extra restrictions:

$$
\operatorname{sgn}(\pi(2i+1) - \pi(2i)) = \operatorname{sgn}(\pi(2j+1) - \pi(2j)) \quad \text{if} \quad w[2i] = w[2j+1].
$$

We can now incorporate the restrictions $\operatorname{sgn}(t_{2i+1} - yt_{2i}) = \operatorname{sgn}(t_{2j+1} - yt_{2j})$ in the integral as we did in the Hankel case. The even non-generating vertices, except for $\pi(2h)$ which is constrained to be equal to $\pi(0)$, do not have a unique choice once we determine the generating vertices. Instead there are $[y_n] + 1$ choices for each of them, as observed above.

Recall from Theorem 2.3.2 of Chapter 2, that for the Symmetric Reverse Circulant, the contribution of each (symmetric) word is equal to 0 or 1. But now, due to the additional constraints on the signs due to asymmetry, the integrand is not necessarily equal to 1. Hence further simplification is not possible and all the relevant indicators will appear in the integrand. We omit the details. This completes the proof for the Reverse Circulant case.

(iv) *The Circulant case.* As before, we need to prove that for each $h \geq 1$ the following limit exists.

$$\beta_h = \lim_{p \to \infty} \sum_{t=0}^{h-1} \left(\frac{p}{n}\right)^t \sum_{w \in \mathcal{W}_{t,h}} \frac{1}{p^{t+1}n^{h-t}} \#\Pi^*(w).$$

Proceeding as before, we now obtain exactly the same set of equations as in the Toeplitz case but with some added restrictions given below:

$$\pi(i) - \sum_{j:j \leq i,\ j \in S} a_{i,j}\pi(j) \in n\mathbb{Z}, \quad \text{for all}\ \ i \notin S.$$

In the Toeplitz proof, we already argued that for any non-symmetric word, $\sum_{j:j \in S} a_{2h,j}\pi(j) = \pi(0)$ induces a non-trivial restriction on the generating vertices. On the other hand, we can have a bounded ($\leq \lfloor y_n \rfloor + 1$) number of choices for the non-generating vertices. Thus, again, non-symmetric words do not contribute in the limit.

Rest of the proof is similar to the (symmetric) Reverse Circulant case. We omit the repetitive details. The proof of the theorem is thus complete. $\square$

Remark 3.2.2. It can be verified that the following pairs of matrices have identical LSDs: $A_p(W)$ and $A_p(W^s)$; $A_p(T)$ and $A_p(H^{(s)})$; and $A_p(C)$ and $A_p(R^{(s)})$. However, the LSDs of $A_p(H)$ and $A_p(H^{(s)})$ are different.

3.3   Aspect ratio $y = 0$

In this case it can be checked that the LSD of $A_p(X)$ is degenerate. To get a non-degenerate limit, recall its scaled and centered version $N_p(X)$ given in (3.2). Now we will proceed along the familiar three-step process, but the details are now quite involved:

(i) In Step 1, prove the convergence of the EESD under boundedness of the variables.

(ii) To move from convergence of the EESD to that of the ESD, we need a Borel-Cantelli-type result as in Lemma 1.4.3 of Chapter 1. This is done in Lemma 3.3.6.

(iii) In Step (iii), we prove a general truncation result to ensure that we can work, without loss of generality, under the assumption that all moments are finite. Recall that for symmetric matrices, to achieve this, our minimal moment assumption was finiteness of the *second moment*. For XX' matrices in Regime I, our minimal moment assumption was finiteness of the *fourth moment*. In Regime II, under a slightly stronger moment assumption we establish a suitable truncation result in Theorem 3.3.2.

As in Regime I, we first deal with the S matrix. To keep the proof transparent, we do *not* do the details of Steps (ii) and (iii) for this particular case and establish only (i).

Steps (ii) and (iii) are done for general XX' matrices in the next section. Those general results are in particular applicable to the S matrix, even though it entails slight strengthening of the moment condition.

3.3.1   Sample variance-covariance matrix

As mentioned earlier, in this case the LSD is the semi-circle law. Bai and Yin (1988)[6] gave a proof of this using the Stieltjes transform. We present a moment method proof that is taken from Bose and Sen (2008)[29].

Theorem 3.3.1. Suppose $\{x_{ij}\}$ satisfy Assumption I or Assumption II with bounded fourth moment or, Assumption III. If $p \to \infty$, $p/n \to 0$, then $F^{\sqrt{\frac{n}{p}}(A_p(W)-I_p)}$ converges to the semi-circle law $\mathcal{L}_W$ almost surely.   ◇

Proof. As already mentioned, we shall give a proof of the convergence of the EESD. Thus, assume that the input sequence satisfies Assumption I.
      Define

$$
\begin{aligned}
\delta_{ij} &= \mathbb{I}\{i = j\} \\
\tilde{x}_j &= x_{\pi(j-1),\pi(j)} x_{\pi(j+1),\pi(j)} - \delta_{\pi(j-1),\pi(j+1)}, \quad 1 \le j \le 2k-1.
\end{aligned}
$$

Then

$$
\mathrm{E}\,\beta_k(N_p) = \mathrm{E}\left[\beta_k\left(\left(\frac{n}{p}\right)^{1/2}\!\left(A_p(W) - I_p\right)\right)\right] = \frac{1}{n^{k/2}p^{1+k/2}} \sum_\pi \mathrm{E}[X_\pi], \quad (3.27)
$$

where,

$$
X_\pi = \prod_{j=1}^{2k-1} \tilde{x}_j. \tag{3.28}
$$

Now $\mathrm{E}[X_\pi] = 0$ whenever the value $(\pi(2i), \pi(2i-1))$ or $(\pi(2i), \pi(2i+1))$ occurs only once in the product or when for some j, the value of $\pi(2j+1)$ does not occur at least twice among $\pi(1), \pi(3), \ldots, \pi(2k-1)$.

Define a graph $G = (V, E)$ with

$$
\begin{aligned}
V_1 &= \{\pi(2j), 0 \leq j \leq k\}, \\
V_2 &= \{\pi(2j-1), 1 \leq j \leq k\} \\
V &= V_1 \cup V_2 \\
E &= \{(\pi(2l), \pi(2l+1)), (\pi(2l+2), \pi(2l+1)) : 0 \leq l \leq k-1\}
\end{aligned}
$$

where multiple edges count as one.

Fix a match (out of finitely many choices) among the even vertices and one among the odd vertices, such that $E[X_\pi] \neq 0$. There are at most $p^{\#V_1} n^{\#V_2}$ corresponding circuits. So the contribution of that term is

$$
O\left(\frac{p^{|V_1|} n^{|V_2|}}{p^{k/2+1} n^{k/2}}\right) = O\left(\left(\frac{p}{n}\right)^{k/2-|V_2|}\right). \tag{3.29}
$$

First consider k to be odd. Then $\#V_2 < k/2$. Since $p/n \to 0$, (3.29) immediately implies that $E[\beta_k(N_p)] \to 0$.

Now consider $k = 2m$ to be even. We look for π which produces a non-trivial contribution. Using (3.29), for a non-trivial contribution, we must have

$$
\#V = 2m+1, \#E = 2m, \#V_2 = m \text{ and } \#V_1 = m+1. \tag{3.30}
$$

Observe that:

(i) $\#V_2 = m$ implies a pair-partition of odd vertices. Denote it by a word w of length k. So, $\pi(2i-1) = \pi(2j-1)$ iff $w[i] = w[j]$.

(ii) Each pair in E must occur exactly twice.

(iii) If $(\pi(2l), \pi(2l+1)) = (\pi(2l+2), \pi(2l+1))$ or equivalently $\pi(2l) = \pi(2l+2)$, then $E[X_\pi] = 0$. So, consecutive even vertices cannot be equal.

(iv) Note that (ii) and (iii) together imply that $E[X_\pi] = 1$. Suppose

$$
w[i] = w[j], \text{ i.e., } \pi(2i-1) = \pi(2j-1) \tag{3.31}
$$

and they are different from the rest of the odd vertices. For every fixed w, the total number of choices of odd vertices that satisfy the pairing imposed by w is exactly equal to $N_1(n) = n(n-1)\cdots(n-m+1)$.

Consider the four pairs of vertices from E, $(\pi(2i-2), \pi(2i-1))$, $(\pi(2i), \pi(2i-1))$, $(\pi(2j-2), \pi(2j-1))$ and $(\pi(2j), \pi(2j-1))$.

By (3.31) and (ii), they have to be matched in pairs among themselves. Also, (iii) rules out the possibility that the first pair is matched with the second and the third is matched with the fourth. So the other two combinations are the only possibilities. It is easy to verify that this is the same as saying that

$$
L(\pi(2i-2), \pi(2i)) = L(\pi(2j-2), \pi(2j)) \tag{3.32}
$$

where L is the Wigner link function. Let $\pi^*(i) = \pi(2i)$. Equation (3.32) implies that π^* is a matched circuit of length k. Let $\Pi^*(w)=$ all circuits π^* satisfying the Wigner link function. From the results of Chapter 2 (see 2.4)), we already know that, $\lim \frac{1}{p^{m+1}} \#\Pi^*(w) = 1$ or 0, according when, w is or is not Catalan. Hence, the following equalities hold and (M1) is established.

$$\lim_{p \to \infty} E[\beta_{2m}(N_p)] = \lim_{p \to \infty} \frac{1}{p^{m+1} n^m} \sum_{w:\text{matched}, \ |w|=m} N_1(n) \#\Pi^*(w)$$

$$= \lim_{p \to \infty} \frac{1}{p^{m+1}} \sum_{w:\text{matched}, \ |w|=m} \#\Pi^*(w) = \frac{(2m)!}{(m+1)!m!}$$

which is the $(2m)$-th moment of the semi-circle law. This proves that the EESD converges to the semi-circle law under Assumption I.

The Borel-Cantelli argument to move from convergence of the EESD to that of the ESD is given for general matrices later in Lemma 3.3.6. When applied to the S matrix, this requires a finite moment of order slightly higher than four. However, by a direct argument, one can show that Lemma 3.3.6 holds for the S matrix solely under the finiteness of the fourth moment. In Remark 3.3.1 in the next section we will also argue that if the fourth moment is finite, then the LSD exists *in probability*. We shall remain satisfied with this convergence in probability.

Finally, the truncation argument to relax the assumption of boundedness of the variables is a bit tricky in Regime II. The reader will see this argument for general XX' in the next section (Theorem 3.3.2). This establishes the almost sure LSD under Assumption III with a finite moment of order just higher than four. Again, with some work, one can get down to finiteness of the fourth moment for the S matrix.

A complete proof for the almost sure LSD when only the fourth moment is assumed to be finite is available in Bai (1988)[6]. $\qquad\qquad\square$

3.3.2 Other XX' matrices

We now move to the other matrices in this regime. Let $X_p = X$ stand for the non-symmetric versions of any of the four matrices Toeplitz, Hankel, Circulant and Reverse Circulant. It turns out that in Regime II, the LSD for $N_p(X)$ for all these matrices exist and equal $\mathcal{L}_T$, the LSD for $n^{-1/2}T_n^{(s)}$ obtained in Section 2.

Straightforward extension of Theorem 3.2.2, that reduces the problem to the bounded case, is not possible here. In Theorem 3.3.2 we show how the matrix N_p defined above may be approximated by the following matrix $B_p = B_p(X)$ with bounded entries under an additional moment assumption.

$$(B_p)_{ij} = \frac{1}{\sqrt{np}} \sum_{l=1}^{n} \hat{x}_{L(i,l)} \hat{x}_{L(j,l)}, \text{ if } i \neq j \text{ and } (B_p)_{ii} = 0, \tag{3.33}$$

where

$$\tilde{x}_\alpha = x_\alpha \mathbb{I}(|x_\alpha| \le \epsilon_p n^{1/4}), \quad \hat{x}_\alpha = \tilde{x}_\alpha - \mathrm{E}\,\tilde{x}_\alpha, \quad \text{and} \quad \epsilon_p \text{ will be chosen later.}$$

Recall that for the Sample variance-covariance matrix, we had assumed the finiteness of the fourth moment in Theorem 3.3.1. Since in this section we wish to work with more general link functions, we assume finiteness of a slightly higher moment as given below in Assumption IV. Note that in this assumption, the order of the moment is related to how fast $p/n \to 0$.

Assumption IV. The input sequence $\{x_i, i \in \mathcal{Z}\}$ *is independent with mean zero and variance 1. There exists $\lambda > 1$, such that $p = \mathrm{O}(n^{1/\lambda})$ and* $\sup_i \mathrm{E}(|x_i|^{4(1+1/\lambda)+\delta}) < \infty$ *for some $\delta > 0$.*

Theorem 3.3.2. Suppose $p, n \to \infty$ and Assumption IV holds. Suppose L_p satisfies Assumptions A and B. Then there exists a non-random sequence $\{\epsilon_p\}$ with the property that $\epsilon_p \downarrow 0$ but $\epsilon_p p^{1/4} \uparrow \infty$ as $p \uparrow \infty$ such that $W_2(F^{B_p}, F^{N_p}) \to 0$ a.s.   ◇

Proof. Let

$$\tilde{X}_p = ((\tilde{x}_{L(i,j)}))_{1 \le i \le p, 1 \le j \le n} \quad \text{and} \quad \tilde{N}_p = \frac{1}{\sqrt{np}}(\tilde{X}_p \tilde{X}_p' - n I_p).$$

$$\sup_x \left| F^{N_p}(x) - F^{\tilde{N}_p}(x) \right| \quad \le \quad p^{-1} \mathrm{rank}(X_p - \tilde{X}_p)$$

$$\le \quad p^{-1} \sum_{i=1}^{p} \sum_{j=1}^{n} \eta_{L(i,j)} \quad \text{where} \quad \eta_\alpha = \mathbb{I}(|x_\alpha| > \epsilon_p n^{1/4}).$$

The first inequality above follows from Bai (1999)[5, Lemma 2.6].

Write $q_p = \sup_\alpha \mathrm{P}(|x_\alpha| > \epsilon_p n^{1/4})$. We claim that there exists a sequence $\epsilon_p \downarrow 0$ arbitrarily slowly such that

$$q_p \le \epsilon_p n^{-(1+1/\lambda)+\delta/8}. \tag{3.34}$$

To establish the claim, for simplicity, assume $n = n(p)$ is an increasing function of p. Fix any $\epsilon > 0$. We have

$$n^{(1+1/\lambda)+\delta/8} \sup_\alpha \mathrm{P}(|x_\alpha| > \epsilon n^{1/4}) \le \epsilon^{-4(1+1/\lambda)-\delta/2} \times$$

$$\sup_\alpha \mathrm{E}\left[|x_\alpha|^{4(1+1/\lambda)+\delta/2} \mathbb{I}(|x_\alpha| > \epsilon n^{1/4}) \right]$$

$$\to 0 \tag{3.35}$$

since the random variables $\{|x_\alpha|^{4(1+1/\lambda)+\delta/2}\}$ are uniformly integrable.

Given $m \geq 1$, by (3.35) find an integer p_m such that $n \geq n_m := n(p_m)$ implies

$$n^{(1+1/\lambda)+\delta/8} \sup_\alpha \mathrm{P}(|x_\alpha| > m^{-1}n^{1/4}) \leq m^{-1}.$$

Define $\epsilon_p = 1/m$ if $p_m \leq p < p_{m+1}$ and $\epsilon_p = n(p_1)^{(1+1/\lambda)+\delta/8}$ for $p < p_1$. Note that by choosing the integers in the sequence $p_1 < p_2 < \cdots$ as large as we want, we can make ϵ_p go to zero as slowly as we like. Clearly, ϵ_p satisfies the inequality (3.34).

For any $\beta > 0$, and with Y_i independent Bernoulli with $E(Y_i) \leq q_p$,

$$\mathrm{P}\big(\sup_x |F^{N_p}(x) - F^{\tilde{N}_p}(x)| > \beta\big) \leq \mathrm{P}\big(p^{-1} \sum_{i=1}^p \sum_{j=1}^n \eta_{L(i,j)} > \beta\big)$$

$$\leq \mathrm{P}\big(\alpha_p p^{-1} \sum_{i=1}^{k_p} Y_i > \beta\big)$$

$$\leq (\beta p)^{-1} \alpha_p \sum_{i=1}^{k_p} E\, Y_i$$

$$\leq (\beta p)^{-1} \alpha_p k_p q_p$$

$$\leq C n q_p = o(n^{-(1/\lambda+\delta/8)}) = o(p^{-(1+\lambda\delta/8)})$$

where C is such that $\alpha_p k_p \leq C\beta np$. Hence by Borel-Cantelli lemma,

$$\sup_x |F^{N_p}(x) - F^{\tilde{N}_p}(x)| \to 0 \text{ a.s.}$$

Let

$$\hat{X}_p = ((\hat{x}_{L(i,j)}))_{1 \leq i \leq p,\ 1 \leq j \leq n} \quad \text{and} \quad \hat{N}_p = \frac{1}{\sqrt{np}}(\hat{X}_p \hat{X}'_p - nI_p).$$

Using Lemma 3.2.1 (a) and (b),

$$W_2^2(F^{\hat{N}_p}, F^{\tilde{N}_p}) \leq \big(p^{-1} \sum_{i=1}^p |\lambda_i(\hat{N}_p) - \lambda_i(\tilde{N}_p)|\big)^2$$

$$\leq \frac{1}{np^3} \mathrm{Tr}\,(\tilde{X}_p \tilde{X}'_p + \hat{X}_p \hat{X}'_p)\, \mathrm{Tr}\,\big(\mathrm{E}(\tilde{X}_p)(\mathrm{E}(\tilde{X}'_p)\big).$$

Using the conditions on the moments, the truncation level, $\alpha_p k_p = \mathrm{O}(np)$, and an appropriate SLLN for independent random variables, it is easy to show that the above tends to 0 a.s. We omit the tedious details. On the other hand,

$$W_2^2(F^{\hat{N}_p}, F^{B_p}) \leq \frac{1}{p} \mathrm{Tr}(\hat{N}_p - B_p)^2$$

$$\leq \frac{1}{2np^2} \sum_{i=1}^p \big(\sum_{l=1}^n (\hat{x}_{L(i,l)}^2 - \mathrm{E}\,\hat{x}_{L(i,l)}^2)\big)^2$$

$$+ \frac{1}{2np^2} \sum_{i=1}^p \big(\sum_{l=1}^n (1 - \mathrm{E}\,\hat{x}_{L(i,l)}^2)\big)^2 = M + N, \text{ say.}$$

Note that for every i, j, there exists α such that

$$0 \le 1 - \mathrm{E}\,\hat{x}^2_{L(i,j)} = \mathrm{E}\,x^2_\alpha \mathbb{I}(|x_\alpha| > \epsilon_p n^{1/4}) + (\mathrm{E}\,x_\alpha \mathbb{I}(|x_\alpha| > \epsilon_p n^{1/4}))^2$$

$$\le \frac{1}{\epsilon_p^2 n^{1/2}}(\mathrm{E}\,x^4_\alpha + \mathrm{E}^2\,x^2_\alpha).$$

From this it is immediate that

$$N \le \frac{n^2 p}{2np^2}\frac{1}{n\epsilon_p^4}\left(\sup_\alpha(\mathrm{E}\,x^4_\alpha + \mathrm{E}^2\,x^2_\alpha)\right)^2 \to 0 \;\; \text{since} \;\; \epsilon_p p^{1/4} \to \infty.$$

Now we will deal with the first term, M.

$$\sum_{i=1}^{p}\left(\sum_{l=1}^{n}(\hat{x}^2_{L(i,l)} - \mathrm{E}\,\hat{x}^2_{L(i,l)})\right)^2 = \sum_\alpha a_\alpha\,(\hat{x}^2_\alpha - \mathrm{E}\,\hat{x}^2_\alpha)^2$$

$$+ \sum_{\alpha \ne \alpha'} b_{\alpha,\alpha'}\,(\hat{x}^2_\alpha - \mathrm{E}\,\hat{x}^2_\alpha)(\hat{x}^2_{\alpha'} - \mathrm{E}\,\hat{x}^2_{\alpha'})$$

$$= T_{1p} + T_{2p} \;\; \text{(say)}$$

where $a_\alpha, b_{\alpha,\alpha'} \ge 0$. Obviously,

$$\#\{\alpha \in \mathcal{Z} : a_\alpha \ge 1\} \le k_p \quad \text{and} \quad \#\{(\alpha,\alpha') \in \mathcal{Z}^2 : \alpha \ne \alpha', b_{\alpha,\alpha'} \ge 1\} \le k_p^2.$$

Also, $a_\alpha \le \alpha_p$ for all α and $b_{\alpha,\alpha'} \le \Delta^2 \alpha_p$ for all $\alpha \ne \alpha'$. Hence,

$$\sum_p \frac{1}{4n^2 p^4}\,\mathrm{E}\,T^2_{1p} = \sum_p \frac{1}{4n^2 p^4}\sum_\alpha a^2_\alpha\,\mathrm{E}(\hat{x}^2_\alpha - \mathrm{E}\,\hat{x}^2_\alpha)^4$$

$$+ \sum_p \frac{1}{4n^2 p^4}\sum_{\alpha \ne \alpha'} a_\alpha a_{\alpha'}\,\mathrm{E}(\hat{x}^2_\alpha - \mathrm{E}\,\hat{x}^2_\alpha)^2\,\mathrm{E}(\hat{x}^2_{\alpha'} - \mathrm{E}\,\hat{x}^2_{\alpha'})^2$$

$$\le \sum_p \frac{\alpha_p^2 k_p}{4n^2 p^4}\left[\sup_\alpha \mathrm{E}(\hat{x}^2_\alpha - \mathrm{E}\,\hat{x}^2_\alpha)^4 + \sup_\alpha \mathrm{E}^2(\hat{x}^2_\alpha - \mathrm{E}\,\hat{x}^2_\alpha)^2\right]$$

$$\le \sup_\alpha \sum_p \frac{\alpha_p^2 k_p}{4n^2 p^4}(n^{1/4}\epsilon_p)^4\,\mathrm{E}\,x^4_\alpha + \sup_\alpha \sum_p \frac{\alpha_p^2 k_p^2}{4n^2 p^4}\,\mathrm{E}^2\,x^4_\alpha < \infty$$

where we used the relations $\alpha_p k_p = O(np)$ and $\alpha_p \le p^2$. Thus, $T_{1p}/(2np^2) \to 0$ a.s.

It now remains to tackle T_{2p}. Let $y_\alpha = \hat{x}^2_\alpha - \mathrm{E}\,\hat{x}^2_\alpha$, $\alpha \in \mathcal{Z}$. Then $\{y_\alpha\}$ are mean zero independent random variables.

$$\sum_p \frac{1}{4n^2 p^4}\,\mathrm{E}\,T^2_{2p} = \sum_p \frac{1}{4n^2 p^4}\,\mathrm{E}\left(\sum_{\alpha \ne \alpha'} b_{\alpha,\alpha'} y_\alpha y_{\alpha'}\right)^2$$

$$\le \sum_p \frac{1}{4n^2 p^4}\sum_{\alpha \ne \alpha'} b^2_{\alpha,\alpha'}\,\mathrm{E}\,y^2_\alpha y^2_{\alpha'}$$

$$\le \sup_\alpha \sum_p \frac{k_p^2 \alpha_p^2}{4n^2 p^4}\,\mathrm{E}^2\,x^2_\alpha < \infty.$$

Thus, by Borel-Cantelli lemma, $T_{2p}/(np^2) \to 0$ almost surely and this completes the proof. □

Remark 3.3.1. If we carefully follow the above proof, finiteness of the $(4(1 + 1/\lambda) + \delta)$-th moment was only needed to establish that $\sup |F^{N_p}(x)-, F^{\tilde{N}_p}(x)| \overset{a.s.}{\to} 0$. If we impose the weaker assumption $\sup_\alpha E\, x_\alpha^4 < \infty$, then $q_p \leq \epsilon_p/n$ for a suitably chosen sequence $\{\epsilon_p\} \downarrow 0$ and $\sup |F^{N_p}(x) - F^{\tilde{N}_p}(x)| \overset{P}{\to} 0$ holds and hence $W_2(F^{B_p}, F^{N_p}) \to 0$ in probability. ◇

Having approximated N_p by B_p, we now need to establish the behavior of the moments of B_p. This is done through a series of lemmata, finally leading to Theorem 3.3.7. In the subsequent discussion, we will use the following notation for words $w, w_1, w_2, \ldots, w_k$:

$$\Pi_{\neq}(w) = \{\pi \in \Pi(w) : \pi(2i - 2) \neq \pi(2i) \ \forall 1 \leq i \leq h\},$$
$$\Pi_{\neq}^*(w) = \{\pi \in \Pi^*(w) : \pi(2i - 2) \neq \pi(2i) \ \forall 1 \leq i \leq h\}$$
$$\Pi(w_1, \ldots, w_k) = \{(\pi_1, \ldots, \pi_k) : w_i[s] = w_j[\ell] \Leftrightarrow \xi_{\pi_i}(s) = \xi_{\pi_j}(\ell), 1 \leq i, j \leq k\}$$
$$\Pi^*(w_1, \ldots, w_k) = \{(\pi_1, \ldots, \pi_k) : w_i[k] = w_j[\ell] \Rightarrow \xi_{\pi_i}(k) = \xi_{\pi_j}(\ell), 1 \leq i, j \leq k\}$$
$$\Pi_{\neq}(w_1, \ldots, w_k) = \{(\pi_1, \ldots, \pi_k) \in \Pi(w_1, w_2, \ldots, w_k) : \pi_i(2j) \neq \pi_i(2(j + 1))),$$
$$1 \leq i \leq k, 1 \leq j \leq h - 1\}.$$

For Lemmata 3.3.3–3.3.5, we will always assume that the link function L satisfies Assumption $A(ii)$ and A'.

Lemma 3.3.3. Fix $h \geq 1$ and a matched word w of length $2h$. Then we have

$$\#\Pi_{\neq}(w) \leq K_h p^{1 + [\frac{|w| + 1}{2}]} n^{[\frac{|w|}{2}]}, \tag{3.36}$$

where K_h is some constant depending on h. ◇

Proof. Suppose we have k odd generating vertices, $\pi(2i_1 - 1), \pi(2i_2 - 1), \ldots, \pi(2i_k - 1)$ where $1 \leq i_1 < i_2 < \cdots < i_k < h$. We wish to emphasize that $i_k < h$ since $\{\xi_\pi(2h - 1), \xi_\pi(2h)\}$ cannot be a matched edge as $\xi_\pi(2h - 1) \neq \xi_\pi(2h)$ by Assumption A'.

Recall the definition of Type I and Type II generating vertices given in Section 3.1. Let t be the number of Type I odd generating vertices. Since the total number of generating vertices apart from $\pi(0)$ is $|w|$, we have $t \leq [|w|/2]$.

Now, fix a Type II generating vertex $\pi(2i - 1)$. Also, suppose we have already made our choices for the vertices $\pi(j)$, $j < 2i - 1$ which come before $\pi(2i - 1)$. Since $\pi(2i)$ is not a generating vertex,

$$\xi_\pi(2i) = \xi_\pi(j) \quad \text{for some} \quad j < 2i - 1. \tag{3.37}$$

Note that since $\pi(2i - 2) \neq \pi(2i)$, we cannot have $\xi_\pi(2i) = \xi_\pi(2i - 1)$ by Assumption A'. Now that the value of $\xi_\pi(j)$ has been fixed, for each value of

$\pi(2i)$, there can be at most Δ many choices for $\pi(2i-1)$ such that (3.37) is satisfied. Thus, we can only have at most $p\Delta$ many choices for the generating vertex $\pi(2i-1)$. In short, there are only $O(p)$ choices for any Type II odd generating vertex.

With the above crucial observation before us, it is now easy to conclude that

$$
\begin{aligned}
\#\Pi_{\neq}(w) &= O\left(p^{\#\text{even generating vertices}+\#\text{Type II vertices}} \times n^{\#\text{Type I vertices}}\right) \\
&= O(p^{1+[\frac{|w|+1}{2}]}n^{[\frac{|w|}{2}]}).
\end{aligned}
$$

$\square$

Lemma 3.3.4. (a) For every $h \geq 1$ even,

$$
p^{-1}\operatorname{E}\operatorname{Tr}B_p^h - \sum_{w \text{ matched, } |w|=h} p^{-1-h/2}n^{-h/2}\#\Pi_{\neq}^*(w) \to 0.
$$

(b) For every $h \geq 1$ odd, $\lim_{p\to\infty} p^{-1}\operatorname{E}\operatorname{Tr}B_p^h = 0.$ $\diamond$

Proof. Let $\hat{\mathbb{X}}_\pi$ be as defined in (3.6) with x_α replaced by $\hat{x}_\alpha$. From the fact that $\operatorname{E}\hat{x}_\alpha = 0$ and $B_{ii} = 0$ for all i, we have

$$
p^{-1}\operatorname{E}\operatorname{Tr}B_p^h = p^{-(1+h/2)}n^{-h/2} \sum_{w \text{ matched}} \sum_{\substack{\pi\in\Pi(w), \\ \pi(2i-2)\neq\pi(2i),\ \forall\ 1\leq i\leq h}} \operatorname{E}\hat{\mathbb{X}}_\pi. \quad (3.38)
$$

Fix a matched word w of length $2h$. It induces a partition on $2h$ L-values $\xi_\pi(1), \xi_\pi(2), \xi_\pi(3), \ldots, \xi_\pi(2h)$ resulting in $|w|$ many groups (partition blocks) where the values of ξ_π within a group are the same, but across groups they are different. Let C_k be the number of groups of size k. Clearly,

$$
C_2 + C_3 + \cdots + C_{2h} = |w| \text{ and } 2C_2 + 3C_3 + \cdots + 2hC_{2h} = 2h.
$$

Note that

$$
\sup_\alpha |\operatorname{E}\hat{x}_\alpha^2 - 1| = o(1) \text{ and } \sup_\alpha \operatorname{E}|\hat{x}_\alpha|^k \leq (\epsilon_p n^{1/4})^{k-2} \ \forall\ k \geq 2.
$$

Thus, if $\pi \in \Pi(w)$,

$$
\operatorname{E}|\hat{x}_{\xi_\pi(1)}\cdots\hat{x}_{\xi_\pi(2h)}| \leq (\epsilon_p n^{1/4})^{0.C_2+1.C_3+\cdots+(2h-2).C_{2h}} \leq (\epsilon_p n^{1/4})^{2h-2|w|}.
$$

Using Lemma 3.3.3,

$$
\sum_{\substack{\pi\in\Pi(w), \\ \pi(2i-2)\neq\pi(2i)\ \forall\ 1\leq i\leq h}} \operatorname{E}|\hat{x}_{\xi_\pi(1)}\cdots\hat{x}_{\xi_\pi(2h)}| \leq (\epsilon_p n^{1/4})^{2h-2|w|}\left(K_h p^{1+[\frac{|w|+1}{2}]}n^{[\frac{|w|}{2}]}\right)
$$

$$
= (\epsilon_p)^{2h-2|w|}n^{h/2-(|w|/2-[|w|/2])}p^{1+[\frac{|w|+1}{2}]}.
$$

Case I. Either $|w| < h$ or $|w| = h$ with h odd. Then clearly

$$\frac{1}{p^{1+h/2}n^{h/2}} \sum_{\substack{\pi \in \Pi(w), \\ \pi(2i-2) \neq \pi(2i) \ \forall \ 1 \leq i \leq h}} \mathrm{E} \, |\hat{x}_{\xi_\pi(1)} \hat{x}_{\xi_\pi(2)} \cdots \hat{x}_{\xi_\pi(2h)}| \to 0. \qquad (3.39)$$

Case II. $|w| = h$ with h even. Then

$$\lim_p \frac{1}{p^{1+h/2}n^{h/2}} \sum_{\substack{\pi \in \Pi(w), \\ \pi(2i-2) \neq \pi(2i) \ \forall \ 1 \leq i \leq h}} \mathrm{E} \, \hat{x}_{\xi_\pi(1)} \cdots \hat{x}_{\xi_\pi(2h)} = \lim_p \frac{1}{p^{1+h/2}n^{h/2}} \#\Pi_{\neq}(w)(1+o(1))^h$$

$$= \lim_p \frac{1}{p^{1+h/2}n^{h/2}} \#\Pi_{\neq}(w)$$

$$= \lim_p \frac{1}{p^{1+h/2}n^{h/2}} \#\Pi_{\neq}^*(w).$$

The proof now follows by using the above and (3.38). □

Fix jointly- and cross-matched words (w_1, w_2, w_3, w_4) of length $(2h)$ each. Let

$$\kappa = \text{total number of distinct letters in } w_1, w_2, w_3 \text{ and } w_4. \qquad (3.40)$$

Lemma 3.3.5. *For some constants C_h depending on h,*

$$\#\Big\{ (\pi_1, \pi_2, \pi_3, \pi_4) \in \Pi(w_1, w_2, w_3, w_4) :$$

$$\pi_i(0) \neq \pi_i(2), \ldots, \pi_i(2h-2) \neq \pi_i(2h), 1 \leq i \leq 4 \Big\}$$

$$\leq \begin{cases} C_h p^{2+2h} n^{2h} & \text{if } \kappa = 4h \text{ or } 4h-1 \\[2mm] C_h p^{4+[\frac{1+\kappa}{2}]} n^{[\frac{\kappa}{2}]} & \text{if } \kappa \leq 4h-2. \end{cases} \qquad (3.41)$$

◇

Proof. **Case I.** $\kappa = 4h-1$ or $4h$. Since the circuits are cross-matched, upon reordering the circuits, making circular shift and counting anti-clockwise if necessary, we may assume, without loss of generality, that $\xi_{\pi_1}(2h)$ does not match with any L-value in π_1 when $\kappa = 4h-1$. Because of cross-matching, when $\kappa = 4h$, we may further assume that $\xi_{\pi_2}(2h)$ does not match with any L-value in π_1 or π_2.

We first fix the values of the p-vertices $\pi_i(0), 1 \leq i \leq 4$, all of which are even generating vertices. Then we scan all the vertices from left to right, one circuit after another. We will then obtain a total of $(4h+2)$ generating vertices instead of $(4h+4)$ generating vertices which we would have obtained by usual counting.

In our dynamic counting, we scan the L-values in the following order:

$$\xi_{\pi_1}(1), \xi_{\pi_1}(2), \ldots, \xi_{\pi_1}(2h), \xi_{\pi_2}(1), \xi_{\pi_2}(2), \ldots, \xi_{\pi_2}(2h), \xi_{\pi_3}(1), \ldots, \xi_{\pi_4}(2h).$$

From the arguments given in Lemma 3.3.3, it is clear that for an odd vertex $\pi_i(2j - 1)$ to be Type I, both $\xi_{\pi_i}(2j - 1)$ and $\xi_{\pi_i}(2j)$ have to be the first appearances of two distinct L values. So, the total number of Type I n-generating vertices is at most $\kappa/2 \leq 2h$.

Case II. $\kappa \leq 4h - 2$. We again apply the crucial fact that if an odd vertex $\pi_i(2j - 1)$ is Type I, then as we scan the circuits left to right, both $\xi_{\pi_j}(2j - 1)$ and $\xi_{\pi_j}(2j)$ must be the first appearances of two distinct L-values, Since there are exactly κ distinct L-values, the number of Type I odd vertices is not more than $\kappa/2$. Combining this with the fact that the total number of generating vertices equals $(\kappa + 4)$, we get the required bound. □

Lemma 3.3.6. For each fixed $h \geq 1$,

$$\sum_{p=1}^{\infty} \mathrm{E}\left[p^{-1}(\mathrm{Tr}\, B_p^h - \mathrm{E}(\mathrm{Tr}\, B_p^h))\right]^4 < \infty.$$

◇

Proof. As argued earlier and from the fact that the diagonal elements of B_p are all zero, we have

$$\mathrm{E}\left[p^{-1}(\mathrm{Tr}\, B_p^h - \mathrm{E}(\mathrm{Tr}\, B_p^h))\right]^4 = p^{-4-2h}n^{-2h}\sum_{*}^{*}\sum \mathrm{E}\left(\prod_{i=1}^{4}(\hat{\mathbb{X}}_{\pi_i} - \mathrm{E}\hat{\mathbb{X}}_{\pi_i})\right)$$

where the outer sum $\sum_{*}$ is over all quadruples of jointly- and cross-matched words (w_1, w_2, w_3, w_4), each of length $2h$. The inner sum $\sum^{*}$ is over all quadruples of circuits $(\pi_1, \pi_2, \pi_3, \pi_4) \in \Pi(w_1, w_2, w_3, w_4)$ such that

$$\pi_i(0) \neq \pi_i(2), \ldots, \pi_i(2h - 2) \neq \pi_i(2h), i = 1, 2, 3, 4.$$

Note that by definition of κ, $\kappa \leq 4h$ for any jointly-matched quadruple of words (w_1, w_2, w_3, w_4) of total length $8h$. Fix w_1, w_2, w_3, w_4, jointly- and cross-matched.

Case I. $\kappa = 4h$ or $4h - 1$. Then the maximum power with which any $\hat{x}_\alpha$ can occur in $\prod_{i=1}^{4}\hat{\mathbb{X}}_{\pi_i}$ is bounded by 4. But since $\sup_\alpha \mathrm{E}(\hat{x}_\alpha)^4 < \infty$, we immediately have $|\mathrm{E}\left(\prod_{i=1}^{4}(\hat{\mathbb{X}}_{\pi_i} - \mathrm{E}\hat{\mathbb{X}}_{\pi_i})\right)| < \infty$. Thus, by Lemma 3.3.5,

$$p^{-4-2h}n^{-2h}\sum_{*:\ \kappa\in\{4h-1,4h\}}\sum^{*}\mathrm{E}\left(\prod_{i=1}^{4}(\hat{\mathbb{X}}_{\pi_i} - \mathrm{E}\hat{\mathbb{X}}_{\pi_i})\right) = p^{-4-2h}n^{-2h}\,\mathrm{O}(p^{2+2h}n^{2h})$$

$$= \mathrm{O}(p^{-2}).$$

Case II. Suppose now that $\kappa = 4h - k, k \geq 2$. Borrowing notation from Lemma 3.3.4, we have

$$C_2 + C_3 + \cdots + C_{8h} = \kappa \quad \text{and} \quad 2C_2 + 3C_3 + \cdots + 8hC_{8h} = 8h.$$

These two equations immediately give

$$C_3 + 2C_4 + \cdots + (8h - 2)C_{8h} = 8h - 2\kappa = 2k.$$

It is also easy to see that

$$C_{2k+2+i} = 0 \; \forall \; i \geq 1 \quad \text{and} \quad C_5 + 2C_6 + \cdots + (2k - 2)C_{2k+2} \leq 2k - 2.$$

Since the input variables are truncated at $\epsilon_p n^{1/4}$ and since $\sup_\alpha E(x_\alpha^4) < \infty$,

$$\sup_\alpha E \, | \prod_{i=1}^{4} \hat{\mathbb{X}}_{\pi_i} | \leq E(\hat{x}_\alpha)^4 n^{\frac{C_5 + 2C_6 + \cdots + (2k-2)C_{2k+2}}{4}} = O(n^{\frac{2k-2}{4}}).$$

Thus, by Lemma 3.3.5, for any $k \geq 2$,

$$p^{-4-2h} n^{-2h} \sum_{*:\; \kappa \in \{4h-k\}} \sum^{*} E \, \Big(\prod_{i=1}^{4} (\hat{\mathbb{X}}_{\pi_i} - E\,\hat{\mathbb{X}}_{\pi_i}) \Big) \leq p^{-4-2h} n^{-2h} C_h p^{4 + [\frac{4h-k+1}{2}]}$$

$$\times n^{[\frac{4h-k}{2}]} O(n^{\frac{k-1}{2}})$$

$$= O(p^{-(\frac{k}{2}+1)}).$$

For a quick explanation of the first equality above, just note that the total power of n in the previous expression is negative.   □

We are now ready to summarize the results of Theorem 3.3.2 and Lemma 3.3.3–3.3.6 in the following theorem in Regime II.

Theorem 3.3.7. Let $p, n \to \infty$ so that $p/n \to 0$ and Assumption IV holds. Suppose L_p satisfies Assumptions A(ii), A' and B. Suppose $\{\epsilon_p\}$ satisfying $\{\epsilon_p\} \downarrow 0$ and $\epsilon_p p^{1/4} \to \infty$ is appropriately chosen. Then

(a) For every $h \geq 1$ even,

$$p^{-1} E \operatorname{Tr} B_p^h - \sum_{\substack{w \text{ matched,} \\ |w|=h}} p^{-(1+h/2)} n^{-h/2} \# \Pi_{\neq}^{*}(w) \to 0.$$

(b) For every $h \geq 1$ odd, $\lim_{p \to \infty} p^{-1} E \operatorname{Tr} B_p^h = 0$.

(c) For each $h \geq 1$, $p^{-1} \operatorname{Tr} B_p^h - E\, p^{-1} \operatorname{Tr} B_p^h \overset{a.s.}{\to} 0$.

(d) $\beta_{2h} \equiv \limsup\limits_{p} \sum\limits_{\substack{w \text{ matched,} \\ |w|=2h}} p^{-(1+h)}n^{-h}\#\Pi^*_{\neq}(w)$, satisfies Carleman's or Riesz's condition.

Hence, the sequence $\{F^{N_p}\}$ is almost surely tight. Every sub-sequential limit is symmetric about 0 and its moments are dominated by moments of a Gaussian distribution. The LSD of $\{F^{N_p}\}$ exists almost surely, iff

$$\lim \sum_{w \in W_{2h}} p^{-(1+h)}n^{-h}\#\Pi^*_{\neq}(w) \text{ exists.}$$

These give the $(2h)$-th moment of the LSD.   ◇

Now we will deal with two matrices with different link functions $L^{(1)}$ or $L^{(2)}$. Suppose these two link functions satisfy Assumptions A(ii), A′ and B and they agree on the set $\{(i,j) : 1 \leq i \leq p,\ 1 \leq j \leq n \text{ and } i < j\}$. Then we show that the LSD of the corresponding matrices $N_p(X)$ are identical.

To distinguish between the same quantities for the two link functions, we shall add superscripts (1) and (2).

Theorem 3.3.8. *Let $L^{(1)}$ and $L^{(2)}$ be two link functions which satisfy Assumptions A(ii), A′ and B and agree on the set $\{(i,j) : 1 \leq i \leq p,\ 1 \leq j \leq n,\ i < j\}$. Then, for each matched word w of length $(4h)$ with $|w| = 2h$,*

$$\frac{1}{p^{h+1}n^h}\left(\#\Pi^{*\,(1)}_{\neq}(w) - \Pi^{*\,(2)}_{\neq}(w)\right) \to 0.$$

Hence, in Regime II under Assumption IV, $F^{N_p^{(i)}}, i = 1,2$ have identical LSD behavior.   ◇

Proof. Define for each link function $L^{(i)}$, $i = 1,2$

$$\Gamma^{(i)}_j := \{\pi \in \Pi^{*\,(i)}_{\neq}(w) : 1 \leq \pi(2j+1) \leq p\},\ j = 1,2,\ldots,2h.$$

Now, it is enough to prove that for a fixed j,

$$\frac{1}{p^{h+1}n^h}\#\Gamma^{(i)}_j \to 0.$$

Consider the transformation $\pi \mapsto \hat{\pi}$, where $\hat{\pi}$ is also a circuit with

$$\hat{\pi}(0) = \pi(2j),\ \ \hat{\pi}(1) = \pi(2j+1),\ldots,\pi(4h-1) = \pi(2j-1),\ \ \hat{\pi}(4h) = \pi(2j).$$

Then it is easy to show that the map $\pi \mapsto \hat{\pi}$ is a bijection between $\Gamma^{(i)}_j$ and $\Gamma^{(i)}_1$ and therefore, $\#\Gamma^{(i)}_j = \#\Gamma^{(i)}_1$.

Observe that $\pi(1)$ is always a Type I odd generating vertex. By Lemma 3.3.3, we have $|\Pi^{*\,(i)}(w)| = O(p^{h+1}n^h)$. Since, in the definition of $\Gamma^{(i)}_1$ we are restricting one of the Type I odd generating vertices to be a p generating vertex, we are going to lose a factor of n in the bound and pick up a factor of p instead. Thus, $|\Gamma_1^{(i)}| = O(p^{h+2}n^{h-1})$ and hence

$$\frac{1}{p^{h+1}n^h} \#\Gamma_j^{(i)} = O\left(\frac{p}{n}\right) \to 0.$$

$\square$

The Symmetric Toeplitz link function $L(i,j) = |i - j|$ does not satisfy Assumption A' but the non-symmetric Toeplitz link function $L(i,j) = i - j$ does. Hence the above result is not applicable for this pair of link functions. However, under Assumption A, we can directly claim the closeness of their LSD, but only in probability. In Theorem 3.3.9 we work with Assumption A and show the closeness of the LSD in probability when two link functions agree on the above set.

Theorem 3.3.9. Assume $\{x_\alpha\}$ are independent with mean zero and $\sup_\alpha \mathrm{E}\, x_\alpha^4 < \infty$. Suppose $L^{(1)}$ and $L^{(2)}$ are two link functions such that $L^{(1)}(i,j) = L^{(2)}(i,j)$ on the set $\{(i,j) : 1 \leq i \leq p,\ p \leq j \leq n\}$ and both satisfy Assumption A. Set $X = ((x_{L^{(1)}(i,j)}))_{p\times n}$ and $Y = ((x_{L^{(2)}(i,j)}))_{p\times n}$. Then

$$W_2 := W_2\left(F^{\sqrt{\frac{n}{p}}(A_p(X)-I_p)},\ F^{\sqrt{\frac{n}{p}}(A_p(Y)-I_p)}\right) \xrightarrow{\mathrm{P}} 0.$$

$\diamond$

Proof. Let $X = [X_0 : Z]$ and $Y = [Y_0 : Z]$ where X_0 and Y_0 are $p \times p$ sub-matrices of X and Y respectively. Note that

$$\begin{aligned}
\mathrm{E}[W_2^2] &\leq\ n^{-1}p^{-2}\,\mathrm{E}\,\mathrm{Tr}(XX' - YY')^2 \\
&=\ n^{-1}p^{-2}\,\mathrm{E}\,\mathrm{Tr}(X_0X_0' - Y_0Y_0')^2 \\
&\leq\ 2n^{-1}\left(\mathrm{E}\,\mathrm{Tr}(p^{-1}X_0X_0')^2 + \mathrm{E}\,\mathrm{Tr}(p^{-1}Y_0Y_0')^2\right).
\end{aligned}$$

Calculations similar to those done in Regime I now imply that

$$\mathrm{E}\,p^{-1}\mathrm{Tr}(p^{-1}X_0X_0')^2 \leq K \quad \text{and} \quad \mathrm{E}\,p^{-1}\mathrm{Tr}(p^{-1}Y_0Y_0')^2 \leq K,$$

for some constant K. Since $p/n \to 0$, the result follows immediately. $\square$

We are now ready to establish a surprising result on the LSD of $\left(\frac{n}{p}\right)^{1/2}(A_p(X_p) - I_p)$ when X_p is any of the four matrices Toeplitz, Hankel, Circulant and Reverse Circulant.

Theorem 3.3.10. If Assumption IV holds, $p \to \infty$ and $p/n \to 0$, then $F^{\sqrt{\frac{n}{p}}(A_p(X)-I_p)} \to \mathcal{L}_T$ a.s. for X equal to T, H, R, C, $H_p^{(s)}$ or $R_p^{(s)}$. $\diamond$

Proof. We first consider the Toeplitz matrix. In view of Theorem 3.3.7, to show the existence of the LSD, we only need to show that for each matched word of length $(4h)$, with $|w| = 2h$,

$$\lim_{p \to \infty} p^{-(1+h)} n^{-h} \#\Pi^*_{\neq}(w) \quad \text{exists.}$$

Note that if $\pi \in \Pi^*_{\neq}(w)$, then $\xi_\pi(i) \neq \xi_\pi(i+1)$ for all i odd. Hence, there can be only two types of matching between the L-values as listed below:

1. **Double bond.** A matching is said to have a double bond if there exists two consecutive odd-even L-values which match pair-wise with another two consecutive odd-even L-values. There can be again two possibilities,

    (a) **Crossing.** $\xi_\pi(2i+1) = \xi_\pi(2j+2)$ and $\xi_\pi(2i+2) = \xi_\pi(2j+1)$ for some $i < j$.

    (b) **Non-crossing.** $\xi_\pi(2i + 1) = \xi_\pi(2j + 1)$ and $\xi_\pi(2i + 2) = \xi_\pi(2j + 2)$ for some $i < j$.

2. **Single bond.** The remaining types of pairing will be termed a single bond. They give rise to the following types of equations:
    $\xi_\pi(2i+1) = \xi_\pi(s)$ and $\xi_\pi(2i+2) = \xi_\pi(t)$ where $\{s, t\} \neq \{2j+1, 2j+2\}$ for all j.

Claim. Let w be a matched word of length $4h$. If w has a single bond, then

$$\lim_{p \to \infty} p^{-(1+h)} n^{-h} \#\Pi^*_{\neq}(w) = 0.$$

Proof. Recall the definition of a Type I generating vertex. It is clear that if $\pi(2i - 1)$ is Type I, then

$$\xi_\pi(s) = \xi_\pi(2i - 1) \quad \text{or} \quad \xi_\pi(s) = \xi_\pi(2i) \Rightarrow s > 2i.$$

We show that the number of Type I odd generating vertices is strictly less than h. Then the proof will follow immediately since the total number of generating vertices is $(2h + 1)$.

The total number of odd vertices (generating, non-generating together) is $(2h)$. Let us form two mutually exclusive and exhaustive sets U and V where U contains all odd vertices involved in double bonds and V contains the rest of the odd vertices. Quite clearly, if $\{\xi_\pi(2u_1 - 1), \xi_\pi(2u_1)\} = \{\xi_\pi(2u_2 - 1), \xi_\pi(2u_2)\}$ is a double bond with $u_1 < u_2$, then $\pi(2u_1 - 1), \pi(2u_2 - 1) \in U$ and $\pi(2u_1 - 1)$ is a Type I odd generating vertex. Thus, the total number of Type I odd generating vertices in U is $(1/2)\#U$. Next we argue that the total number of Type I generating vertices in V is strictly less than $(1/2)\#V$ and hence, the total number of Type I odd generating vertices is strictly less than h.

Note that exactly half of the odd vertices which are involved in double bond matching are Type I odd vertices. Now, let us count the number of

Type I odd vertices which are involved in single bond matches. We list the Type I odd generating vertices in V as

$$V_1 := \{\pi(2g_1 - 1), \pi(2g_2 - 1), \ldots, \pi(2g_s - 1)\}$$

and the rest of the vertices of V as

$$V_2 := V - V_1 = \{\pi(2d_1 - 1), \pi(2d_2 - 1), \ldots, \pi(2d_t - 1)\}.$$

For $i \neq j$, write $2i - 1 \leftrightarrow 2j - 1$ if $\{\xi_{\pi(2i-1)}, \xi_{\pi(2i)}\} \cap \{\xi_{\pi(2j-1)}, \xi_{\pi(2j)}\} \neq \emptyset$. From the definition of a Type I odd generating vertex, it is clear that $2g_i - 1 \leftrightarrow 2g_j - 1$ is not possible.

We claim that $2d_i - 1 \leftrightarrow 2g_l - 1$ and $2d_i - 1 \leftrightarrow 2g_m - 1, l \neq m$ cannot occur simultaneously. Because if that happens, then we have

$$\begin{aligned}\pi(2d_i - 2) - \pi(2d_i - 1) &= \pi(2g_a) - \pi(2g_a - 1)\\ \pi(2d_i) - \pi(2d_i - 1) &= \pi(2g_b - 2) - \pi(2g_b - 1), \quad \{a, b\} = \{l, m\}.\end{aligned}$$

Subtracting we get,

$$\pi(2g_b - 1) - \pi(2g_a - 1) = \pi(2d_i - 2) - \pi(2d_i) + \pi(2g_b - 2) - \pi(2g_a).$$

Vertices on the right side are all even and hence the number of choices on the right side is $O(p)$. On the other hand, in the left side we have two Type I odd vertices each of which has free choices of the order n. This is an impossibility. So in summary, the relation $\leftrightarrow$ associates a vertex in V_1 with two vertices in V_2 (single bond), but a vertex in V_2 is not associated to two distinct vertices in V_1. Therefore, $\#V_1 < \#V_2$. So the total number of Type I odd generating vertices in V is strictly less than $\#V/2$. Thus, the total number of Type I odd vertices is strictly less than $(\#U + \#V)/2 = h$ which concludes the proof of the claim. □

Reverting to the main proof for the Toeplitz matrix, we may now, for the rest of our calculation, consider only those words which produce no single bond. By the circuit constraint, we have

$$\xi_\pi(1) - \xi_\pi(2) + \cdots + \xi_\pi(2i-1) - \xi_\pi(2i) + \cdots + \xi_\pi(4h-1) - \xi_\pi(4h) = 0. \quad (3.42)$$

Note that if i forms a non-crossing double bond with j then $\xi_\pi(2i-1) - \xi_\pi(2i) = \xi_\pi(2j-1) - \xi_\pi(2j)$. If w has at least one non-crossing double bond then (3.42) leads to a non-trivial restriction on the vertices of the circuit reducing the number of even generating vertices by one and thus $p^{-(1+h)}n^{-h}\#\Pi_{\neq}^*(w) \to 0$. Thus, we may restrict ourselves to those words which give rise to only crossing double bonds. Let us fix one such word w of length $(4h)$, with $|w| = 2h$. Now let us consider a pair of equations forming a crossing double bond:

$$\begin{aligned}\pi(2i) - \pi(2i + 1) &= \pi(2j + 2) - \pi(2j + 1)\\ \pi(2i + 2) - \pi(2i + 1) &= \pi(2j) - \pi(2j + 1), \quad \text{for } i < j.\end{aligned} \quad (3.43)$$

In the above equations $\pi(2i+1)$ is a Type I odd generating vertex and $\pi(2j+1)$ is an odd non-generating vertex which pairs up with $\pi(2i+1)$. Note that,

$$\begin{aligned}
\pi(2j+1) &= \pi(2j+2) - \pi(2i) + \pi(2i+1) \\
&= \pi(2j) - \pi(2i+2) + \pi(2i+1).
\end{aligned}$$

Since $-p < \pi(2j+2) - \pi(2i) < p$, if $\pi(2i+1)$ is chosen freely between p and $(n-p)$, we do not have any restriction on even vertices imposed by odd vertices, that is, even vertices can be chosen independent of the choice of odd vertices satisfying the following restrictions:

$$\pi(2i) - \pi(2i+2) = \pi(2j+2) - \pi(2j). \tag{3.44}$$

But if $\pi(2i+1) \in \{1, 2, \ldots, p-1\} \cup \{n-p+1, n-p+2, \ldots, n\}$, then the choice of even vertices is restricted by the choice of $\pi(2i+1)$ because of the constraint $1 \le \pi(2j+1) \le n$.

Define a new word $\hat{w}$ of length $2h$, so that

$$\hat{w}[i] = \hat{w}[j] \quad \text{iff} \quad w[2i-1] = w[2j] \quad \text{and} \quad w[2i] = w[2j-1].$$

It is easy to see that $|\hat{w}| = h$. Let $\hat{\pi}$ be a circuit of length $2h$ given by $\hat{\pi}(i) = \pi(2i)$. Since $\frac{p}{n} \to 0$,

$$\begin{aligned}
\frac{\#\Pi_{\neq}^*(w)}{n^h p^{1+h}} &= \frac{\#\left[\Pi_{\neq}^*(w) \cap \{\pi : p \le \pi(2i-1) \le n-p, \ \forall 1 \le i \le 2h\}\right]}{p^{1+h} n^h} + \mathrm{o}(1) \\
&= \frac{\#\{\hat{\pi} : \hat{w}[i] = \hat{w}[j] \Rightarrow \hat{\pi}(i) - \hat{\pi}(i+1) = \hat{\pi}(j+1) - \hat{\pi}(j)\}}{p^{1+h}} + \mathrm{o}(1).
\end{aligned}$$

The above restriction on $\hat{\pi}$ is precisely the same restriction on pair-matched circuits of length $2h$ that appeared in the proof of the LSD for the symmetric Toeplitz matrix, see Theorem 2.2.1 in Chapter 2.

Also, note that every word of length $2h$ with h letters can be obtained through this procedure. Therefore, in this case, the LSD of $\left(\frac{n}{p}\right)^{1/2} (A_p(T) - I_p)$ is $\mathcal{L}_T$. This completes the proof for the non-symmetric Toeplitz matrix.

Now consider $X = T_p^{(s)}$. The symmetric link function $L(i,j) = |i-j|$ does not obey Assumption A'. However, since it obeys Assumption A, by Theorem 3.3.9, $\left(\frac{n}{p}\right)^{1/2} (A_p(T^{(s)}) - I_p)$ has the same LSD as for the non-symmetric case, namely $\mathcal{L}_T$. This completes the proof for the symmetric Toeplitz matrix.

We now consider the two Hankel matrices. By Theorem 3.3.8, it is enough to prove the existence of the LSD for the symmetric Hankel case. Now we can simply imitate the argument for the non-symmetric Toeplitz matrix in Regime II. Here also the essential contribution comes from the words that have only crossing double bonds. But we now have a different pair of equations instead of (3.43) that we had in the Toeplitz case. We now have

$$\begin{aligned} \pi(2i) + \pi(2i+1) &= \pi(2j+2) + \pi(2j+1), \\ \pi(2i+2) + \pi(2i+1) &= \pi(2j) + \pi(2j+1), \qquad \text{for } i < j. \end{aligned} \tag{3.45}$$

But once we cancel the odd vertices as we did in the Toeplitz case, we are again reduced to the Toeplitz-type restrictions on even vertices. We omit the details. Hence we conclude that the LSD exists and is $\mathcal{L}_T$, concluding the proof for Hankel matrices.

Now consider the Reverse Circulant matrices. Since we are in Regime II, invoking Theorem 3.3.8, it suffices to work with the symmetric link function $L(i, j) = (i + j - 2) \bmod n$. We again imitate the proof of the Toeplitz case. Here a typical restriction in a word containing only crossing double bonds reads as,

$$(\pi(2i) + \pi(2i+1) - 2) \bmod n = (\pi(2j+2) + \pi(2j+1) - 2) \bmod n \tag{3.46}$$
$$(\pi(2i+2) + \pi(2i+1) - 2) \bmod n = (\pi(2j) + \pi(2j+1) - 2) \bmod n \text{ for } i < j.$$

As in the case of Toeplitz and Hankel matrices, we choose generating odd vertices between p and $(n - p)$; the only restriction that even vertices need to satisfy is

$$(\pi(2i) - \pi(2i+2) - 2) \bmod n = (\pi(2j+2) - \pi(2j) - 2) \bmod n.$$

But since p is negligible compared to n, and an even vertex can take values between 1 and p, this is equivalent to the usual Toeplitz restriction

$$\pi(2i) - \pi(2i+2) = \pi(2j+2) - \pi(2j).$$

Hence, the LSD exists and the limit is exactly $\mathcal{L}_T$. This completes the proof for the Reverse Circulant matrices.

The proof for the Circulant matrices is now easy. All we need to do is note that on and above the diagonal, the circulant link function exactly matches with the link function $L(i, j) = j - i$. But the link function $L(i, j) = j - i$ is nothing but the non-symmetric Toeplitz link function once we index input random variables as $\{x_{-i} : i \in \mathbb{Z}\}$. We may now invoke Theorem 3.3.8 to conclude the proof. $\qquad\square$

Remark 3.3.2. The LSD results in Regimes I and II are summarized in Table 3.1. The histogram for the ESD of $A_p(T)$ is given in Figure 3.2, illustrating Theorem 3.2.7 for the Toeplitz matrix. It is not too difficult to show theoretically that the support is unbounded. The more interesting observation is that the support of the LSD excludes a neighborhood of zero. Recall that for the S matrix with $y < 1$, the infimum of the support is $(1 - \sqrt{y})^2$. It will be interesting to establish the infimum of the support in the Toeplitz case. Apparently, such results are of interest due to the numerical technique of "premultiplication" by patterned matrices which is used to solve large systems of sparse equations, see for example Kaltofen (1995)[68]. ◇

TABLE 3.1
Words and moments for XX' matrices.

MATRIX	w Cat.	Other w	β_k and LSD
$p/n \to 0$ $\sqrt{\frac{n}{p}}(S - I_p)$ $(S = n^{-1}W_pW_p')$	1 (Cat. in p)	0	$\frac{(2k)!}{k!(k+1)!}$, $\mathcal{L}_W$
$\sqrt{\frac{n}{p}}(A_p(X) - I_p)$ $(X = T,\ H,\ R,\ C)$			$\mathcal{L}_T$
$p/n \to y \neq 0, \infty$ $S = n^{-1}W_pW_p'$	1	0	$\sum_{t=0}^{k-1} \frac{1}{t+1}\binom{k}{t}\binom{k-1}{t}y^t$, $\mathcal{L}_{MPy}$
$A_p(X)$ $(X = T, H, R_p, C_p)$			different, but universal

3.4 Exercises

1. Complete the proof of Theorem 3.2.2 when Assumption III holds.

2. Complete the proof of Theorem 3.2.3 when Assumption III holds.

3. Show that except for the S matrix, all the other LSDs obtained in this chapter have unbounded support.

4. Show that in Regime I, the LSD of the Sample variance-covariance matrix adjusted for the mean is also MP_y.

5. Suppose $p = yn$ where y is an integer. Show that the LSD of $A_p(C)$ in Regime I is the distribution of ξR^2 where ξ is a Bernoulli random variable with $P(\xi = 0) = 1 - P(\xi = 1) = (1 - 1/y)$, R is distributed

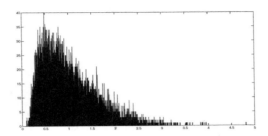

FIGURE 3.2
Histogram for the ESD of $(1/n)TT'$ for $p = 300$, $n = 900$, 50 replications. Input is i.i.d. $U(-\sqrt{3}, \sqrt{3})$.

as the Reverse Circulant LSD given in Chapter 2 and they are independent. Hint: Using the expressions developed in the proofs, show that the h-th moment of the LSD of $A_p(C)$ is given by

$$\beta_h^C = \sum_{t=0}^{h-1} y^t |\mathcal{W}_{t,h}^0| y^{h-t-1} = y^{h-1} \sum_{t=0}^{h-1} |\mathcal{W}_{t,h}^0| = y^{h-1} h!.$$

6. Show that $A_p(W)$ and $A_p(W^s)$ have identical LSD in Regime I.

7. Suppose $p/n \to 1$. Show that the LSD of $A_p(W^s)$ is MP_1.

8. Show that the LSD of $A_p(T)$ and $A_p(H^{(s)})$ are identical in Regime I. Hint: In the spirit of Bryc, Dembo and Jiang (2006, Remark 2.1)[34], let $P_n := ((\mathbb{I}\{i + j = n + 1\}))_{i,j=1}^n$ be a symmetric permutation matrix. Verify that $\tilde{H}_p^{(s)} := T_p P_n$ is a $p \times n$ Hankel matrix, with symmetric link, for the input sequence $\{x_{n+1+k} : k \geq 0\}$.

9. Show that the LSD of $A_p(C)$ and $A_p(R^{(s)})$ are identical in Regime I. Hint: Let $\tilde{C}_p$ be the $p \times n$ Circulant matrix with the link function $L(i, j) = (n + i - j) \bmod n$. Verify that $R_p^{(s)} := \tilde{C}_p P_n$ is the $p \times n$ symmetric Reverse Circulant matrix for the input sequence $\{x_{1+k} : k \geq 0\}$ where P_n is as in the previous exercise.

10. Show that the LSD of $A_p(H)$ and $A_p(H^{(s)})$ are different.

11. Simulate the matrices of this chapter and check how fast the LSDs take hold in Regimes I and II. In particular, check that some of these LSDs are identical.

12. Show that all results of this chapter remain valid if we replace Δ by Δ_p in Assumption A and impose the restriction $\Delta_p^k/p \to 0$ for every integer k.

4

k-Circulant matrices

For positive integers k and n, the $n \times n$ k-Circulant matrix with the input sequence $\{x_k\}$, is defined as

$$
A_{k,n} = \begin{bmatrix}
x_0 & x_1 & x_2 & \cdots & x_{n-2} & x_{n-1} \\
x_{n-k} & x_{n-k+1} & x_{n-k+2} & \cdots & x_{n-k-2} & x_{n-k-1} \\
x_{n-2k} & x_{n-2k+1} & x_{n-2k+2} & \cdots & x_{n-2k-2} & x_{n-2k-1} \\
\vdots & \vdots & \vdots & \vdots & \vdots & \vdots \\
x_k & x_{k+1} & x_{k+2} & \cdots & x_{k-2} & x_{k-1}
\end{bmatrix}_{n \times n}.
$$

The subscript of the entries are to be read modulo n. For $1 \leq j < n - 1$, its $(j + 1)$-th row is obtained by giving its j-th row a right circular shift by k positions (equivalently, $k \bmod n$ positions). In this chapter we study the LSD of this class of matrices.

When $k = 1$, we obtain the usual *Circulant* matrix, given as

$$
C_n = \begin{bmatrix}
x_0 & x_1 & x_2 & \cdots & x_{n-2} & x_{n-1} \\
x_{n-1} & x_0 & x_1 & \cdots & x_{n-3} & x_{n-2} \\
x_{n-2} & x_{n-1} & x_0 & \cdots & x_{n-4} & x_{n-3} \\
\vdots & \vdots & \vdots & \vdots & \vdots & \vdots \\
x_1 & x_2 & x_3 & \cdots & x_{n-1} & x_0
\end{bmatrix}_{n \times n}.
$$

We have already mentioned the connection between the usual Circulant matrix and the Toeplitz matrix. Moreover, the eigenvalues of the Circulant matrix also arises crucially in time series analysis. For instance, the periodogram of a sequence $\{x_l\}_{l \geq 0}$ is a straightforward function of the eigenvalues of the corresponding Circulant matrix. The study of the properties of the periodogram is fundamental in the spectral analysis of time series. See for instance Fan and Yao (2003)[54].

We have already seen in Chapter 2 that the *Symmetric Circulant matrix*, which is the symmetric version of C_n with $x_k = x_{n-k}$ $\forall k \geq 1, 1 \leq k \leq n-1$, has the Gaussian LSD.

The k-Circulant matrix with $k = n - 1$ is the *Reverse Circulant matrix* that we encountered in Chapter 2. This matrix has gained importance recently due to its connection with *half independence* which we shall describe later in Chapter 9. It can be checked that for an arbitrary input sequence, $A_{k,n}$ is symmetric if and only if $k = n - 1$.

General k-Circulant matrices and their block versions appear in multi-level supersaturated design of experiments (see Georgiou and Koukouvinos (2006) [58]) and time series analysis (see Pollock (2002)[84]). If A is an $n \times n$ matrix with entries 0 or 1, then the digraph of A is defined as the graph with vertices $\{1, \ldots, n\}$ and there is an edge between i and j if the (i, j)th entry of A is 1. Now suppose A is a k-Circulant matrix of order n with entries 0 or 1. An important question one studies here is, what are the conditions on k and n which ensure that there exists an integer m such that $A^m = J_n$ where J_n is a matrix all of whose elements are 1? The study of such solutions are generally achieved by considering the digraphs of A. These are also closely related in graph theory to the spectra of De Bruijn digraphs. The adjacency matrix of a De Bruijn graph is a k-Circulant matrix. For results of this type we refer to Strock (1992)[99] and Wu, Jia and Li (2002)[109]. See also Davis (1979)[47] and Pollock (2002)[84].

Clearly, the parity of k and n determines the number of zero eigenvalues. Moreover, as we shall see later, there is a formula solution known for the eigenvalues of $A_{k,n}$. This formula exhibits the differing behavior of the eigenvalues depending on the common prime factors between n and p and helps to a large extent in obtaining the LSDs for suitable combinations of k and n. If k is fixed and n increases to infinity, then different LSDs are possible for different sub-sequences. We shall also consider some scenarios where both k and n tend to ∞ with suitable restrictions.

As is evident from the simulation results given in Figures 4.1–4.4 (see later), a variety of LSDs are possible. For instance, suppose $\{x_i\}$ are i.i.d. $N(0,1)$. Then the LSD of $F^{n^{-1/2}A_{1,n}}$ is the bivariate normal distribution. On the other hand, if $k = n^{o(1)}$ (≥ 2) and $gcd(k, n) = 1$, then the LSD of $F^{n^{-1/2}A_{k,n}}$ is degenerate at zero, in probability. We shall work with some specific sub-sequences of k and n to obtain non-trivial limits.

In this chapter we shall find it convenient to use the *in probability* notion of convergence of the ESD. However, since the eigenvalues of these matrices are not in general real, the method of moments is not applicable. Instead, we shall use the method of *normal approximation*.

4.1   Normal approximation

The method of our choice in this chapter is *normal approximation*, so ubiquitous in statistics. The distribution of a statistic is often asymptotically normal and it has been of much interest to study how good the normal approximation is for a given statistic. The most basic normal approximation results are the *Berry-Esseen* bounds. One version of the bound which we shall heavily use in this chapter is given in Lemma 4.1.1 below. Its proof, which we omit, follows easily from Corollary 18.1, page 181 and Corollary 18.3, page 184 of Bhattacharya and Rao (1976)[22].

For a set $B \subseteq \mathbb{R}^d, d \geq 1$, let ∂B and $(\partial B)^\eta$ denote, respectively, the boundary and the "η-boundary" of the set B. That is,

$$\partial B = \{y \in \mathbb{R}^d : \text{ any neighborhood of } y \text{ contains an element of } B \cap B^c\},$$
$$(\partial B)^\eta = \{y \in \mathbb{R}^d : \|y - z\| \leq \eta \text{ for some } z \in \partial B\}.$$

By $\Phi_d(\cdot)$ we always mean the probability distribution of a d-dimensional standard normal vector. We drop the dimension subscript d and write just $\Phi(\cdot)$. The dimension will be clear from the context.

Let $\mathcal{C}$ denote the class of all Borel measurable *convex* subsets of $\mathbb{R}^d$.

Lemma 4.1.1. Let $X_1, \ldots, X_k$ be independent $\mathbb{R}^d$-valued random vectors which and have zero mean and an average positive-definite covariance matrix $V_k = k^{-1} \sum_{j=1}^k \mathrm{Cov}(X_j)$. Let G_k denote the distribution of $k^{-1/2} T_k (X_1 + \cdots + X_k)$, where T_k is the symmetric, positive-definite matrix satisfying $T_k^2 = V_k^{-1}$, $n \geq 1$. Suppose for some $\delta > 0$,

$$\rho_{2+\delta} = k^{-1} \sum_{j=1}^k \mathrm{E} \, \|X_j\|^{(2+\delta)} < \infty.$$

Then

(a) There exists $C > 0$, depending only on d, such that

$$\sup_{B \in \mathcal{C}} |G_k(B) - \Phi_d(B)| \leq Ck^{-\delta/2} [\lambda_{\min}(V_k)]^{-(2+\delta)} \rho_{2+\delta}.$$

(b) There exists $C > 0$, depending only on d, such that for $\eta = C\rho_{2+\delta} n^{-\delta/2}$ and any Borel set A,

$$|G_k(A) - \Phi_d(A)| \leq Ck^{-\delta/2} [\lambda_{\min}(V_k)]^{-(2+\delta)} \rho_{2+\delta} + 2 \sup_{y \in \mathbb{R}^d} \Phi_d((\partial A)^\eta - y).$$

$\diamond$

The $(2+\delta)$ moment condition in the above lemma necessitates the following assumption on the input sequence and will be in force throughout this chapter.

Assumption III(δ) $\{x_i\}$ are independent, $\mathrm{E}(x_i) = 0$, $\mathrm{Var}(x_i) = 1$ and $\sup_i \mathrm{E} \, |x_i|^{2+\delta} < \infty$ for some $\delta > 0$.

As we shall soon see, the normal approximation method is most suited for the k-Circulant matrices. What makes this work is the explicit formula for the eigenvalues of $A_{k,n}$ derived later. Bose and Mitra (2002)[27] first used this method to find the LSD of Reverse Circulant and Symmetric Circulant matrices with i.i.d. entries. Later Bose, Mitra and Sen (2012)[28] used this to establish the LSD for some specific type of k-Circulant matrices with i.i.d. entries.

As mentioned earlier, we will be satisfied with the *in probability* convergence of the ESD in this chapter. It is easy to see that the ESD of A_n converges to F in probability if for $\epsilon > 0$ and at all continuity points (x, y) of F,

$$\mathrm{P}\big(|F^{A_n}(x,y) - F(x,y)| > \epsilon\big) \to 0 \text{ as } n \to \infty. \qquad (4.1)$$

Since the distribution functions are always bounded by 1, the above convergence is equivalent to *convergence in L_2*: at all continuity points (x, y) of F,

$$\int_\Omega \big[F^{A_n}(x,y) - F(x,y)\big]^2 d\mathrm{P}(\omega) \to 0 \text{ as } n \to \infty.$$

Note that this holds if, at all continuity points (x, y) of F,

$$\mathrm{E}\big[F^{A_n}(x,y)\big] \to F(x,y) \text{ and } \mathrm{Var}\big[F^{A_n}(x,y)\big] \to 0. \qquad (4.2)$$

This provides us a with a simple but valuable tool to establish LSD in probability.

4.2   Circulant matrix

Let C_n be a Circulant matrix with input sequence $\{x_i\}$. Fix n and let $\{\lambda_t\}$ denote the eigenvalues of C_n. Then it is easy to see that $(1, \dots, 1)$ is an eigenvector with the eigenvalue $\sum_{i=0}^{n-1} x_i = \lambda_n$ (say). For $1 \le t < n$, the other eigenvalues λ_t may be written as $\lambda_t = a_{t,n} + ib_{t,n}$, where,

$$a_{t,n} = \sum_{l=0}^{n-1} x_l \cos\Big(\frac{2\pi t l}{n}\Big), \quad b_{t,n} = \sum_{l=0}^{n-1} x_l \sin\Big(\frac{2\pi t l}{n}\Big). \qquad (4.3)$$

Recall the following trigonometric identities:

$$\sum_{l=0}^{n-1} \cos\Big(\frac{2\pi t l}{n}\Big) \sin\Big(\frac{2\pi t' l}{n}\Big) = 0 \ \forall t, t' \qquad (4.4)$$

$$\sum_{l=0}^{n-1} \cos^2\Big(\frac{2\pi t l}{n}\Big) = \sum_{l=0}^{n-1} \sin^2\Big(\frac{2\pi t l}{n}\Big) = n/2, \ t \ne n/2, n. \qquad (4.5)$$

$$\sum_{l=0}^{n-1} \cos\Big(\frac{2\pi t l}{n}\Big) \cos\Big(\frac{2\pi t' l}{n}\Big) = 0 \ t \ne t'; \qquad (4.6)$$

$$\sum_{l=0}^{n-1} \sin\Big(\frac{2\pi t l}{n}\Big) \sin\Big(\frac{2\pi t' l}{n}\Big) = 0 \ \forall t \ne t' \pmod{n}. \qquad (4.7)$$

Now suppose $\{x_l\}_{l \geq 0}$ are independent, mean zero and variance one random variables. For $z \in \mathbb{C}$, by $\bar{z}$ we mean, as usual, the complex conjugate of z. For all $0 < t, t' < n$, the following identities can easily be verified using the above orthogonality relations:

$$\mathrm{E}\,(a_{t,n} b_{t,n}) = 0, \quad \mathrm{E}\,(a_{t,n}^2) = \mathrm{E}\,(b_{t,n}^2) = n/2,$$

$$\bar{\lambda}_t = \lambda_{n-t}, \quad \mathrm{E}\,(\lambda_t \lambda_{t'}) = n\mathbb{I}(t + t' = n), \quad \text{and} \quad \mathrm{E}\,(|\lambda_t|^2) = n.$$

Theorem 4.2.1. If Assumption III(δ) is satisfied then the ESD of $\frac{1}{\sqrt{n}} C_n$ converges in L_2 to the two-dimensional normal distribution given by $\mathbf{N}(0, D)$ where D is a 2×2 diagonal matrix with diagonal entries $1/2$.   ◇

Incidentally, Sen (2006)[92] proved the above result when $\delta = 1$. Meckes (2009)[76] allowed independent complex entries in C_n. He showed that if $\mathrm{E}(x_j) = 0$, $\mathrm{E}\,|x_j|^2 = 1$ and for every $\epsilon > 0$,

$$\lim_{n \to \infty} \frac{1}{n} \sum_{j=0}^{n-1} \mathrm{E}\,\left(|x_j|^2 \mathbb{I}_{|x_j| > \epsilon \sqrt{n}}\right) = 0,$$

then the ESD converges in L_2 to the standard complex normal distribution.

Proof of Theorem 4.2.1. For notational simplicity, let F_n denote the ESD of $n^{-1/2} C_n$. Observe that we may ignore the eigenvalue λ_n, and also $\lambda_{n/2}$ whenever n is even, since they contribute at most $2/n$ to the ESD $F_n(x, y)$.

Gaussian input. First suppose that $\{x_i\}$ are Gaussian. Then by the orthogonality relations stated above, $a_{t,n}$ and $b_{t,n}$ are i.i.d. Gaussian. Thus, as n varies, F_n is essentially the *empirical distribution function* of i.i.d. bivariate Gaussian random vectors. By a Glivenko-Cantelli argument, F_n converges to the required row-wise bivariate Gaussian distribution.

Non-Gaussian input. Now we move to non-Gaussian inputs. Let $\Phi_{0,D}(\cdot, \cdot)$ denote the distribution function of the bivariate normal variable with mean 0 and dispersion matrix D. Recall from (4.2) that it is enough to show that

$$\mathrm{E}\,[F_n(x, y)] \to \Phi_{0,D}(x, y) \quad \text{and} \quad \mathrm{Var}\,[F_n(x, y)] \to 0.$$

Now, for $x, y \in \mathbb{R}$,

$$\mathrm{E}\,[F_n(x, y)] \sim n^{-1} \sum_{k=1, k \neq n/2}^{n-1} \mathrm{P}\,(b_k \leq x, c_k \leq y) + o(1)$$

where

$$b_k = \frac{1}{\sqrt{n}} \sum_{j=0}^{n-1} x_j \cos(\omega_k j), \quad c_k = \frac{1}{\sqrt{n}} \sum_{j=0}^{n-1} x_j \sin(\omega_k j), \quad \omega_k = \frac{2\pi k}{n}. \qquad (4.8)$$

To show $\mathrm{E}\left[F_n(x,y)\right] \to \Phi_{0,D}(x,y)$, define for $1 \leq k \leq n-1$, (except for $k = n/2$) and $0 \leq l \leq n-1$,

$$X_{l,k} = \left(\sqrt{2}x_l \cos(\omega_k l), \quad \sqrt{2}x_l \sin(\omega_k l)\right)'.$$

Note that

$$\mathrm{E}\left(X_{l,k}\right) = 0 \tag{4.9}$$

$$n^{-1} \sum_{l=0}^{n-1} \mathrm{Cov}\left(X_{l,k}\right) = I \tag{4.10}$$

$$\sup_n \sup_{1 \leq k \leq n} \left[n^{-1} \sum_{l=0}^{n-1} \mathrm{E} \parallel X_{lk} \parallel^{(2+\delta)} \right] \leq C < \infty. \tag{4.11}$$

For $k \neq n/2$

$$\{b_k \leq x, c_k \leq y\} = \left\{n^{-1/2} \sum_{l=0}^{n-1} X_{l,k} \leq \left(\sqrt{2}x, \sqrt{2}y\right)'\right\}.$$

Note that $\left\{(r,s) : (r,s) \leq \left(\sqrt{2}x, \sqrt{2}y\right)\right\}$ is a *convex* set in $\mathbb{R}^2$ and further, $\{X_{l,k}, \; l = 0, 1, \ldots (n-1)\}$ satisfies (4.9)–(4.11). Hence, we can apply Part (a) of Lemma 4.1.1 for $k \neq n/2$ to get

$$\left|\mathrm{P}\left(\frac{\sum_{l=0}^{n-1} X_{l,k}}{\sqrt{n}} \leq \left(\sqrt{2}x, \sqrt{2}y\right)'\right) - \mathrm{P}\left((N_1, N_2) \leq \left(\sqrt{2}x, \sqrt{2}y\right)\right)\right|$$

$$\leq C\frac{\displaystyle\sum_{l=0}^{n-1} \mathrm{E} \parallel X_{lk} \parallel^{(2+\delta)}}{n^{1+\delta/2}} \leq Cn^{-\delta/2} \to 0.$$

Therefore

$$\begin{aligned}
\lim_{n\to\infty} \mathrm{E}\left[F_n(x,y)\right] &= \lim_{n\to\infty} \frac{1}{n} \sum_{k=1,k\neq n/2}^{n-1} \mathrm{P}\left(b_k \leq x, c_k \leq y\right) \\
&= \lim_{n\to\infty} \frac{1}{n} \sum_{k=1,k\neq n/2}^{n-1} P\left((N_1, \; N_2)' \leq \left(\sqrt{2}x, \sqrt{2}y\right)'\right) \\
&= \Phi_{0,D}(x,y). \tag{4.12}
\end{aligned}$$

Now, to show $\mathrm{Var}\left[F_n(x,y)\right] \to 0$, it is enough to show that

$$\frac{1}{n^2} \sum_{k\neq k';k,k'=1}^{n} \mathrm{Cov}\left(J_k, J_{k'}\right) = \frac{1}{n^2} \sum_{k\neq k';k,k'=1}^{n} \left[\mathrm{E}\left(J_k, J_{k'}\right) - \mathrm{E}(J_k)\,\mathrm{E}(J_{k'})\right] \to 0 \tag{4.13}$$

where for $1 \leq k \leq n$, J_k is the indicator that $\{b_k \leq x, c_k \leq y\}$. Now as $n \to \infty$,

$$\frac{1}{n^2} \sum_{k\neq k';k,k'=1}^{n} \mathrm{E}\left(J_k\right)\mathrm{E}\left(J_{k'}\right) = \left[\frac{1}{n}\sum_{k=1}^{n} \mathrm{E}\left(J_k\right)\right]^2 - \frac{1}{n^2}\sum_{k=1}^{n}\left[\mathrm{E}\left(J_k\right)\right]^2 \to \left[\Phi_{0,D}(x,y)\right]^2.$$

So to show (4.13), it is enough to show as $n \to \infty$,

$$\frac{1}{n^2} \sum_{k \neq k'; k, k'=1}^{n} \mathrm{E}\left(J_k, J_{k'}\right) \to \left[\Phi_{0,D}(x, y)\right]^2.$$

Along the lines of the proof used to show (4.12) one may now extend the vectors of two coordinates defined above to ones with four coordinates and proceed exactly as above to verify this. We omit the routine details. This completes the proof of Theorem 4.2.1. □

4.3   *k*-Circulant matrices

4.3.1   Eigenvalues

A formula solution for the eigenvalues of the *k*-Circulant matrix is given in Zhou (1996)[114]. We first describe this result and establish some related auxiliary properties of the solution that will be useful to us. Let

$$\omega = \omega_n := \cos(2\pi/n) + i\sin(2\pi/n), \ i^2 = -1 \ \text{and} \ \lambda_t = \sum_{l=0}^{n-1} x_l \omega^{tl}, \ \ 0 \leq t < n.$$

$$(4.14)$$

Note that $\{\lambda_t, 0 \leq t < n\}$ are eigenvalues of the usual Circulant matrix $A_{1,n}$. Let $p_1 < p_2 < \cdots < p_c$ be all the common prime factors of n and k. Then we may write,

$$n = n' \prod_{q=1}^{c} p_q^{\beta_q} \quad \text{and} \quad k = k' \prod_{q=1}^{c} p_q^{\alpha_q}.$$

$$(4.15)$$

Here α_q, $\beta_q \geq 1$ and n', k', p_q are pairwise relatively prime. It will turn out that $A_{k,n}$ has $(n - n')$ zero eigenvalues. The remaining n' eigenvalues are non-zero functions of $\{\lambda_t\}$.

To identify the non-zero eigenvalues of $A_{k,n}$, we need some preparation. For any positive integer m, let

$$\mathbb{Z}_m = \{0, 1, 2, \ldots, m - 1\}.$$

We introduce the following family of sets

$$S(x) := \{xk^b \bmod n' : b \geq 0\}, \quad x \in \mathbb{Z}_{n'}.$$

$$(4.16)$$

We observe the following facts about the family of sets $\{S(x)\}_{x \in \mathbb{Z}_{n'}}$.

(I) Let $g_x = \#S(x)$. We call g_x the *order* of x. Note that $g_0 = 1$. It is easy to see that

$$S(x) = \{xk^b \bmod n' : 0 \leq b < g_x\}.$$

An alternative description of g_x, which we will use later extensively, is the following. For $x \in \mathbb{Z}_{n'}$, let

$$\mathcal{O}_x = \{b > 0 \ : b \text{ is an integer and } xk^b = x \mod n'\}.$$

Then $g_x = \min \mathcal{O}_x$, that is, g_x is the smallest positive integer b such that $xk^b = x \mod n'$.

(II) The distinct sets from the collection $\{S(x)\}_{x \in \mathbb{Z}_{n'}}$ forms a partition of $\mathbb{Z}_{n'}$. To see this, first note that $x \in S(x)$ and hence, $\bigcup_{x \in \mathbb{Z}_{n'}} S(x) = \mathbb{Z}_{n'}$. Now suppose $S(x) \cap S(y) \neq \emptyset$. Then, $xk^{b_1} = yk^{b_2} \mod n'$ for some integers $b_1, b_2 \geq 1$. Multiplying both sides by $k^{g_x - b_1}$ we see that, $x \in S(y)$ so that, $S(x) \subseteq S(y)$. Hence, reversing the roles, $S(x) = S(y)$.

We call the distinct sets in $\{S(x)\}_{x \in \mathbb{Z}_{n'}}$ the *eigenvalue partition* of $\mathbb{Z}_{n'}$ and denote the partitioning sets and their sizes by

$$\mathcal{P}_0 = \{0\}, \mathcal{P}_1, \ldots, \mathcal{P}_{\ell-1} \text{ and } n_j = \#\mathcal{P}_j, \ 0 \leq j < \ell. \qquad (4.17)$$

Define

$$\Pi_j := \prod_{t \in \mathcal{P}_j} \lambda_{tn/n'}, \quad j = 0, 1, \ldots, \ell - 1. \qquad (4.18)$$

The following theorem of Zhou (1996)[114] provides the formula solution for the eigenvalues of $A_{k,n}$. Its proof is available in Bose, Mitra and Sen (2012)[28].

Theorem 4.3.1. The characteristic polynomial of $A_{k,n}$ is given by

$$\chi(A_{k,n})(\lambda) = \lambda^{n-n'} \prod_{j=0}^{\ell-1} (\lambda^{n_j} - \Pi_j). \qquad (4.19)$$

◇

4.3.2 Eigenvalue partition

We collect some simple but useful properties about the eigenvalue partition $\{\mathcal{P}_j, 0 \leq j < \ell\}$ in the following lemma.

Lemma 4.3.2. (a) Let $x, y \in \mathbb{Z}_{n'}$. If $n' - t_0 \in S(y)$ for some $t_0 \in S(x)$, then for every $t \in S(x)$, we have $n' - t \in S(y)$.

(b) Fix $x \in \mathbb{Z}_{n'}$. Then g_x divides g for every $g \in \mathcal{O}_x$. Furthermore, g_x divides g_1 for each $x \in \mathbb{Z}_{n'}$.

(c) Suppose g divides g_1. Set $m := \gcd(k^g - 1, n')$. Let $X(g)$ and $Y(g)$ be defined as

$$
\begin{aligned}
X(g) &:= \{x : x \in \mathbb{Z}_{n'} \text{ and } x \text{ has order } g\}, \\
Y(g) &:= \{bn'/m : 0 \le b < m\}.
\end{aligned} \tag{4.20}
$$

Then

$$
X(g) \subseteq Y(g), \quad \#Y(g) = m \quad \text{and} \quad \bigcup_{h:h|g} X(h) = Y(g).
$$

◇

Proof. (a) Since $t \in S(x) = S(t_0)$, we can write $t = t_0 k^b \mod n'$ for some $b \ge 0$. Therefore, $n' - t = (n' - t_0)k^b \mod n' \in S(n' - t_0) = S(y)$.

(b) Fix $g \in \mathcal{O}_x$. Since g_x is the smallest element of $\mathcal{O}_x$, it follows that $g_x \le g$. Suppose, if possible, $g = qg_x + r$ where $0 < r < g_x$. By the fact that $xg_x = x \mod n'$, it then follows that

$$
\begin{aligned}
x &= xk^g \mod n' \\
&= xk^{qg_x + r} \mod n' \\
&= xk^r \mod n'.
\end{aligned}
$$

This implies that $r \in \mathcal{O}_x$ and $r < g_x$ which is a contradiction to the fact that g_x is the smallest element in $\mathcal{O}_x$. Hence, we must have $r = 0$ proving that g divides g_1.

Note that $k^{g_1} = 1 \mod n'$. This implies that $xk^{g_1} = x \mod n'$. Therefore $g_1 \in \mathcal{O}_x$ proving the assertion.

(c) Clearly, $\#Y(g) = m$. Fix $x \in X(h)$ where h divides g. Then, $xk^g = x(k^h)^{g/h} = x \mod n'$, since g/h is a positive integer. Therefore n' divides $x(k^g - 1)$. So, n'/m divides $x(k^g - 1)/m$. But n'/m is relatively prime to $(k^g - 1)/m$ and hence, n'/m divides x. So, $x = bn'/m$ for some integer $b \ge 0$. Since $0 \le x < n'$, we have $0 \le b < m$, and $x \in Y(g)$, proving $\bigcup_{h:h|g} X(h) \subseteq Y(g)$ and in particular, $X(g) \subseteq Y(g)$.

On the other hand, take $0 \le b < m$. Then $(bn'/m) k^g = (bn'/m) \mod n'$. Hence, $g \in \mathcal{O}_{bn'/m}$ which implies, by part (b) of the lemma, that $g_{bn'/m}$ divides g. Therefore, $Y(g) \subseteq \bigcup_{h:h|g} X(h)$ which completes the proof. □

Lemma 4.3.3. Let $g_1 = q_1^{\gamma_1} q_2^{\gamma_2} \cdots q_m^{\gamma_m}$ where $q_1 < q_2 < \cdots < q_m$ are primes. Define, for $1 \le j \le m$,

$$
\begin{aligned}
L_j &:= \{q_{i_1} q_{i_2} \cdots q_{i_j} : 1 \le i_1 < \cdots < i_j \le m\}, \\
G_j &:= \sum_{l_j \in L_j} \#Y(g_1/\ell_j) = \sum_{l_j \in L_j} \gcd\left(k^{g_1/\ell_j} - 1, n'\right).
\end{aligned}
$$

Then we have

(a) $\#\{x \in \mathbb{Z}_{n'} : g_x < g_1\} = G_1 - G_2 + G_3 - G_4 + \cdots.$

(b) $G_1 - G_2 + G_3 - G_4 + \cdots \leq G_1.$ ◇

Proof. Fix $x \in \mathbb{Z}_{n'}$. By Lemma 4.3.2 (b), g_x divides g_1 and hence, we can write $g_x = q_1^{\eta_1} \ldots q_m^{\eta_m}$ where, $0 \leq \eta_b \leq \gamma_b$ for $1 \leq b \leq m$. Since $g_x < g_1$, there is at least one b so that $\eta_b < \gamma_b$. Suppose that exactly h-many η's are equal to the corresponding γ's where $0 \leq h < m$. To keep notation simple, we will assume that, $\eta_b = \gamma_b$, $1 \leq b \leq h$ and $\eta_b < \gamma_b$, $h + 1 \leq b \leq m$.

(a) Then $x \in Y(g_1/q_b)$ for $h + 1 \leq b \leq m$ and $x \notin Y(g_1/q_b)$ for $1 \leq b \leq h$. So, x is counted $(m - h)$ times in G_1. Similarly, x is counted $\binom{m-h}{2}$ times in G_2, $\binom{m-h}{3}$ times in G_3, and so on. Hence, the total number of times x is counted in $(G_1 - G_2 + G_3 - \cdots)$ is

$$\binom{m-h}{1} - \binom{m-h}{2} + \binom{m-h}{3} - \cdots = 1.$$

(b) Note that $m - h \geq 1$. Further, each element in the set $\{x \in \mathbb{Z}_{n'} : g_x < g_1\}$ is counted once in $G_1 - G_2 + G_3 - \cdots$ and $(m - h)$ times in G_1. The result follows immediately.   □

4.3.3   Lower-order elements

We will now consider the elements in $\mathbb{Z}_{n'}$ with order less than that of $1 \in \mathbb{Z}_{n'}$ which has the highest order g_1. We will need the proportion of such elements in $\mathbb{Z}_{n'}$. So, we define

$$\upsilon_{k,n'} := \frac{1}{n'} \#\{x \in \mathbb{Z}_{n'} : g_x < g_1\}. \tag{4.21}$$

The following two lemmata establish upper bounds on $\upsilon_{k,n'}$ and will be crucially used later.

Lemma 4.3.4. (a) If $g_1 = 2$, then $\upsilon_{k,n'} = \gcd(k - 1, n')/n'$.

(b) Suppose $g_1 \geq 4$ is even, and $k^{g_1/2} = -1 \bmod n'$. Then

$$\upsilon_{k,n'} \leq n'^{-1}[1 + \sum_{b|g_1,\ b \geq 3} \gcd(k^{g_1/b} - 1, n')].$$

(c) If $g_1 \geq 2$ and q_1 is the smallest prime divisor of g_1, then $\upsilon_{k,n'} < 2n'^{-1}k^{g_1/q_1}$.   ◇

Proof. Part (a) is immediate from Lemma 4.3.3 which asserts that $n'\upsilon_{k,n'} = \#Y(1) = \gcd(k - 1, n')$.

(b) Fix $x \in \mathbb{Z}_{n'}$ with $g_x < g_1$. Since g_x divides g_1 and $g_x < g_1$, g_x must be of the form g_1/b for some integer $b \geq 2$ provided g_1/b is an integer. If $b = 2$, then $xk^{g_1/2} = xk^{g_x} = x \mod n'$. But $k^{g_1/2} = -1 \mod n'$ and so, $xk^{g_1/2} = -x \mod n'$. Therefore, $2x = 0 \mod n'$ and x can be either 0 or $n'/2$, provided, of course, that $n'/2$ is an integer. But $g_0 = 1 < 2 \leq g_1/2$, so x cannot be 0. So, there is at most one element in the set $X(g_1/2)$. Thus, we have,

$$\#\{x \in \mathbb{Z}_{n'} : g_x < g_1\} = \#X(g_1/2) + \sum_{b|g_1, \, b \geq 3} \#\{x \in \mathbb{Z}_{n'} : g_x = g_1/b\}$$

$$= \#X(g_1/2) + \sum_{b|g_1, \, b \geq 3} \#X(g_1/b)$$

$$\leq 1 + \sum_{b|g_1, \, b \geq 3} \#Y(g_1/b) \quad ((\text{by Lemma 4.3.2(c)})$$

$$= 1 + \sum_{b|g_1, \, b \geq 3} \gcd(k^{g_1/b} - 1, n') \quad (\text{by Lemma 4.3.2(c)}).$$

(c) As in Lemma 4.3.3, let $g_1 = q_1^{\gamma_1} q_2^{\gamma_2} \cdots q_m^{\gamma_m}$ where $q_1 < q_2 < \cdots < q_m$ are primes. Then by Lemma 4.3.3,

$$n' \times v_{k,n'} = G_1 - G_2 + G_3 - G_4 + \cdots \leq G_1$$

$$= \sum_{b=1}^{m} \gcd(k^{g_1/q_b} - 1, n')$$

$$< \sum_{b=1}^{m} k^{g_1/q_b} \leq 2k^{g_1/q_1},$$

where the last inequality follows from the observation

$$\sum_{b=1}^{m} k^{g_1/q_b} \leq k^{g_1/q_1} \sum_{b=1}^{m} k^{-g_1(q_b - q_1)/q_1 q_b}$$

$$\leq k^{g_1/q_1} \sum_{b=1}^{m} k^{-(q_b - q_1)}$$

$$\leq k^{g_1/q_1} \sum_{b=1}^{m} k^{-(b-1)} \leq 2k^{g_1/q_1}.$$

$\square$

Lemma 4.3.5. Let b and c be two fixed positive integers. Then for any integer $k \geq 2$, the following inequality holds in each of the four cases,

$$\gcd(k^b \pm 1, k^c \pm 1) \leq k^{\gcd(b,c)} + 1.$$

$\diamond$

Proof. The assertion trivially follows if one of b and c divides the other. So, we assume, without loss, that $b < c$ and b does not divide c. Since $k^c \pm 1 = k^{c-b}(k^b + 1) + (-k^{c-b} \pm 1)$, we can write

$$\gcd(k^b + 1, k^c \pm 1) = \gcd(k^b + 1, k^{c-b} \mp 1).$$

Similarly,

$$\gcd(k^b - 1, k^c \pm 1) = \gcd(k^b - 1, k^{c-b} \pm 1).$$

Moreover, if we write $c_1 = c - \lfloor c/b \rfloor b$, then by repeating the above step $\lfloor c/b \rfloor$ times, we can see that $\gcd(k^b \pm 1, k^c \pm 1)$ is equal to one of $\gcd(k^b \pm 1, k^{c_1} \pm 1)$. Now if c_1 divides b, then $\gcd(b, c) = c_1$ and we are done. Otherwise, we can now repeat the whole argument with $b = c_1$ and $c = b$ to deduce that $\gcd(k^b \pm 1, k^{c_1} \pm 1)$ is one of $\gcd(k^{b_1} \pm 1, k^{c_1} \pm 1)$ where $b_1 = b - \lfloor b/c_1 \rfloor c_1$. We continue in a similar fashion, reducing each time one of the two exponents of k in the gcd and the lemma follows once we recall Euclid's recursive algorithm for computing the gcd of two numbers. $\square$

Lemma 4.3.6. (a) Fix $g \geq 1$. Suppose $k^g = -1 + sn$, $n \to \infty$ with $s = 1$ if $g = 1$ and $s = o(n^{p_1 - 1})$ if $g > 1$ where p_1 is the smallest prime divisor of g. Then $g_1 = 2g$ for all but finitely many n and $v_{k,n} \to 0$.

(b) Suppose $k^g = 1 + sn$, $g \geq 1$ fixed, $n \to \infty$ with $s = 1$ if $g = 1$ and $s = o(n^{p_1 - 1})$ if $g > 1$, where p_1 is the smallest prime divisor of g. Then $g_1 = g$ for all but finitely many n and $v_{k,n} \to 0$. $\diamond$

Proof. (a) First note that $\gcd(n, k) = 1$ and therefore $n' = n$. When $g = 1$, it is easy to check that $g_1 = 2$ and by Lemma 4.3.4(a), $v_{k,n} \leq 2/n$.

Now assume $g > 1$. Since $k^{2g} = (sn - 1)^2 = 1 \mod n$, g_1 divides $2g$. Observe that $g_1 \neq g = 2g/2$ because $k^g = -1 \mod n$.

If $g_1 = 2g/b$, where b divides g and $b \geq 3$, then by Lemma 4.3.5,

$$\gcd(k^{g_1} - 1, n) = \gcd\left(k^{2g/b} - 1, (k^g + 1)/s\right) \leq \gcd\left(k^{2g/b} - 1, k^g + 1\right)$$
$$\leq k^{\gcd(2g/b,\, g)} + 1.$$

Note that since $\gcd(2g/b, g)$ divides g and $\gcd(2g/b, g) \leq 2g/b < g$, we have $\gcd(2g/b, g) \leq g/p_1$. Consequently,

$$\gcd(k^{2g/b} - 1, n) \leq k^{g/p_1} + 1 \leq (sn - 1)^{1/p_1} + 1 = o(n), \qquad (4.22)$$

which is a contradiction to the fact that $k^{g_1} = 1 \mod n$, which implies that $\gcd(k^{g_1} - 1, n) = n$. Hence, $g_1 = 2g$. Now by Lemma 4.3.4(b) it is enough to show that for any fixed $b \geq 3$ so that b divides g_1,

$$\gcd(k^{g_1/b} - 1, n)/n = o(1) \quad \text{as } n \to \infty,$$

which we have already proved in (4.22).

(b) Again $\gcd(n, k) = 1$ and $n' = n$. The case when $g = 1$ is trivial as then we have $g_x = 1$ for all $x \in \mathbb{Z}_n$ and $v_{k,n} = 0$.

Now assume $g > 1$. Since $k^g = 1 \mod n$, g_1 divides g. If $g_1 < g$, then $g_1 \leq g/p_1$ which implies that $k^{g_1} \leq k^{g/p_1} = (sn + 1)^{1/p_1} = o(n)$, which is a contradiction. Thus, $g = g_1$.

Now Lemma 4.3.4(c) immediately yields,

$$v_{k,n} < \frac{2k^{g_1/p_1}}{n} \leq \frac{2(1 + sn)^{1/p_1}}{n} = o(1). \qquad \square$$

4.3.4 Degenerate limit

The next theorem, due to Bose, Mitra and Sen (2008)[28], implies that the radial component of the LSD of k-Circulants with $k \geq 2$ is always degenerate, at least when the input sequence is i.i.d. normal, as long as $k = n^{o(1)}$ and $\gcd(k, n) = 1$. Observe that, in this case also n tends to infinity along a subsequence and it is determined by the condition $\gcd(k, n) = 1$.

Theorem 4.3.7. Suppose $\{x_i\}_{i \geq 0}$ is an i.i.d. sequence of $N(0, 1)$ random variables. Let $k \geq 2$ be such that $k = n^{o(1)}$ and $n \to \infty$ with $\gcd(n, k) = 1$. Then $F^{n^{-1/2}A_{k,n}}$ converges weakly in probability to the uniform distribution over the circle with center at $(0, 0)$ and radius $r = \exp(\mathrm{E}\left[\log \sqrt{Y}\right])$, Y being an exponential random variable with mean one. $\diamond$

Remark 4.3.1. $-\log Y$ has the standard Gumbel distribution with mean

$$\gamma = \lim_{n \to \infty} \left[1 + \frac{1}{2} + \cdots + \frac{1}{n} - \log n\right] \approx 0.57721 \text{ (the Euler-Mascheroni constant)}.$$

It follows that $r = e^{-\gamma/2} \approx 0.74930$.

It is thus natural to consider the case when k^g is of the order n and $\gcd(k, n) = 1$ where g is a fixed integer. In the next two theorems, we consider two special cases of the above scenario, namely when n divides $k^g \pm 1$.

Proof of Theorem 4.3.7. We shall need the following lemma.

Lemma 4.3.8. Fix k and n. Suppose that $\{x_l\}_{0 \leq l < n}$ are i.i.d. standard normal random variables. Then

(a) For every n, $n^{-1/2}a_{t,n}, n^{-1/2}b_{t,n}$, $0 \leq t \leq n/2$ are i.i.d. normal with mean zero and variance $1/2$. Any sub-collection $\{\Pi_{j_1}, \Pi_{j_2}, \ldots\}$ of $\{\Pi_j\}_{0 \leq j < \ell}$, so that none of them are conjugates of each other, are mutually independent.

(b) If $1 \leq j < \ell$ where $\mathcal{P}_j \cap (n - \mathcal{P}_j) = \emptyset$, then $n^{-n_j/2}\Pi_j$ is distributed as the product of n_j i.i.d. copies of $E^{1/2}U$ where E and U are independent, E is exponential with mean one and U is uniform over the unit circle in $\mathbb{R}^2$.

(c) For $1 \leq j < \ell$ where $\mathcal{P}_j = n - \mathcal{P}_j$ and $n/2 \notin \mathcal{P}_j$, $n^{-n_j/2}\Pi_j$ are distributed as the product of $(n_j/2)$i.i.d. exponentials with mean one. $\diamond$

Proof. (a) $n^{-1/2}a_{t,n}, n^{-1/2}b_{t,n}, 0 \le t \le n/2$ are linear combinations of $\{x_l\}$, and by (4.4), they have mean zero, variance $1/2$ and are independent.

(b) By part (a), $n^{-1/2}\lambda_t = n^{-1/2}a_{t,n} + in^{-1/2}b_{t,n}$ is a complex normal random variable with mean zero and variance $1/2$ for every $0 < t < n$. Moreover, they are independent by the given restriction on $\mathcal{P}_j$. Now observe that such a complex normal is distributed as $E^{1/2}U$.

(c) If $t \in \mathcal{P}_j$ then $n - t \in \mathcal{P}_j$ too and $t \ne n - t$. By part (a), $n^{-1}\lambda_t\lambda_{n-t} = n^{-1}\left(a_{t,n}^2 + b_{t,n}^2\right)$ which is distributed as $Y/2$ where Y is Chi-square with two degrees of freedom. This is the same distribution as that of an exponential random variable with mean one. The proof is complete once we observe that n_j is necessarily even and the λ_t's associated with $\mathcal{P}_j$ can be grouped into $n_j/2$ disjoint pairs like above, which are mutually independent.   □

Now we get back to the proof of the theorem. Recall $\lambda_j, \ell, \mathcal{P}_j, n_j$ and g_x from Section 4.3.1. By Theorem 4.3.1, the eigenvalues of $n^{-1/2}A_{k,n}$ are

$$\exp\left(\frac{2\pi i(s + \Theta_j)}{n_j}\right) \times \left(\prod_{t \in \mathcal{P}_j} |n^{-1/2}\lambda_t|\right)^{1/n_j}, 1 \le s \le n_j, \ 0 \le j < \ell,$$

where $2\pi\Theta_j = \arg(\prod_{t \in \mathcal{P}_j} \lambda_t), \Theta_j \in [0, 1)$ where $\arg(z)$ is the usual argument of z between 0 and 2π. Fix any $\epsilon > 0$ and $0 < \theta_1 < \theta_2 < 2\pi$. Define

$$B(\theta_1, \theta_2, \epsilon) = \{(x, y) \in \mathbb{R}^2 : r - \epsilon < \sqrt{x^2 + y^2} < r + \epsilon, \tan^{-1}(y/x) \in [\theta_1, \theta_2]\}.$$

Clearly, it is enough to prove that as $n \to \infty$,

$$\frac{1}{n}\sum_{j=0}^{\ell-1}\sum_{s=1}^{n_j} \mathbb{I}\left(\exp\left(\frac{2\pi i(s + \Theta_j)}{n_j}\right) \times \left(\prod_{t \in \mathcal{P}_j} \left|\frac{\lambda_t}{\sqrt{n}}\right|\right)^{1/n_j} \in B(\theta_1, \theta_2, \epsilon)\right) \xrightarrow{P} \frac{(\theta_2 - \theta_1)}{2\pi}.$$

$$(4.23)$$

Note that for a fixed positive integer C, we have

$$n^{-1}\sum_{1 \le j < \ell : n_j \le C} n_j \le n^{-1}\sum_{u=2}^{C} \#\{1 \le x < n : g_x = u\}$$

$$\le n^{-1}\sum_{u=2}^{C} \#\{1 \le x < n : xk^u = x \bmod n\}$$

$$= n^{-1}\sum_{u=2}^{C} \#\{1 \le x < n : x(k^u - 1) = sn \text{ for some } s \ge 1\}$$

$$\le n^{-1}\sum_{u=2}^{C} (k^u - 1) \le n^{-1}Ck^C \to 0, \text{ as } n \to \infty.$$

Therefore, letting $N_C = \sum_{j=0: n_j \le C}^{\ell-1} n_j$, then the above bound, along with the fact $\mathcal{P}_0 = \{0\}$ yields $N_C/n \to 0$. With $C > (2\pi)/(\theta_2 - \theta_1)$, the left side of (4.23) can be written as

$$\frac{1}{n}\sum_{j=0}^{\ell-1} \#\left\{s : \frac{2\pi(s+\Theta_j)}{n_j} \in [\theta_1, \theta_2], s = 1, 2, \ldots, n_j\right\}$$

$$\times \mathbb{I}\left(\left(\prod_{t \in \mathcal{P}_j} |n^{-1/2}\lambda_t|\right)^{1/n_j} \in (r - \epsilon, r + \epsilon)\right)$$

$$= \frac{n - N_C}{n}\frac{1}{n - N_C}\sum_{j=0,\, n_j > C}^{\ell-1} n_j \times n_j^{-1}\#\left\{s : \frac{s+\Theta_j}{n_j} \in \frac{1}{2\pi}[\theta_1,\ \theta_2], s = 1, \ldots, n_j\right\}$$

$$\times \mathbb{I}\left(\left(\prod_{t \in \mathcal{P}_j} |n^{-1/2}\lambda_t|\right)^{1/n_j} \in (r - \epsilon, r + \epsilon)\right) + O\left(\frac{N_C}{n}\right)$$

$$= \frac{1}{n - N_C}\sum_{j=0,\, n_j > C}^{\ell-1} n_j\left(\frac{(\theta_2 - \theta_1)}{2\pi} + O(C^{-1})\right)\mathbb{I}\left(\left(\prod_{t \in \mathcal{P}_j} \left|\frac{\lambda_t}{\sqrt{n}}\right|\right)^{1/n_j} \in (r - \epsilon, r + \epsilon)\right)$$

$$+ O\left(\frac{N_C}{n}\right)$$

$$= \frac{1}{n - N_C}\sum_{j=0,\, n_j > C}^{\ell-1} n_j \times \frac{(\theta_2 - \theta_1)}{2\pi} \times \mathbb{I}\left(\left(\prod_{t \in \mathcal{P}_j} |n^{-1/2}\lambda_t|\right)^{1/n_j} \in (r - \epsilon, r + \epsilon)\right)$$

$$+ O(C^{-1}) + O\left(\frac{N_C}{n}\right)$$

$$= \frac{(\theta_2 - \theta_1)}{2\pi} + \frac{1}{n - N_C}\sum_{j=0,\, n_j > C}^{\ell-1} n_j \times \mathbb{I}\left(\left(\prod_{t \in \mathcal{P}_j} |n^{-1/2}\lambda_t|\right)^{1/n_j} \notin (r - \epsilon, r + \epsilon)\right)$$

$$+ O(C^{-1}) + O\left(\frac{N_C}{n}\right). \tag{4.24}$$

To show that the second term in the above expression converges to zero in L_1, and hence in probability, it remains to prove,

$$P\left(\left(\prod_{t \in \mathcal{P}_j} |n^{-1/2}\lambda_t|\right)^{1/n_j} \notin (r - \epsilon, r + \epsilon)\right) \tag{4.25}$$

is uniformly small for all j such that $n_j > C$ and for all but finitely many n, provided we take C sufficiently large.

By Lemma 4.3.8, for each $1 \le t < n$, $|n^{-1/2}\lambda_t|^2$ is an exponential random variable with mean one, and λ_t is independent of $\lambda_{t'}$ if $t' \ne n-t$ and $|\lambda_t| = |\lambda_{t'}|$ otherwise. Let $E, E_1, E_2, \ldots$ be i.i.d. exponential random variables with mean one. Observe that depending on whether or not $\mathcal{P}_j$ is conjugate to itself, (4.25) equals respectively,

$$P\left(\left(\prod_{t=1}^{n_j/2} E_t\right)^{1/n_j} \notin (r - \epsilon, r + \epsilon)\right) \quad \text{or} \quad P\left(\left(\prod_{t=1}^{n_j} \sqrt{E_t}\right)^{1/n_j} \notin (r - \epsilon, r + \epsilon)\right).$$

The theorem now follows by letting first $n \to \infty$ and then $C \to \infty$ in (4.24) and by observing that the SLLN implies that

$$
(\prod_{t=1}^{C} \sqrt{E_t})^{1/C} \to r = \exp\left(\mathrm{E}\left[\log \sqrt{E}\right]\right) \quad \text{almost surely, as} \quad C \to \infty. \quad \square
$$

4.3.5   Non-degenerate limit

Establishing the LSD for general k-Circulant matrices appears to be a difficult problem. From the formula solution given in Section 4.3.1, it is clear that for many combinations of k and n, a lot of eigenvalues are zero. For example, if k is prime and $n = m \times k$ where $\gcd(m, k) = 1$, then 0 is an eigenvalue with multiplicity $(n - m)$. To avoid this degeneracy we primarily restrict our attention to the case when $\gcd(k, n) = 1$.

We have already seen that the LSD of the usual Circulant matrix $n^{-1/2}A_{1,n}$ is bivariate normal. If we keep k (other than 1) fixed and let n tend to infinity, then LSD may not exist. The ESD of $n^{-1/2}A_{2,n}$ for large n has mass at zero if n is even, and looks like a solar ring with no mass at zero if n is odd (see Figures 4.1 and 4.2). Similarly, if $k = 3$ then the behavior of the ESD depends on whether n is a multiple of 3 (see Figure 4.2). So, for a fixed value of $k(\neq 1)$ the LSD may exist if we let n tend to infinity only along a sub-sequence depending on k. LSD in a few special cases of k and n were derived in Bose, Mitra and Sen (2008)[28] for i.i.d. inputs. These two cases are essentially (i) $k^g = 1 + sn$ and (ii) $k^g = -1 + sn$. We present these results now. See Figure 4.3 for illustrative simulations in two specific cases.

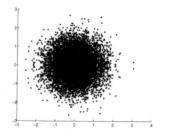 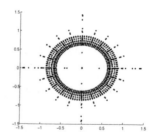

FIGURE 4.1
Eigenvalues of $n^{-1/2}A_{k,n}$, x_i i.i.d. $N(0, 1)$,
(i) $k = 1$, $n = 2000$ (left) (ii) $k = 2$, $n = 2000$ (right). 10 replications.

Let $\{U_i\}$, $\{E_i\}$ be mutually independent where $\{E_i\}$ are i.i.d. $Exp(1)$, U_1 is uniformly distributed over $(2g)$th roots of unity and U_2 is uniformly distributed over the unit circle. The following result is due to Bose, Mitra and Sen (2008)[28].

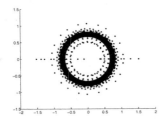 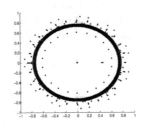

FIGURE 4.2
Eigenvalues of 10 replications of $n^{-1/2}A_{k,n}$, x_ii.i.d.$N(0,1)$,
(i) $k=2$, $n=2001$ (left) (ii) $k=3$, $n=2001$ (right). 10 replications.

Theorem 4.3.9 (Bose, Mitra and Sen (2008)[28]). Suppose $\{x_l\}_{l\geq0}$ satisfies Assumption III(δ). Fix $g\geq1$ and let p_1 be the smallest prime divisor of g. Suppose $k^g=-1+sn$ where $s=1$ if $g=1$ and $s=o\left(n^{p_1-1}\right)$ if $g>1$. Then $F^{n^{-1/2}A_{k,n}}$ converges weakly in probability to $U_1\left(\prod_{j=1}^g E_j\right)^{1/2g}$ as $n\to\infty$ where $\{E_j\}_{1\leq j\leq g}$ are i.i.d. exponentials with mean one and U_1 is uniformly distributed over the $(2g)$th roots of unity, independent of $\{E_j\}_{1\leq j\leq g}$. ◇

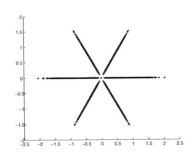 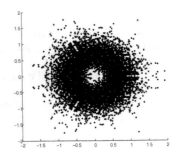

FIGURE 4.3
Eigenvalues of $n^{-1/2}A_{k,n}$, x_ii.i.d.$N(0,1)$,
(i) $k=10$, $n=k^3+1$ (left) (ii) $k=10$, $n=k^3-1$ (right). 10 replications.

Proof. Since $\gcd(k,n)=1$, $n'=n$ in Theorem 4.3.1, and hence there are no zero eigenvalues. By Lemma 4.3.6(a), $v_{k,n}\to0$ and hence the corresponding eigenvalues do not contribute to the LSD. It remains to consider only the eigenvalues corresponding to the sets $\mathcal{P}_j$ of size *exactly* equal to g_1. From Lemma 4.3.6(a), $g_1=2g$ for n sufficiently large.

Recall the quantities $n_j = \#\mathcal{P}_j$, $\Pi_j = \Pi_{t \in \mathcal{P}_l} \lambda_t$, where $\lambda_t = \sum_{l=0}^{n-1} x_l \omega^{tl}$, $0 \le j < n$. Also, for every integer $t \ge 0$, $tk^g = (-1 + sn)t = -t \mod n$, so that, λ_t and λ_{n-t} belong to the same partition block $S(t) = S(n-t)$. Thus, each Π_j is a non-negative real number. Let us define

$$J_n = \{0 \le j < \ell : \#\mathcal{P}_j = 2g\},$$

so that $n = 2g\#J_n + n v_{k,n}$. Since, $v_{k,n} \to 0$, $(\#J_n)^{-1} n \to 2g$. Without any loss, we denote the index set of such j as $J_n = \{1, 2, \ldots, \#J_n\}$.

Let $1, \varrho, \varrho^2, \ldots, \varrho^{2g-1}$ be all the $(2g)$th roots of unity. Since $n_j = 2g$ for every $j \in J_n$, the eigenvalues corresponding to the set $\mathcal{P}_j$ are:

$$\Pi_j^{1/2g}, \Pi_j^{1/2g} \varrho, \ldots, \Pi_j^{1/2g} \varrho^{2g-1}.$$

Hence, it suffices to consider only the empirical distribution of $\Pi_j^{1/2g}$ as j varies over the index set J_n: if this sequence of empirical distributions has a limiting distribution F, say, then the LSD of the original sequence $n^{-1/2} A_{k,n}$ will be (r, θ) in polar coordinates where r is distributed according to F and θ is distributed uniformly across all the $2g$ roots of unity, and r and θ are independent. With this in mind, and remembering the scaling $\sqrt{n}$, we consider

$$F_n(x) = (\#J_n)^{-1} \sum_{j=1}^{\#J_n} \mathbb{I}\big(\big[n^{-g} \Pi_j \big]^{\frac{1}{2g}} \le x \big).$$

Since the set of λ values corresponding to any $\mathcal{P}_j$ is closed under conjugation, there exists a set $\mathcal{A}_j \subset \mathcal{P}_j$ of size g such that

$$\mathcal{P}_j = \{x : x \in \mathcal{A}_j \text{ or } n - x \in \mathcal{A}_j\}.$$

Combining each λ_t with its conjugate, and recalling the definition of $\{a_{t,n}\}$ and $\{b_{t,n}\}$ in (4.3), we may write Π_j as

$$\Pi_j = \prod_{t \in \mathcal{A}_j} \left(a_{t,n}^2 + b_{t,n}^2 \right).$$

Gaussian input: First assume that the random variables $\{x_l\}_{l \ge 0}$ are i.i.d. standard normal. Then by Lemma 4.3.8(c), F_n is the usual empirical distribution of $\#J_n$ observations on $\left(\prod_{j=1}^g E_j \right)^{1/2g}$ where $\{E_j\}_{1 \le j \le g}$ are i.i.d. exponentials with mean one. Thus, by the Glivenko-Cantelli lemma, this converges to the distribution of $\left(\prod_{j=1}^g E_j \right)^{1/2g}$. Note that though the variables involved in the empirical distribution form a triangular sequence, the convergence is still almost sure due to the specific bounded nature of the indicator functions involved in the ESD. This may be proved easily by applying Hoeffding's inequality and the Borel-Cantelli lemma. We omit the details.

As mentioned earlier, all eigenvalues corresponding to any partition block

$\mathcal{P}_j$ are all the $(2g)$th roots of the product Π_j. Thus, the limit claimed in the statement of the theorem holds. So we have proved the result when the random variables $\{x_l\}_{l\geq 0}$ are i.i.d. standard normal.

Non-Gaussian input: Now suppose that the variables $\{x_l\}_{l\geq 0}$ are not necessarily normal. This case is tackled by normal approximation arguments similar to the case $k = 1$ (and hence $g = 1$) given earlier in the proof of Theorem 4.2.1. We now sketch some of the main steps. The basic idea remains the same but in this general case, a technical complication arises. Unlike in the $k = 1$ case, while applying normal approximation, the relevant set is *not* convex. Thus, we need to control the Gaussian measure of the η-boundaries of non-convex sets. We achieve this by a suitable compactness argument.

We start by defining

$$F(x) = \mathrm{P}\left(\left(\prod_{j=1}^{g} E_j\right)^{1/2g} \leq x\right), \quad x \in \mathbb{R}.$$

To show that the ESD converges to the required LSD in probability, we show that for every $x > 0$,

$$\mathrm{E}\left[F_n(x)\right] \to F(x) \quad \text{and} \quad \mathrm{Var}\left[F_n(x)\right] \to 0.$$

Note that for $x > 0$,

$$\mathrm{E}\left[F_n(x)\right] = (\#J_n)^{-1} \sum_{j=1}^{\#J_n} \mathrm{P}\left(n^{-g}\Pi_j \leq x^{2g}\right).$$

Lemma 4.3.8 motivates us to use the normal approximation given in Lemma 4.1.1. Towards this, define $2g$-dimensional random vectors

$$X_{l,j} = 2^{1/2}\left(x_l \cos\left(\frac{2\pi t l}{n}\right), \ x_l \sin\left(\frac{2\pi t l}{n}\right) : \ t \in \mathcal{A}_j\right) \quad 0 \leq l < n, 1 \leq j \leq \#J_n.$$

Note that

$$\mathrm{E}(X_{l,j}) = 0 \quad \text{and} \quad n^{-1}\sum_{l=1}^{n-1} \mathrm{Cov}(X_{l,j}) = I_{2g} \ \forall \ l, \ j.$$

Fix $x > 0$. Define the set $A \subseteq \mathbb{R}^{2g}$ as

$$A := \left\{(x_j, y_j : 1 \leq j \leq g) : \prod_{j=1}^{g}\left[2^{-1}(x_j^2 + y_j^2)\right] \leq x^{2g}\right\}.$$

Note that

$$\left\{n^{-g}\Pi_j \leq x^{2g}\right\} = \left\{n^{-1/2}\sum_{l=0}^{n-1} X_{l,j} \in A\right\}.$$

We want to prove

$$\mathrm{E}\left[F_n(x)\right] - \Phi_{2g}(A) = (\#J_n)^{-1} \sum_{l=1}^{\#J_n} \left(\mathrm{P}\left(n^{-g}\Pi_j \leq x^{2g}\right) - \Phi_{2g}(A)\right) \to 0.$$

However, now A is not convex. It suffices to show that for every $\epsilon > 0$ there exists $N = N(\epsilon)$ such that for all $n \geq N$,

$$\sup_{j \in J_n} \left| \mathrm{P}\left(n^{-1/2} \sum_{l=0}^{n-1} X_{l,j} \in A\right) - \Phi_{2g}(A) \right| \leq \epsilon.$$

Fix $\epsilon > 0$. Find $M_1 > 0$ large such that $\Phi([-M_1, M_1]^c) \leq \epsilon/(8g)$. By Assumption III($\delta$), $\mathrm{E}\left(n^{-1/2}a_{t,n}\right)^2 = \mathrm{E}\left(n^{-1/2}b_{t,n}\right)^2 = 1/2$ for any $n \geq 1$ and $0 < t < n$. Now by Chebyshev bound, we can find $M_2 > 0$ such that for each $n \geq 1$ and for each $0 < t < n$,

$$\mathrm{P}(|n^{-1/2}a_{t,n}| \geq M_2) \leq \epsilon/(8g) \quad \text{and} \quad \mathrm{P}(|n^{-1/2}b_{t,n}| \geq M_2) \leq \epsilon/(8g).$$

Set $M = \max\{M_1, M_2\}$. Define

$$B := \left\{(x_j, y_j : 1 \leq j \leq g) \in \mathbb{R}^{2g} : |x_j|, |y_j| \leq M \;\; \forall j\right\}.$$

Then for all sufficiently large n,

$$\left| \mathrm{P}\left(\frac{\sum_{l=0}^{n-1} X_{l,j}}{\sqrt{n}} \in A\right) - \Phi_{2g}(A) \right| \leq \left| \mathrm{P}\left(\frac{\sum_{l=0}^{n-1} X_{l,j}}{\sqrt{n}} \in A \cap B\right) - \Phi_{2g}(A \cap B) \right|$$
$$+ \epsilon/2.$$

We now apply Lemma 4.1.1 to $A \cap B$ to conclude that the first term in the right side of the above inequality is bounded above by

$$C_1 n^{-\delta/2}\rho_{2+\delta} + 2 \sup_{z \in \mathbb{R}^{2g}} \Phi_{2g}\left((\partial(A \cap B))^\eta - z\right)$$

where

$$\rho_{2+\delta} = \sup_{0 \leq l < n, j \in J_n} n^{-1} \sum_{l=0}^{n-1} \mathrm{E}\|X_{l,j}\|^{2+\delta} \quad \text{and} \quad \eta = \eta(n) = C_2\rho_{2+\delta}n^{-\delta/2}.$$

Note that $\rho_{2+\delta}$ is uniformly bounded in n by Assumption III(δ). It thus remains to show that for all sufficiently large n,

$$\Delta := \sup_{z \in \mathbb{R}^{2g}} \Phi_{2g}\left((\partial(A \cap B))^\eta - z\right) \leq \epsilon/8.$$

Observe that

$$\Delta \leq \sup_{z \in \mathbb{R}^{2g}} \int_{(\partial(A \cap B))^\eta} \phi(x_1 - z_1) \ldots \phi(y_g - z_{2g}) dx_1 \ldots dy_g$$
$$\leq \int_{(\partial(A \cap B))^\eta} dx_1 \ldots dy_g.$$

Finally, note that $\partial(A \cap B)$ is a *compact* $(2g - 1)$-dimensional manifold which has zero measure under the $2g$-dimensional Lebesgue measure. By compactness of $\partial(A \cap B)$, we have

$$(\partial(A \cap B))^\eta \downarrow \partial(A \cap B) \qquad \text{as } \eta \to 0,$$

and the claim follows upon using the Dominated Convergence Theorem.

This proves that for $x > 0$, $\mathrm{E}[F_n(x)] \to F(x)$. It remains to show that $\mathrm{Var}[F_n(x)] \to 0$. Since the variables involved are all bounded, it is enough to show that

$$n^{-2} \sum_{j \neq j'} \mathrm{Cov}\left(\mathbb{I}\left(n^{-g}\Pi_j \leq x^{2g}\right), \mathbb{I}\left(n^{-g}\Pi_{j'} \leq x^{2g}\right)\right) \to 0.$$

Along the lines of the proof used to show $\mathrm{E}[F_n(x)] \to F(x)$, one may now extend the vectors with $2g$ coordinates defined above to ones with $4g$ coordinates and proceed exactly as above, to verify this. We omit the routine details. This completes the proof of Theorem 4.3.9. $\qquad\square$

The following result is due to Bose, Mitra and Sen (2008)[28] for the case $k^g = 1 + sn$.

Theorem 4.3.10. Suppose $\{x_l\}_{l \geq 0}$ satisfies Assumption III(δ). Fix $g \geq 1$ and let p_1 be the smallest prime divisor of g. Suppose $k^g = 1 + sn$ where $s = 0$ if $g = 1$ and $s = o(n^{p_1 - 1})$ if $g > 1$. Then $F^{n^{-1/2}A_{k,n}}$ converges weakly in probability to $U_2\left(\prod_{j=1}^g E_j\right)^{1/2g}$ as $n \to \infty$ where $\{E_j\}_{1 \leq j \leq g}$ are i.i.d. exponentials with mean one and U_2 is uniformly distributed over the unit circle in $\mathbb{R}^2$, independent of $\{E_j\}_{1 \leq j \leq g}$. $\qquad\diamond$

Proof of Theorem 4.3.10. Since this proof is similar to the proof of the previous theorem, we will only sketch the main idea. First of all, note that $\gcd(k, n) = 1$ under the given hypothesis. When $g = 1$, then $k = 1$ and the eigenvalue partition is the trivial partition which consists of only singletons and clearly the partition sets $\mathcal{P}_j$, unlike the previous theorem, are not self-conjugate.

For $g \geq 2$, by Lemma 4.3.6 (b), it follows that $g_1 = g$ for n sufficiently large and $v_{k,n} \to 0$. In this case also, the partition sets $\mathcal{P}_j$ are not necessarily self-conjugate. Indeed, we will show that the number of indices j such that $\mathcal{P}_j$ is self-conjugate is asymptotically negligible compared to n. For that, we need to bound the cardinality of the following sets for $1 \leq b < g_1 = g$:

$$W_b := \left\{0 < t < n : tk^b = -t \mod n\right\} = \left\{0 < t < n : n|t(k^b + 1)\right\}.$$

Note that $t_0(b) := n/\gcd(n, k^b + 1)$ is the minimum element of W_b and every other element of the set W_b is a multiple of $t_0(b)$. Thus, the cardinality of the set W_b can be bounded by

$$\#W_b \leq n/t_0(b) = \gcd(n, k^b + 1).$$

Let us now estimate $\gcd(n, k^b + 1)$. For $1 \le b < g$,

$$
\begin{aligned}
\gcd(n, k^b + 1) &\le \gcd(k^g - 1, k^b + 1) \\
&\le k^{\gcd(g,b)} + 1 \\
&\le k^{g/p_1} + 1 = (1 + sn)^{1/p_1} + 1 = o(n),
\end{aligned}
$$

which implies

$$
n^{-1} \sum_{1 \le b < g} \#W_b = o(1),
$$

as desired. So, we can ignore the partition sets which are self-conjugate.

Let J_n denote the set of all those indices j for which $\#\mathcal{P}_j = g$ and $\mathcal{P}_j \cap (n - \mathcal{P}_j) = \emptyset$. Without loss, we assume that $J_n = \{1, 2, \ldots, \#J_n\}$.

Let $1, \varrho, \varrho^2, \ldots \varrho^{g-1}$ be all the gth roots of unity. The eigenvalues corresponding to the set $\mathcal{P}_j, j \in J_n$ are:

$$
\Pi_j^{1/g}, \Pi_j^{1/g} \varrho, \ldots, \Pi_j^{1/g} \varrho^{g-1}.
$$

Note that for $j \in J_n$, unlike in the previous theorem, now $\Pi_j = \prod_{t \in \mathcal{P}_j} (a_{t,n} + ib_{t,n})$ will be complex.

Hence, we need to consider the empirical distribution:

$$
G_n(x, y) = \frac{1}{g \# J_n} \sum_{j=1}^{\#J_n} \sum_{r=1}^{g} \mathbb{I}(n^{-1/2} \Pi_j^{1/g} \varrho^{r-1} \le x + iy), \quad x, y \in \mathbb{R},
$$

where for two complex numbers $w = x_1 + iy_1$ and $z = x_2 + iy_2$, by $w \le z$, we mean $x_1 \le x_2$ and $x_2 \le y_2$.

Now first suppose that $\{x_l\}_{l \ge 0}$ are i.i.d. $N(0, 1)$. Then by Lemma 4.3.8, $n^{-1/2} \Pi_j^{1/g}, j \in \mathcal{P}_j$ are independent and each of them is distributed as $\left(\prod_{t=1}^{g} E_t \right)^{1/2g} U_2$ as given in the statement of the theorem. This coupled with the fact that $\varrho^{r-1} U_2$ has the same distribution as that of U_2 for each $1 \le r \le g$ implies that $\{G_n\}_{n \ge 1}$ converges to the desired LSD (say G) as described in the theorem.

When $\{x_l\}_{l \ge 0}$ are not necessarily normals but only satisfy Assumption III(δ), we show that $\mathrm{E}\, G_n(x, y) \to G(x, y)$ and $\mathrm{Var}\,(G_n(x, y)) \to 0$ using the same line of argument as given in the proof of Theorem 4.3.9. For that, we again define $2g$-dimensional random vectors,

$$
X_{l,j} = 2^{1/2} \left(x_l \cos\left(\frac{2\pi t l}{n} \right), \quad x_l \sin\left(\frac{2\pi t l}{n} \right) : t \in \mathcal{P}_j \right) \quad 0 \le l < n, 1 \le j \le \#J_n,
$$

which satisfy

$$
\mathrm{E}\,(X_{l,j}) = 0 \quad \text{and} \quad n^{-1} \sum_{l=1}^{n-1} \mathrm{Cov}\,(X_{l,j}) = I_{2g} \quad \forall\ l, j.
$$

Fix $x, y \in \mathbb{R}$. Define the set $A \subseteq \mathbb{R}^{2g}$ as

$$A := \left\{ (x_j, y_j : 1 \leq j \leq g) : \left(\prod_{j=1}^{g} \left[2^{-1/2}(x_j + iy_j) \right] \right)^{1/g} \leq x + iy \right\}$$

so that

$$\left\{ n^{-1/2} \Pi_j^{1/g} \varrho^{r-1} \leq x + iy \right\} = \left\{ n^{-1/2} \sum_{l=0}^{n-1} X_{l,j} \in \varrho^{g+1-r} A \right\}.$$

The rest of the arguments is similar to the of proof of Theorem 4.3.9, once we realize that for each $1 \leq r \leq g$, $\partial(\varrho^{g+1-r}A)$ is again a $(2g-1)$-dimensional manifold which has zero measure $2g$-dimensional Lebesgue measure. $\square$

Remark 4.3.2. (a) From Theorems 4.3.9 and 4.3.10 we can recover the LSD of k-Circulants for $k = n-1$ (Reverse Circulant) and $k = 1$ (Circulant).

(b) While the radial coordinates of the LSDs described in Theorems 4.3.9 and 4.3.10 are the same, their angular coordinates differ. One puts its mass only at $e^{i2\pi j/2g}, 1 \leq j \leq 2g$ on the unit circle, while the other spreads its mass uniformly over this circle. This is also evident from Figure 4.3 given earlier.

(c) The restriction on $s = (k^g \pm 1)/n$ in the above two theorems seems to be a natural one. Suppose g is a prime and so $g = p_1$. In this case if $s \geq n^{p_1-1}$, then k becomes greater than or equal to n violating the assumption that $k < n$.

(d) Theorems 4.3.9 and 4.3.10 gives the LSDs only for $k^g = \pm 1 + n$. But simulations suggest the possibility of more varied limits when $k^g = r + n, r \neq \pm 1$. See Figure 4.4 for the case $r = \pm 3$ and $g = 2$.

4.4 Exercises

1. Suppose $\{F_n\}$ is a sequence of random distribution functions on a fixed probability space $(\Omega, \mathcal{A}, \mathrm{P})$ and F is a non-random distribution function. Show that the following are equivalent:

   (i) F_n converges weakly in probability to F;

   (ii) For every $\epsilon > 0$ and at all continuity points (x, y) of F,

   $$\mathrm{P}\big(|F_n(x, y) - F(x, y)| > \epsilon\big) \to 0 \text{ as } n \to \infty. \qquad (4.26)$$

   (iii) At all continuity points (x, y) of F,

   $$\int_{\Omega} [F_n(x, y) - F(x, y)]^2 \, d\mathrm{P}(\omega) \to 0 \text{ as } n \to \infty.$$

   (iv) At all continuity points (x, y) of F, condition (4.2) holds.

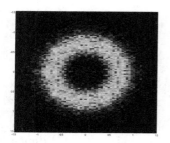 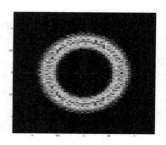

FIGURE 4.4

Eigenvalues of $n^{-1/2}A_{k,n}$ when $a_l \sim N(0,1)$, (i) $k = 16, n = -3 + k^2$ (left) (ii) $k = 16, n = 3 + k^2$ (right). 100 replications. Color represents the height of the histogram, from red (high) to blue (low). (Color figure appears following page 120)

2. Replicate the simulations of this chapter.

3. Verify that $\{\lambda_t\}$ in equation (4.14) are indeed the eigenvalues of the $n \times n$ Circulant matrix with input $\{x_i\}$.

4. Verify Remark 4.3.1.

5. Fix $g > 1$ and a positive integer m. Also, fix m primes $q_1, q_2, \ldots, q_m$ and m positive integers $\beta_1, \beta_2, \ldots, \beta_m$. Suppose the sequences k and $n \to \infty$ are such that

   (i) $k = q_1 q_2 \cdots q_m \hat{k}$ and $n = q_1^{\beta_1} q_2^{\beta_2} \cdots q_m^{\beta_m} \hat{n}$, $\hat{k}$ and $\hat{n} \to \infty$,

   (ii) $k^g = -1 + s\hat{n}$ where $s = o\left(\hat{n}^{p_1 - 1}\right) = o\left(n^{p_1 - 1}\right)$ where p_1 is the smallest prime divisor of g.

   Then show that $F^{n^{-1/2}A_{k,n}}$ converges weakly in probability to the distribution which has mass at $1 - \prod_{j=1}^{m} q_j^{-\beta_j}$ at zero, and the rest of the probability mass is distributed as $U_1 \left(\prod_{j=1}^{g} E_j\right)^{1/2g}$ where U_1 and $\{E_j\}_{1 \leq j \leq g}$ are as in Theorem 4.3.9.

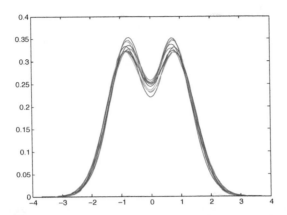

FIGURE 2.5
Smoothed ESD of $R_n^{(s)}$ for $n = 400$, 15 replications. Input is i.i.d. $N(0, 1)$.

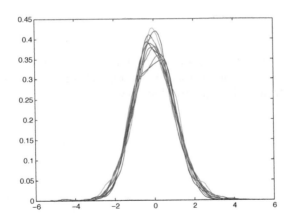

FIGURE 2.6
Smoothed ESD of $C_n^{(s)}$ for $n = 400$, 15 replications. Input is $N(0, 1)$.

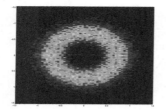 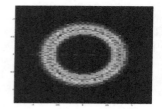

FIGURE 4.5
Eigenvalues of $n^{-1/2}A_{k,n}$ when $a_l \sim N(0,1)$, (i) $k = 16, n = -3 + k^2$ (left) (ii) $k = 16, n = 3 + k^2$ (right). 100 replications. Color represents the height of the histogram, from red (high) to blue (low).

5

Wigner-type matrices

The convergence of the spectral distribution of the Wigner matrix to the semi-circle law may be considered as the "central limit theorem" for random matrices. We have also seen the role of Catalan words in the convergence of the Wigner matrix. It is thus natural to ask the following questions for non-Wigner matrices:

(i) When is there a non-zero contribution *only* from Catalan words, and $p(w) = 1$ for such words, so that we get the semi-circle law as the limit?

(ii) When does each Catalan word contribute one (with possible non-zero contribution from non-Catalan words)?

(iii) For what matrix models do Catalan words not necessarily contribute one each and non-semi-circle limits arise, even when non-Catalan words have zero contribution?

(iv) In what ways may the assumption of the Wigner pattern be relaxed while the semi-circle LSD still holds?

The goal of this chapter is to investigate these issues. In particular we shall see that in a general sense, the semi-circle law serves as a *lower bound* for possible limits. Further, there is a large class of non-Wigner matrices whose limit is also the semi-circle law.

5.1 Wigner-type matrix

Recall the three assumptions from Chapter 1:

Assumption I $\{x_i\}$ or $\{x_{i,j}\}$ are independent and uniformly bounded with mean zero and variance 1.

Assumption II $\{x_i\}$ or $\{x_{i,j}\}$ are i.i.d. with mean zero and variance one.

Assumption III $\{x_{i,j}^{(n)}\}(1 \leq i, j \leq k_n)$ are independent and satisfy

$$\mathrm{E}[x_{i,j}^{(n)}] = 0 \quad \mathrm{E}[(x_{i,j}^{(n)})^2] = 1 \text{ and for all } m \in \mathbb{N}, \quad \sup_{1 \leq i,j \leq k_n} \mathrm{E}[|x_{i,j}^{(n)}|^m] \leq c_m < \infty.$$

We assume similar conditions if we have one sequence $\{x_j^{(n)}\}$.

We shall have occasion to deal with $(2k + 1)$ tuples $\{\pi(j)\}$ that satisfy all the constraints imposed by a pair-matched word w of length $(2k)$ except that they are not necessarily circuits (that is, $\pi(0)$ is not necessarily equal to $\pi(2k)$). Hence we define

$$
\begin{aligned}
C(w) &= \{\pi \text{ (not necessarily circuit)} : w[i] = w[j] \\
&\quad \Rightarrow L(\pi(i-1), \pi(i)) = L(\pi(j-1), \pi(j))\}, \\
\Gamma_{i,j}(w) &= \{\pi \in C(w) : \pi(0) = i, \pi(2k) = j\} \quad (1 \le i, j \le n), \\
\gamma_{i,j}(w) &= \#\Gamma_{i,j}(w) \quad (1 \le i, j \le n).
\end{aligned}
$$

Clearly,

$$\#\Pi^*(w) = \sum_{i=1}^{n} \gamma_{i,i}(w). \tag{5.1}$$

Recall Definitions 1.4.1 and 1.4.2 for Property B, k_n and α_n from Chapter 1.

Theorem 5.1.1. Consider a sequence of patterned random matrices $\{A_n\}$ with the link function L that satisfies Property B.

(a) Suppose L satisfies $\alpha_n = o(n)$. If w is a pair-matched non-Catalan word of length $(2k)$, then

$$\#\Pi^*(w) = O(n^k \alpha_n).$$

Hence
$$p(w) = 0 \text{ if } w \text{ is not Catalan.}$$

Further, if the input sequence satisfies Assumption I, II or III with $k_n^d \to \infty$ and $\alpha_n k_n^d = O(n^2)$ then any sub-sequential limit of $F^{n^{-1/2}A_n}$ has essential support contained in $[-2\sqrt{\Delta}, 2\sqrt{\Delta}]$.

(b) Suppose L satisfies $\Delta = 1$. Then, for any Catalan word w of length $(2k)$, and any n,
$$\#\Pi^*(w) = n^{k+1}.$$

Consequently,
$$p(w) = 1 \text{ if } w \text{ is Catalan.}$$

If in addition, the input sequence satisfies Assumption I, II or III with $k_n \to \infty$ and $\alpha_n k_n^d = O(n^2)$ and L satisfies $\alpha_n = o(n)$, then the LSD of $n^{-1/2}A_n$ is the semi-circle law almost surely. ◇

Proof. (a) Without loss of generality, we assume that the input sequence is uniformly bounded. Suppose w is not a Catalan word. Then there is $i < j < t < l$ such that

$$w[i] = w[t] = a \text{ (say) and } w[j] = w[l] = b \text{ (say).} \tag{5.2}$$

Without loss of generality, let i be the minimum such choice (for which the above condition is satisfied for some j, t, l), and for this i, let j be the maximum such choice (for some t, l). Thus, all letters of w in $\{w[j+1], ..., w[t-1]\}$ have both copies at the positions $\{i+1, ..., t-1\}$. We use the following algorithm to obtain the count of all circuits π corresponding to w:

(1) First fill up $\pi(0), ..., \pi(j-1)$ from *left to right* honoring the word constraints.

(2) Now fill up $\pi(j), ..., \pi(t)$ from *right to left*, honoring the word constraints. Since

$$L(\pi(i-1), \pi(i)) = L(\pi(t-1), \pi(t)),$$

$(\pi(t-1), \pi(t))$ can be chosen in at most α_n ways. If $w[s]$ $(j < s < t)$ is the second occurrence of a letter in the filling sequence, then, having chosen up to $\pi(s)$, $\pi(s-1)$ has at most Δ choices. Let

$$w[i_1], ..., w[i_y] \quad (i_1 < i_2 < .. < i_y)$$

be the first occurrences of the letters in the filling sequence occurring in $\{t, .., j+1\}$. Note that $i_1 > j+1$, and each vertex in $\{t-2, ..., j\} \cap \{i_y - 1, ..., i_1 - 1\}^c$ can be chosen in at most Δ ways, given all previous vertices in the filling sequence.

(3) Finally, fill $\pi(t+1), ..., \pi(2k)$ from *left to right*. If the number of first occurrences in $w[t+1], ..., w[2k]$ is z, then the number of generating vertices is z.

Now, the circuit is completely specified. The number of vertices other than $\pi(t)$ and $\pi(t-1)$ with choices more than Δ, given previously filled vertices in the filling sequence, is at most $(x+1) + y + z$. The first occurrence of letter b (defined in (5.2)) is never counted among the y first occurrences in step (2). Therefore

$$x + (y+1) + z = k.$$

As a consequence,

$$\#\Pi^*(w) = O(n^k \alpha_n)$$

and hence $p(w) = 0$.

Now, for any sub-sequential limit G,

$$\beta_{2G}(G) \leq \Delta^k \frac{(2k)!}{k!(k+1)!} = \Delta^k \frac{1}{k+1} \binom{2k}{k} \leq \Delta^k 2^{2k}.$$

Therefore

$$\limsup_{k \to \infty} (\beta_{2k}(G))^{1/2k} \leq 2\Delta^{1/2}.$$

The claim on the support then follows as the L^p norm converges to the essential sup when $p \to \infty$.

(b) The essential implication of $\Delta = 1$ is

$$L(\pi(i-1), \pi(i)) = L(\pi(i), \pi(i+1)) \Rightarrow \pi(i-1) = \pi(i+1). \tag{5.3}$$

For $k = 1$, choose $\pi(0)$ and $\pi(1)$ in n ways each. For all such choices, $\pi(2)$ has exactly one choice by (5.3), i.e., $\pi(2) = \pi(0)$. Therefore $\#\Pi^*(w) = n^2$.

Now we apply induction on k. Suppose the statement is true for some $k \geq 1$. Take any Catalan word w of length $(2k + 2)$. Then

$$w[i] = w[i + 1] \quad \text{for some} \quad i \leq 2k + 1.$$

Let w' be the Catalan word obtained from w by deleting $w[i]$ and $w[i + 1]$. Then any circuit π' in $\Pi^*(w')$ yields n distinct circuits $\pi_1, ..., \pi_n$ in $\Pi^*(w)$, where π_s is obtained by putting

$$\begin{aligned}
\pi_s(j) &= \pi'(j), \; \forall \, j \leq i - 1 \,, \\
\pi_s(i) &= s \; \text{and} \\
\pi_s(j) &= \pi'(j - 2), \; \forall \, j \geq i + 1 \; \text{(in view of (5.3))}.
\end{aligned}$$

Also, this process is applied to all circuits in $\Pi^*(w')$ and yields the entire $\Pi^*(w)$. Therefore

$$\#\Pi^*(w) = n \times \#\Pi^*(w').$$

By induction hypothesis,

$$\#\Pi^*(w) = n \times n^{k+1} = n^{k+2}.$$

Now applying (a), first part of (b) is proved. The second part follows by using the first part and (a). Hence (b) is proved completely. □

Let

$$M_{i,j}^{(n)} = \{1 \leq k \leq n : L(k, i) = L(k, j)\}$$

be the *"match set"* between columns (rows) i and j at stage n. Consider the following property:

Property P $M^* \equiv \sup_n \sup_{i,j \leq n} \#M_{i,j}^{(n)} < \infty.$ ◇

In the next theorem and its proof, since we will be dealing with more than one link function, we will use an appropriate subscript for link function L and the class $\Pi^*(w)$.

Theorem 5.1.2. (a) Suppose the matrices $\{A_n\}$ have the link function L which satisfies Property B. Then, for any Catalan word w of length $(2k)$,

$$\#\Pi_{A_n}^*(w) \geq n^{k+1}.$$

Hence if the input sequence satisfies Assumption I, II or III and $k_n \to \infty$, $\alpha_n k_n^d = O(n^2)$, then the even moments of any sub-sequential limit of $F^{n^{-1/2} A_n}$ satisfy

$$\beta_{2k} \geq \frac{(2k)!}{k!(k + 1)!}.$$

(b) Suppose L satisfies Properties B and P. Then

$$p(w) = 1 \quad \text{if } w \text{ is Catalan.}$$

(c) Suppose the input sequence satisfies Assumption I, II or III and $k_n \to \infty$, $\alpha_n k_n^d = O(n^2)$. Further, suppose that the link function L satisfies Properties B and P, and $\alpha_n = o(n)$. Then the LSD of $\{n^{-1/2}A_n\}$ is the semi-circle law. ◇

Note that the Toeplitz, Hankel, Symmetric Circulant and Reverse Circulant link functions all satisfy the assumptions of part (b) of the above theorem and hence $p(w) = 1$ for all Catalan words for these matrices. The above theorem motivates the following definition. We discuss this idea more later after the proof.

Definition 5.1.1. Those patterned matrices which satisfy $\alpha_n = o(n)$, Property B and Property P will be called *Wigner-type* matrices.

Proof of Theorem 5.1.2. (a) To ease notation, suppress dependency on n and write, for example, A for A_n. As mentioned earlier, let L_W and L_A be the link functions for the Wigner matrix and the matrix A and let $\Pi_W^*(w)$ be the $\Pi^*(w)$ for the link function L_W. If $\pi \in \Pi_W^*(w)$, then

$$
\begin{aligned}
w[i] = w[j] \quad &\Rightarrow \quad L_W(\pi(i-1), \pi(i)) = L_W(\pi(j-1), \pi(j)) \\
&\Rightarrow \quad L_A(\pi(i-1), \pi(i)) = L_A(\pi(j-1), \pi(j)) \quad (\text{as } A \text{ is symmetric}).
\end{aligned}
$$

Hence $\pi \in \Pi_A^*(w)$. Therefore

$$\Pi_W^*(w) \subseteq \Pi_A^*(w).$$

As $\#\Pi_W^*(w) = n^{k+1}$, the result (a) follows.

(b) We will prove this part by induction. Consider the following induction statement:

S_k: For any Catalan w of length $(2k)$, there exists $M_k > 0$ such that

$$\gamma_{i,j}(w) \leq M_k n^{k-1} \text{ for all } i \neq j \quad \text{and} \quad \frac{1}{n}\sum_{i=1}^{n}\Big|\frac{\gamma_{i,i}(w)}{n^k} - 1\Big| = O(n^{-1}). \quad (5.4)$$

Note that if S_k is true, then $p(w) = 1$ for any Catalan word w of length $(2k)$ (see (5.1)).

For $k = 1$, $\gamma_{i,i}(w) = n$ and by Property P,

$$\gamma_{i,j}(w) = \#M_{i,j}^{(n)} \leq M^* \quad \text{for all } i \neq j$$

and therefore S_1 is true.

Let S_l be true for all $l \leq k-1$ ($k \geq 2$). We shall now prove that S_k is true. Let w be a Catalan word of length $(2k)$. Then the following two cases arise:

Case (*i*). Assume $w = w_1 w_2$, where w_1 and w_2 are Catalan words of length $(2k_1)$ and $(2k_2)$, respectively, (say) where $k_1 \leq k-1$ and $k_2 \leq k-1$. Then for $i \neq j$, using the induction hypothesis and Property B:

$$
\begin{aligned}
\gamma_{i,j}(w) &= \sum_{l=1}^{n} \gamma_{i,l}(w_1)\gamma_{l,j}(w_2) \\
&= \sum_{l \neq i,j} \gamma_{i,l}(w_1)\gamma_{l,j}(w_2) + \gamma_{i,i}(w_1)\gamma_{i,j}(w_2) + \gamma_{i,j}(w_1)\gamma_{j,j}(w_2) \\
&\leq (M_{k_1} n^{k_1-1})(M_{k_2} n^{k_2-1})(n) \\
&\quad + (\Delta^{k_1-1} n^{k_1})(M_{k_2} n^{k_2-1}) + (\Delta^{k_2-1} n^{k_2})(M_{k_1} n^{k_1-1}) \\
&= (M_{k_1} M_{k_2} + \Delta^{k_1-1} M_{k_2} + \Delta^{k_2-1} M_{k_1}) n^{k-1}.
\end{aligned}
$$

This proves the first part of the induction statement (5.4) in Case (i). Now we prove the other part of the induction statement S_k:

$$
\begin{aligned}
\frac{1}{n}\sum_{i=1}^{n}\left|\frac{\gamma_{i,i}(w)}{n^k} - 1\right| &= \frac{1}{n}\sum_{i=1}^{n}\left|\frac{\sum_{l\neq i}\gamma_{i,l}(w_1)\gamma_{l,i}(w_2) + \gamma_{i,i}(w_1)\gamma_{i,i}(w_2)}{n^k} - 1\right| \\
&\leq \frac{1}{n}\sum_{i=1}^{n}\left|\frac{\gamma_{i,i}(w_1)}{n^{k_1}}\frac{\gamma_{i,i}(w_2)}{n^{k_2}} - 1\right| + \frac{1}{n}\sum_{i,l=1,l\neq i}^{n}\frac{\gamma_{i,l}(w_1)\gamma_{l,i}(w_2)}{n^k} \\
&= T_1 + T_2 \quad \text{(say)}.
\end{aligned}
$$

By using the induction hypothesis,

$$
\begin{aligned}
T_2 &\leq \frac{1}{n}\sum_{i=1}^{n}\sum_{l\neq i}\frac{M_{k_1} M_{k_2} n^{k-2}}{n^k} \\
&\leq M_{k_1} M_{k_2}\frac{n^k}{n^{k+1}} = \frac{M_{k_1} M_{k_2}}{n} = \mathrm{O}(n^{-1}).
\end{aligned}
$$

By the induction hypothesis and Property B,

$$
\begin{aligned}
T_1 &\leq \frac{1}{n}\sum_{i=1}^{n}\frac{\gamma_{i,i}(w_1)}{n^{k_1}}\left|\frac{\gamma_{i,i}(w_2)}{n^{k_2}} - 1\right| + \frac{1}{n}\sum_{i=1}^{n}\left|\frac{\gamma_{i,i}(w_1)}{n^{k_1}} - 1\right| \\
&\leq \frac{1}{n}\sum_{i=1}^{n}\Delta^{k_1-1}\left|\frac{\gamma_{i,i}(w_2)}{n^{k_2}} - 1\right| + \frac{1}{n}\sum_{i=1}^{n}\left|\frac{\gamma_{i,i}(w_1)}{n^{k_1}} - 1\right| = \mathrm{O}(n^{-1}).
\end{aligned}
$$

This establishes the second part of (5.4) and completes the proof in Case (i).

Case (*ii*). Assume $w = a w_1 a$, where w_1 is a Catalan word. Let $i \neq j$. Then

$$
\gamma_{i,j}(w) = \sum_{l \in M_{i,j}^{(n)}} \gamma_{l,l}(w_1) + \sum_{l=1}^{n}\sum_{t\neq l:L(i,l)=L(j,t)} \gamma_{l,t}(w_1) = A_1 + A_2 \quad \text{(say)}.
$$

The first sum is over

$$\{\pi \in C(w): \ \pi(0) = i, \pi(2k) = j \ \text{ for which } \ \pi(1) = \pi(2k-1)\}.$$

The second sum is over

$$\{\pi \in C(w): \ \pi(0) = i, \pi(2k) = j \ \text{ for which } \ \pi(1) \neq \pi(2k-1)\}.$$

Observe that

$A_1 \leq \Delta^{k-2} n^{k-1} M^*$ (Property P),

$A_2 \leq M_{k-1} n^{k-2} \cdot n\Delta = M_{k-1} \Delta n^{k-1}$ (induction hypothesis and Property B).

This proves the first part of (5.4) in Case (ii).

To prove the second part of (5.4), we proceed as in the previous case. By Property P and the induction hypothesis:

$$\frac{1}{n} \sum_{i=1}^{n} |\frac{\gamma_{i,i}(w)}{n^k} - 1| = \frac{1}{n} \sum_{i=1}^{n} |\frac{\sum_{l=1}^{n} \gamma_{l,l}(w_1) + \sum_{l,t:l\neq t, i\in M_{l,t}^{(n)}} \gamma_{l,t}(w_1)}{n^k} - 1|$$

$$\leq \frac{1}{n} \sum_{i=1}^{n} |\frac{\sum_{l=1}^{n} \gamma_{l,l}(w_1)}{n \cdot n^{k-1}} - 1| + \frac{1}{n} \sum_{i=1}^{n} \sum_{l,t:l\neq t, i\in M_{l,t}^{(n)}} \frac{\gamma_{l,t}(w_1)}{n^k}$$

$$\leq \frac{1}{n} \sum_{i=1}^{n} \frac{1}{n} \sum_{l=1}^{n} |\frac{\gamma_{l,l}(w_1)}{n^{k-1}} - 1| + \frac{1}{n^{k+1}} \sum_{l,t:l\neq t} \sum_{i:i\in M_{i,j}^{(n)}} \frac{\gamma_{l,t}(w_1)}{n^k}$$

$$\leq \frac{1}{n} \sum_{l=1}^{n} |\frac{\gamma_{l,l}(w_1)}{n^{k-1}} - 1| + \frac{1}{n^{k+1}} \sum_{l,t:l\neq t} M^* \cdot M_{k-1} n^{k-2}$$

$$\leq \frac{1}{n} \sum_{l=1}^{n} |\frac{\gamma_{l,l}(w_1)}{n^{k-1}} - 1| + \frac{M^* M_{k-1}}{n} = \mathrm{O}(n^{-1}).$$

This proves the second part of (5.4) in Case (ii) and hence (b) is completely proved.

(c) Theorem 5.1.1(a) and Theorem 5.1.2(b) together imply this. $\square$

Example 1. Suppose

$$L(i,j) = |P(i,j)|,$$

where P is a symmetric polynomial in two variables with integer coefficients. Then, conditions of Theorem 5.1.2(b) are satisfied. However, the condition $\alpha_n = o(n)$ may not be satisfied for such link functions. The Toeplitz link function $L(i,j) = |i - j|$ is an example of the latter. $\diamond$

Example 2. Let us now consider a natural subclass of the above. Let

$$L(i,j) = |P(ij)|,$$

where P is a polynomial of degree d (say) in one variable with integer coefficients. Let

$$u_l = \#\{(i,j) : L(i,j) = l\}.$$

Then

$$\alpha_n = \sup_{l \in \{|P(ij)|: 1 \le i, j \le n\}} u_l.$$

Clearly, for each fixed l, the equation $|P(x)| = l$ has at most $2d$ non-negative integer solutions. For each such solution z (say) in $\{1, 2, ..., n^2\}$,

$$\#\{(i,j) : 1 \le i, j \le n, ij = z\} \le d(z),$$

where $d(z)$ is the number of positive divisors of z. It is well known that (see Apostol (1976)[3])

$$\limsup_n \frac{\log d(n)}{\log n / \log \log n} = \log 2. \tag{5.5}$$

In particular, for any $\epsilon > 0$, $d(n) = o(n^\epsilon)$. Choose $\epsilon < \frac{1}{2}$ and $C > 0$ such that $d(m) < Cm^\epsilon$ for all positive integers m. Therefore

$$\begin{aligned}
\alpha_n &= \sup_{l \in \{|P(ij)|: 1 \le i, j \le n\}} u_l \\
&\le (2d)[\sup_{m \le n^2} d(m)] \\
&\le (2d)Mn^{2\epsilon} = o(n).
\end{aligned}$$

Hence by part (c) of Theorem 5.1.2, if the input sequence satisfies Assumption III, the LSD is the semi-circle law. The same LSD continues to hold (under Assumption III) if $L(i,j) = |P(ij)| \mod m_n$ where m_n is a sequence of positive integers that satisfies $\sup \dfrac{n^{2d}}{m_n} < \infty$.

Incidentally, the condition $\alpha_n k_n^d = O(n^2)$ is not satisfied for all polynomials. For example, let

$$L(i,j) = ij.$$

While counting the distinct entries, we count all the integers of the form pq, where p and q are primes and $p \le q$. Hence

$$k_n \ge \frac{1}{2} \left(\frac{n}{\log n} \right)^2.$$

Now,

$$\alpha_n \ge d(n), \quad \text{the number of divisors of } n.$$

On the other hand, by (5.5) we get a sub-sequence $\{n_j\}$ which satisfies

$$\frac{\log d(n_j)}{(\log n_j / \log \log n_j)} \ge C > 0$$

for all j. This implies that

$$d(n_j) \geq \exp\left\{\frac{C \log n_j}{\log \log n_j}\right\}.$$

Thus,

$$\frac{\alpha_{n_j} k_{n_j}}{n_j^2} \geq \frac{1}{2}\exp\left\{C \log n_j/(\log \log n_j)\right\} - (2\log\log n_j)\}$$

$$= \frac{1}{2}\exp\left\{[\frac{C \log n_j}{(\log\log n_j)^2} - 2]\log\log n_j\right\} \to \infty.$$

$\diamond$

Example 3. Wigner-type matrices are easy to construct by *coloring*: take a triangular input sequence $\{x_{i,j}\}_{(1\leq i\leq r_n,\ 0\leq j\leq n-1)}$ which satisfies Assumption III. Consider the *Toeplitz link function* L_T and a *color link function*

$$L_c(i,j) = k \text{ if } l_k \leq i+j < l_{k+1} \ (1 \leq k \leq r_n) \text{ such that}$$

$$2 = l_1 < l_2 < \cdots < l_{r_n+1} = 2n+1 \quad \text{and} \quad \sup_{k\leq r_n}(l_{k+1} - l_k) = o(n).$$

Consider the *colored Toeplitz matrix* A_n with link function

$$L(i,j) = (L_c(i,j), L_T(i,j)).$$

Then it is easy to see that

$$\alpha_n = \sup_{k\leq r_n}(l_{k+1} - l_k) = o(n)$$

and Properties B and P are satisfied. Hence this is a *Wigner-type* matrix. $\diamond$

Non-semi-circle limits can arise if matches between rows are increased, say by *blocking*. Unfortunately, block matrices are not discussed in this book.

5.2   Exercises

1. Look up the references Basu et al. (2012)[16], Bannerjee and Bose [9] and Oraby (2007)[80] for some results on the LSDs of block matrices.

6

Balanced Toeplitz and Hankel matrices

For the Wigner matrix and the group of circulant matrices, the number of times each random variable appears in the matrix is the same across most variables but for asymptotically negligible exceptions. We may call them *balanced* matrices. For them, the LSDs exist after the eigenvalues are scaled by $n^{-1/2}$. However, the Toeplitz and Hankel matrices are unbalanced. It seems natural to consider their balanced versions where each entry is scaled by a constant multiple of the square root of the number of times that entry appears in the matrix instead of the uniform scaling by $n^{-1/2}$. Define the (symmetric) balanced Hankel and Toeplitz matrices BH_n and BT_n with input $\{x_i\}$ as follows:

$$BH_n = \begin{bmatrix} \frac{x_1}{\sqrt{1}} & \frac{x_2}{\sqrt{2}} & \frac{x_3}{\sqrt{3}} & \cdots & \frac{x_{n-1}}{\sqrt{n-1}} & \frac{x_n}{\sqrt{n}} \\ \frac{x_2}{\sqrt{2}} & \frac{x_3}{\sqrt{3}} & \frac{x_4}{\sqrt{4}} & \cdots & \frac{x_n}{\sqrt{n}} & \frac{x_{n+1}}{\sqrt{n-1}} \\ \frac{x_3}{\sqrt{3}} & \frac{x_4}{\sqrt{4}} & \frac{x_5}{\sqrt{5}} & \cdots & \frac{x_{n+1}}{\sqrt{n-1}} & \frac{x_{n+2}}{\sqrt{n-2}} \\ & & & \vdots & & \\ \frac{x_n}{\sqrt{n}} & \frac{x_{n+1}}{\sqrt{n-1}} & \frac{x_{n+2}}{\sqrt{n-2}} & \cdots & \frac{x_{2n-2}}{\sqrt{2}} & \frac{x_{2n-1}}{\sqrt{1}} \end{bmatrix}. \qquad (6.1)$$

$$BT_n = \begin{bmatrix} \frac{x_0}{\sqrt{n}} & \frac{x_1}{\sqrt{n-1}} & \frac{x_2}{\sqrt{n-2}} & \cdots & \frac{x_{n-2}}{\sqrt{2}} & \frac{x_{n-1}}{\sqrt{1}} \\ \frac{x_1}{\sqrt{n-1}} & \frac{x_0}{\sqrt{n}} & \frac{x_1}{\sqrt{n-1}} & \cdots & \frac{x_{n-3}}{\sqrt{3}} & \frac{x_{n-2}}{\sqrt{2}} \\ \frac{x_2}{\sqrt{n-2}} & \frac{x_1}{\sqrt{n-1}} & \frac{x_0}{\sqrt{n}} & \cdots & \frac{x_{n-4}}{\sqrt{4}} & \frac{x_{n-3}}{\sqrt{3}} \\ & & & \vdots & & \\ \frac{x_{n-1}}{\sqrt{1}} & \frac{x_{n-2}}{\sqrt{2}} & \frac{x_{n-3}}{\sqrt{3}} & \cdots & \frac{x_1}{\sqrt{n-1}} & \frac{x_0}{\sqrt{n}} \end{bmatrix}. \qquad (6.2)$$

Strictly speaking, BT_n is not completely balanced, with the main diagonal being unbalanced compared to the rest of the matrix. The main diagonal has all identical elements and making BT_n balanced will shift its eigenvalues by $x_0/\sqrt{2n}$, which in the limit will go to zero and hence this does not affect the asymptotic behavior of the eigenvalues. We use the above version because of the convenience in writing out the calculations later. The goal of this chapter is to investigate the LSD of these two matrices. We use the moment method as has been done so far. However, curiously, it is not clear that the limit moments define a unique probability distribution. Nevertheless, in a roundabout way, we are able to show the existence of the LSD with these moments.

6.1   Main results

Simulation for the two matrices are given in Figures 6.1 and 6.2.

FIGURE 6.1
Histogram of the ESD of Toeplitz (left) and balanced Toeplitz (right) matrices,
$n = 400$, 15 replications with $N(0, 1)$ entries.

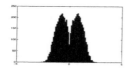 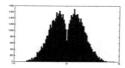

FIGURE 6.2
Histograms of the ESD of Hankel (left) and balanced Hankel (right) matrices,
$n = 400$, 15 replications, with standardized Bernoulli(0.5) entries.

The following result is due to Basak and Bose (2010)[12].

Theorem 6.1.1. Suppose $\{x_i\}$ satisfies Assumption I, II or III of Chapter 1.
Then almost surely the LSD, say BT and BH of the matrices BT_n and BH_n
respectively, exist and are free of the underlying distribution of the $\{x_i\}$.   ⋄

Remark 6.1.1. BT and BH have unbounded support, are symmetric about
zero and have all moments finite. Both LSDs are non-Gaussian. The integral
formulae for the moments are given in (6.8) and (6.11) after we develop the
requisite notation to write them out. It does not seem to be clear if these
moments define a distribution uniquely. Establishing further properties of the
limits is a difficult problem.   ⋄

The theorem will be proved in several steps. The main difference from the
proofs we have seen so far is that while we can verify the (M1) condition, it is
not clear if the limit moments determine a distribution. So, we take a roundabout route. We first look at a truncated (banded) version of the matrix which
is well behaved and for which the limit moments do determine a distribution.

Then we show that the original matrices are close to these matrices when the level of banding shrinks. Hence the proof proceeds in several steps.

Step I. *Reduction of the problem to the uniformly bounded case.* The following truncation lemma is similar to Lemma 1.4.1 in Chapter 1. Its proof is also along the same lines and uses the SLLN and the W_2 metric. The details are left to the reader.

Lemma 6.1.1. Suppose for every bounded, mean zero and variance one i.i.d. input sequence $\{x_i\}$, $\{F^{BT_n}\}$ converges to a non-random distribution F a.s. Then the same limit continues to hold if the $\{x_i\}$ are not bounded. All the above hold for $\{F^{BH_n}\}$ as well. $\diamond$

Step II. *Verification of (M1) condition.*

Step II (a). *Trace formula, etc.* Define

$$\phi_T(i,j) = n - |i-j| \quad \text{and} \quad \phi_H(i,j) = \min(i+j-1, 2n-i-j+1), \quad (6.3)$$

$$\phi_T^n(x,y) = \phi_T^\infty(x,y) = 1 - |x-y|, \quad (6.4)$$

$$\phi_H^n(x,y) = \min(x+y-\frac{1}{n}, 2-x-y+\frac{1}{n}), \quad \phi_H^\infty(x,y) = \lim_{n\to\infty} \phi_H^n(x,y). \quad (6.5)$$

Then note that the *trace formula* implies

$$\frac{1}{n}\operatorname{Tr}[BT_n]^h = \frac{1}{n}\sum_{1\le i_1,\ldots,i_h \le n}\prod_{1\le j\le h-1} \frac{x_{L^T(i_j,i_{j+1})}}{\sqrt{\phi_T(L^T(i_j,i_{j+1}))}}\frac{x_{L^T(i_h,i_1)}}{\sqrt{\phi_T(L^T(i_h,i_1))}}$$

$$\operatorname{E}[\beta_h(BT_n)] = \operatorname{E}[\frac{1}{n}\operatorname{Tr}(BT_n)^h]$$

$$= \frac{1}{n}\sum_{\pi:\pi \text{ circuit}} \frac{\operatorname{E}X_\pi}{\prod_{1\le i\le h}\sqrt{\phi_T(\pi(i-1),\pi(i))}}. \quad (6.6)$$

$$\frac{1}{n}\operatorname{Tr}[BH_n]^h = \frac{1}{n}\sum_{1\le i_1,\ldots,i_h \le n}\prod_{1\le j\le h-1} \frac{x_{L^H(i_j,i_{j+1})}}{\sqrt{\phi_H(L^H(i_j,i_{j+1}))}}\frac{x_{L^H(i_h,i_1)}}{\sqrt{\phi_H(L^H(i_h,i_1))}}$$

$$\operatorname{E}[\beta_h(BH_n)] = \operatorname{E}[\frac{1}{n}\operatorname{Tr}(BH_n)^h]$$

$$= \frac{1}{n}\sum_{\pi:\pi \text{ circuit}} \frac{\operatorname{E}X_\pi}{\prod_{1\le i\le h}\sqrt{\phi_H(\pi(i-1),\pi(i))}}. \quad (6.7)$$

Step II (b). *Only pair-matched words survive.* Fix an integer h. Define

$$\Pi_h^{3+} = \{\pi : l(\pi) = h, \text{ with at least one edge of order greater than equal to 3}\}.$$

$$S_h^{A_n} = \frac{1}{n}\sum_{\pi:\pi\in\Pi_h^{3+}} \frac{1}{\prod_{i=1}^h \sqrt{\phi_A(\pi(i-1),\pi(i))}}, \quad A_n = BH_n \text{ or } BT_n.$$

Lemma 6.1.2. $S_h^{A_n} \to 0$ as $n \to \infty$ for $A_n = BT_n$ or BH_n. Hence, only pair-matched circuits are relevant while calculating $\lim E[\beta_h(A_n)]$.   ◇

Proof. We provide the proof only for BT_n. The proof for BH_n is similar and the details are omitted. Note that

$$S_h^{A_n} = \sum_w \frac{1}{n} \sum_{\pi \in \Pi(w) \cap \Pi_h^{3+}} \frac{1}{\prod_{i=1}^h \sqrt{n - |\pi(i-1) - \pi(i)|}} = \sum_w S_{h,w}, \quad \text{say.}$$

It is enough to prove that for each w, $S_{h,w} \to 0$. We first restrict attention to w which have only one edge of order 3 and all other edges of order 2. Note that this forces h to be odd. Let $h = 2t + 1$ and $|w| = t$. Fix the L-values at say $k_1, k_2, \ldots, k_t$ where k_1 is the L-value corresponding to the order 3 edge. Let i_0 be such that $L(\pi(i_0 - 1), \pi(i_0)) = k_1$. We start counting the number of possible π's from the edge $(\pi(i_0 - 1), \pi(i_0))$. Clearly the number of possible choices of such an edge is at most $2(n - k_1)$. Having chosen the value of the vertex i_0, the number of possible choices of the vertex $(i_0 + 1)$ is at most 2. Carrying on with this argument, we may conclude that the total number of π's having L-values $k_1, k_2, \ldots, k_t$ is at most $C \times (n - k_1)$. Hence for some generic constant C,

$$
\begin{aligned}
S_{h,w} &= \frac{1}{n} \sum_{0 \le k_i \le n-1} \sum_{\substack{\pi:\pi \text{ has } L\text{-values} \\ k_1, k_2, \ldots, k_t}} \frac{1}{(n - k_1)^{3/2}} \prod_{i=2}^t (n - k_i) \\
&\le \frac{1}{n} \sum_{0 \le k_i \le n-1} \frac{C \times (n - k_1)}{(n - k_1)^{3/2} \prod_{i=2}^t (n - k_i)} \\
&= \frac{\mathrm{O}(\sqrt{n})\, \mathrm{O}((\log n)^{t-1})}{n} \to 0 \text{ as } n \to \infty.
\end{aligned}
$$

In the above calculations, we have used the facts that $\sum_{k=1}^n \frac{1}{k} = \mathrm{O}(\log n)$ and for $0 < s < 1$, $\sum_{k=1}^n \frac{1}{k^s} = \mathrm{O}(n^{1-s})$.

It is easy to see that when w contains more than one edge of order 3 or more, the order of the sum will be even smaller. This completes the proof of the first part. The second part is immediate since $E(X_\pi) = 1$ for every pair-matched circuit and $E(|X_\pi|) < \infty$ uniformly over all π.   □

Step II (c). *Slope in BT_n.* We have seen in Chapter 2 that for the standard Toeplitz matrices, out of two possible choices of slopes, only one choice counts. We show now that the same is true for the balanced matrices. Let

$$
\begin{aligned}
\Pi_{h,+} &= \{\pi \text{ pair-matched} : \exists (i_0, j_0) \text{ with} \\
&\qquad \pi(i_0 - 1) - \pi(i_0) + \pi(j_0 - 1) - \pi(j_0) \ne 0\}, \\
\Pi_{h,+}(w) &= \Pi_{h,+} \cap \Pi(w)
\end{aligned}
$$

and recall the slopes $\pi(i - 1) - \pi(i)$.

Lemma 6.1.3. Let $\pi \in \Pi_{h+}$ and $k_1, \ldots, k_h$ be the L-values of π. Then there exists $j_0 \in \{1, 2, \ldots, h\}$ such that $k_{j_0} = \Lambda(k_1, k_2, \ldots, k_{j_0-1}, k_{j_0+1}, \ldots, k_h)$ for some linear function Λ. ◇

Proof. Note that the sum of all the slope values of π is zero. Now the sum of the slope values from the jth matched pair is 0 if the L-values have opposite signs, while it is $2k_j$ or $-2k_j$ if the L-values have the same sign. Hence we have $f(k_1, k_2, \ldots, k_h) = 0$ for some linear function f where the coefficient of k_j equals 0 if the L-values corresponding to k_j have opposite signs while it is ± 2 if the L-values corresponding to k_j are of the same sign and the slope values are positive (negative). Since $\pi \in \Pi_{h,+}$, $\exists\, k_j \neq 0$ such that the L-values corresponding to the jth pair have the same sign. Let

$$\{i_1, \ldots, i_l\} = \{j : \text{coefficient of } k_j \neq 0\} \text{ and}$$
$$j_0 = \max\{j : \text{coefficient of } k_j \neq 0\}.$$

Then clearly $k_{j_0} = \Lambda(k_{i_1}, k_{i_2}, \ldots, k_{i_l})$ for some linear function Λ. □

Lemma 6.1.4. $S_{h+} \stackrel{def}{=} \lim_{n\to\infty} \frac{1}{n} \sum_{\pi \in \Pi_{h,+}} \dfrac{1}{\prod_{i=1}^{h} \sqrt{n - |\pi(i-1) - \pi(i)|}} = 0.$

Hence, to calculate $\lim E[\beta_h(BT_n)]$ we may restrict attention to pair-matched circuits where each edge has an oppositely signed L-value. ◇

Proof. As in Lemma 6.1.2, write $S_{h+} = \sum_w S_{h+,w}$ where $S_{h+,w}$ is the sum restricted to $\pi \in \Pi_{h,+}(w)$. It is enough to show that this tends to zero for each w. Let the L-values corresponding to this w be $k_1, k_2, \ldots, k_h$. Hence

$$S_{h+,w} = \frac{1}{n} \sum_{\substack{k_1, \ldots, k_h \\ \in \{0, 1, \ldots, n-1\}}} \frac{\#\{\pi \in \Pi_{h+}(w) : L \text{ values of } \pi \text{ are } \{k_1, \ldots, k_h\}}{\prod_{i=1}^{h}(n - k_i)}.$$

For this fixed set of L-values, there are at most 2^{2h} sets of slope values. It is enough to prove the result for any one such set of slope values.

Now we start counting the number of possible π's having those slope values. By the previous lemma there exists j_0 such that $k_{j_0} = \Lambda(k_{i_1}, k_{i_2}, \ldots, k_{i_l})$. We start counting the number of possible π from the edge corresponding to the L-value k_{j_0}, say $(\pi(i_* - 1), \pi(i_*))$. Clearly the number of ways to choose vertices $\pi(i_* - 1)$ and $\pi(i_*)$ is $(n - k_{j_0})$. Having chosen $\pi(i_*)$, there is only one choice of $\pi(i_* + 1)$ (since the slope values have been fixed). We continue this procedure to choose all the vertices of the circuit π and hence the number of π's having the fixed set of slope values is at most $(n - k_{j_0})$. Note that since w and the slope signs are fixed, the linear function Λ and the index j_0 are determined as well. Thus, the count $S_{h+,w}^{subset}$ for that fixed subset satisfies

$$S_{h+,w}^{subset} \leq \frac{1}{n} \sum_{0 \leq k_i \leq n-1} \frac{n - k_{j_0}}{\prod_{i=1}^{h}(n - k_i)} = \frac{1}{n} \sum_{\substack{0 \leq k_i \leq n-1 \\ i \neq j_0}} \frac{1}{\prod_{\substack{i=1 \\ i \neq j_0}}^{h}(n - k_i)}.$$

As $k_{j_0} = \Lambda(k_{i_1}, k_{i_2}, \ldots, k_{i_l})$, in the above sum k_{j_0} should be kept fixed; this implies that

$$S_{h+,w} \leq \frac{\mathrm{O}\left((\log n)^{h-1}\right)}{n} \to 0 \text{ as } n \to \infty,$$

proving the first part. The second part now follows immediately.   □

Step II (d) *(M1) condition for BT_n and BH_n.* We need to first establish a few results on moments of a truncated uniform random variable. For a given random variable X (to be chosen), define (whenever it is finite)

$$g_T(x) = \mathrm{E}[\phi_T^n(X, x)^{-(1+\alpha)}] \text{ and } g_H(x) = \mathrm{E}[\phi_H^n(X, x)^{-(1+\alpha)}].$$

Lemma 6.1.5. *Let X be discrete uniform on $\mathbb{N}_n = \{1/n, 2/n, \ldots, 1\}$. Let $x \in \mathbb{N}_n$ and $\alpha > 0$. Then, for some constants C_1, C_2,*

$$\max\{g_T(x), g_H(x)\} \leq C_1 x^{-\alpha} + C_2(1 - x + 1/n)^{-\alpha} + 1/n.$$

                                                                                          ◇

Proof. Fix $x = j/n$ and $1 < j < n$. For $j = 1$ or n we can give an argument similar to the one below. Note that

$$
\begin{aligned}
g_T(x) &= \frac{1}{n} \sum_{y=1}^{n} \frac{1}{[1 - |x - \frac{y}{n}|]^{1+\alpha}} \\
&= \frac{1}{n} \sum_{y<j} \frac{1}{(1 - \frac{j-y}{n})^{1+\alpha}} + \frac{1}{n} \sum_{y>j} \frac{1}{(1 - \frac{y-j}{n})^{1+\alpha}} + \frac{1}{n}.
\end{aligned}
$$

Now,

$$
\begin{aligned}
\frac{1}{n} \sum_{y<j} \left(1 - \frac{j-y}{n}\right)^{-(1+\alpha)} &= \frac{1}{n} \sum_{t=1}^{j-1} \left(1 - \frac{t}{n}\right)^{-(1+\alpha)} \\
&= n^{\alpha} \sum_{t=n-j+1}^{n-1} t^{-(1+\alpha)} \\
&\leq n^{\alpha} \times \frac{C_1}{(n-j+1)^{\alpha}} = C_1(1 - x + 1/n)^{-\alpha}.
\end{aligned}
$$

By similar arguments,

$$\frac{1}{n} \sum_{y>j} \frac{1}{(1 - \frac{y-j}{n})^{1+\alpha}} \leq C_2 x^{-\alpha}$$

and hence the bound holds for $g_T(x)$. By similar calculations the same bound holds for $g_H(x)$.   □

Lemma 6.1.6. Suppose $U_{i,n}$ are i.i.d. discrete uniform on $\mathbb{N}_n$. Let $a_i \in \mathbb{Z}, 1 \leq i \leq m$ be fixed and $0 < \beta < 1$. Let $Y_n = \sum_{i=1}^{m} a_i U_{i,n}$ and $Z_n = 1 - Y_n + 1/n$. Then

$$\sup_n \mathrm{E}[|Y_n|^{-\beta} I(|Y_n| \geq 1/n)] + \sup_n \mathrm{E}[|Z_n|^{-\beta} I(|Z_n| \geq 1/n)] < \infty.$$

$\diamond$

Proof. First note that

$$
\begin{aligned}
\mathbb{P}(|Y_n| \leq \frac{M}{n}) &= \mathrm{E}\left[\mathbb{P}\left(-\frac{M}{n} \leq \sum_{i=1}^{m} a_i U_{i,n} \leq \frac{M}{n} \Big| U_{i,n}, j \neq i_0\right)\right] \\
&= \mathrm{E}\left[\mathbb{P}\left(\frac{M}{n} - \sum_{\substack{i=1 \\ i \neq i_0}}^{m} a_i U_{i,n} \leq a_{i_0} U_{i_0,n} \leq \frac{M}{n} - \sum_{\substack{i=1 \\ i \neq i_0}}^{m} a_i U_{i,n}\right)\right] \\
&\leq (2M+1)/n.
\end{aligned}
$$

Let $U_1, U_2, \ldots, U_m$ be m i.i.d $U(0,1)$ random variables. We note that

$$(U_{1,n}, U_{2,n}, \ldots, U_{m,n}) \overset{\mathcal{D}}{=} \left(\frac{\lceil nU_1 \rceil}{n}, \frac{\lceil nU_2 \rceil}{n}, \ldots, \frac{\lceil nU_m \rceil}{n}\right)$$

where $\lceil x \rceil$ denotes the smallest integer larger than or equal to x. Define

$$\hat{Y}_n = \sum_{i=1}^{m} a_i \frac{\lceil nU_i \rceil}{n} \quad, \quad Y = \sum_{i=1}^{m} a_i U_i \quad \text{and} \quad K = \sum_{i=1}^{m} |a_i|.$$

Then

$$\hat{Y}_n \overset{\mathcal{D}}{=} Y_n \quad \text{and} \quad |\hat{Y}_n - Y| \leq K/n.$$

$$
\begin{aligned}
\mathrm{E}[|Y_n|^{-\beta} I(|Y_n| \geq 1/n)] &= \mathrm{E}[|Y_n|^{-\beta} I(1/n \leq |Y_n| \leq 2K/n)] \\
&\quad + \mathrm{E}[|\hat{Y}_n|^{-\beta} I(|\hat{Y}_n| > 2K/n)] \\
&\leq n^{\beta} \frac{4K+1}{n} + \mathrm{E}[(|Y| - K/n)^{-\beta} I(|\hat{Y}_n| > 2K/n)] \\
&\leq o(1) + \mathrm{E}[(|Y| - K/n)^{-\beta} I(|Y| > K/n)] \\
&\leq o(1) + \int_{x > K/n} (x - K/n)^{-\beta} f_{|Y|}(x) dx.
\end{aligned}
$$

Now

$$\int_{x > K/n} (x - K/n)^{-\beta} f_{|Y|}(x) dx = \int_0^{\infty} x^{-\beta} f_{|Y|}(x + K/n) dx.$$

It is easy to see that f_Y vanishes outside $[-K, K]$. Using induction one can also prove that $f_Y(x) \leq 1$ for all x. These two facts yield

$$\int_0^{\infty} x^{-\beta} f_{|Y|}(x + K/n) dx \leq \int_0^{K + K/n} x^{-\beta} 2 dx = \mathrm{O}(1).$$

Hence
$$\sup_n \mathrm{E}[|Y_n|^{-\beta} I(|Y_n| \geq 1/n)] < \infty.$$

Proof for the other supremum is similar. We omit the details. □

Recall the notation $\mathcal{W}_{2k}(2)$ for the set of pair-matched words of length $2k$.

Lemma 6.1.7. Suppose $\{x_i\}$ are i.i.d. bounded with mean zero and variance 1. Then the (M1) condition holds for BT_n and BH_n. ◇

Proof. Lemma 6.1.2 implies that $\mathrm{E}[\beta_{2k+1}(BA_n)] \to 0$ as $n \to \infty$ where $A_n = BT_n$ or BH_n. From Lemma 6.1.2 and Lemma 6.1.4 (if limit exists) we have

$$\lim_{n\to\infty} \mathrm{E}[\beta_{2k}(BT_n)] = \sum_{w\in\mathcal{W}_{2k}(2)} \lim_{n\to\infty} \frac{1}{n} \sum_{\pi\in\Pi^*(w)} \frac{\mathrm{E}\,X_\pi}{\prod_{i=1}^h \sqrt{\phi_T(\pi(i-1),\pi(i))}}$$

$$= \sum_{w\in\mathcal{W}_{2k}(2)} \lim_{n\to\infty} \frac{1}{n} \sum_{\pi\in\Pi^{**}(w)} \frac{1}{\prod_{i=1}^h \sqrt{\phi_T(\pi(i-1),\pi(i))}}$$

where $\Pi^{**}(w) = \{\pi : w[i] = w[j] \Rightarrow \pi(i-1) - \pi(i) + \pi(j-1) - \pi(j) = 0\}$. Denote $x_i = \pi(i)/n$. Let $S = \{0\} \cup \{\min(i,j) : w[i] = w[j], i \neq j\}$ be the set of all independent vertices of the word w and let $x_S = \{x_i : i \in S\}$. Each x_i can be expressed as a unique linear combination $L_i^T(x_S)$. L_i^T depends on the word w but for notational convenience we suppress its dependence. Note that $L_i^T(x_S) = x_i$ for $i \in S$ and also summing k equations we get $L_{2k}^T(x_S) = x_0$. If $w[i] = w[j]$ then $|L_{i-1}^T(x_S) - L_i^T(x_S)| = |L_{j-1}^T(x_S) - L_j^T(x_S)|$. Thus, using this equality and proceeding as in the proof of Theorem 2.2.1, Chapter 2

$$\lim_{n\to\infty} \mathrm{E}[\beta_{2k}(BT_n)] = \sum_{w\in\mathcal{W}_{2k}(2)} \lim_{n\to\infty} \mathrm{E}\left[\frac{\mathbb{I}(L_i^T(U_{n,S}) \in \mathbb{N}_n, i \notin S \cup \{2k\})}{\prod_{i\in S\setminus\{0\}} \phi_T^n(L_{i-1}(U_{n,S}),U_i)}\right],$$

where for each $i \in S$, $U_{n,i}$ is discrete uniform on $\mathbb{N}_n$ and $U_{n,S}$ is the random vector on $\mathbb{R}^{k+1}$ whose co-ordinates are $U_{n,i}$ and $U_{n,i}$'s are independent of each other. We claim that

$$\lim_{n\to\infty} \mathrm{E}[\beta_{2k}(BT_n)] = m_{2k}^T \tag{6.8}$$

$$= \sum_{w\in\mathcal{W}_{2k}(2)} m_{2k,w}^T$$

$$= \sum_{w\in\mathcal{W}_{2k}(2)} \mathrm{E}\left[\frac{\mathbb{I}(L_i^T(U_S) \in (0,1), i \notin S \cup \{2k\})}{\prod_{i\in S\setminus\{0\}} \phi_T^\infty(L_{i-1}(U_S),U_i)}\right], \tag{6.9}$$

where for each $i \in S$, $U_i \sim U(0,1)$ and U_S is an $\mathbb{R}^{k+1}$-dimensional random vector whose co-ordinates are U_i and they are independent of each other. Note that to prove (6.8) it is enough to show that for each pair-matched word w and for each k there exists $\alpha_k > 0$ such that

$$\sup_n \mathrm{E}\left[\left(\frac{\mathbb{I}(L_i^T(U_{n,S}) \in \mathbb{N}_n, i \notin S \cup \{2k\})}{\prod_{i\in S\setminus\{0\}} \phi_T^n(L_{i-1}(U_{n,S}),U_i)}\right)^{1+\alpha_k}\right] < \infty. \tag{6.10}$$

We will prove that for each pair-matched word w

$$\sup_n \mathrm{E}\left[\left(\frac{\mathbb{I}(L_i^T(U_{n,S}) \in \mathbb{N}_n, i \notin S \cup \{2k\}, i < \max S)}{\prod_{i \in S \setminus \{0\}} \phi_T^n(L_{i-1}(U_{n,S}), U_{n,i})}\right)^{1+\alpha_k}\right] < \infty.$$

We prove the above by induction on k. For $k = 1$ the expression reduces to

$$\mathrm{E}\left[\left(\frac{1}{1 - |U_{n,0} - U_{n,1}|}\right)^{1+\alpha}\right] = \mathrm{E}\left[\mathrm{E}\left\{\left(\frac{1}{1 - |U_{n,0} - U_{n,1}|}\right)^{1+\alpha} | U_{n,0}\right\}\right]$$

$$= \mathrm{E}[g_T(U_{n,0})]$$

$$\leq C_1 \mathrm{E}[U_{n,0}^{-\alpha}] + C_2 \mathrm{E}[(1 - U_{n,0})^{-\alpha}]$$

$$+ 1/n \text{ by Lemma 6.1.5.}$$

Hence Lemma 6.1.6 implies that $\sup_n \mathrm{E}\left[\left(\frac{1}{1 - |U_{n,0} - U_{n,1}|}\right)^{1+\alpha}\right] < \infty$ for all $0 < \alpha < 1$.

Suppose that the result is true for $k = 1, 2, \ldots, t$. We then show that it is true for $k = t + 1$. Fix any pair-matched word w_0. Note that the random variable corresponding to the generating vertex of the last letter appears only once and hence we can do the following calculations. Let

$$B_{t+1} = \left[\frac{I\left(L_i^T(U_{n,S}) \in \mathbb{N}_n, i \notin S \cup \{2(t+1)\}, i < \max S\right)}{\prod_{i \in S \setminus \{0\}} (1 - |L_{i-1}^T(U_{n,S}) - U_{n,i}|)}\right]^{1+\alpha}.$$

Then

$$\mathrm{E}[B_{t+1}] = \mathrm{E}\left[\mathrm{E}[B_{t+1} | U_{n,i}, i \in S \setminus \{i_{t+1}\}]\right]$$

$$= \mathrm{E}\left[\underbrace{\left(\frac{I\left(L_i^T(U_{n,S}) \in \mathbb{N}_n, i \notin S, i < \max S \setminus \{i_{t+1}\}\right)}{\prod_{i \in S \setminus \{0, i_{t+1}\}} (1 - |L_{i-1}^T(U_{n,S}) - U_{n,i}|)}\right)^{1+\alpha}}_{\Phi_n}\right.$$

$$\left.\times \underbrace{g_T(U_{n,i_{t+1}-1}) I[L_{i_{t+1}-1}^T(U_{n,S}) \in \mathbb{N}_n]}_{\Psi_n}\right].$$

Letting $\| \cdot \|_q$ denote the L_q norm, by Lemma 6.1.5 and Lemma 6.1.6, $\sup_n \|\Psi_n\|_q < \infty$ whenever $\alpha q < 1$.

Let us now consider the word w_0^* obtained from w_0 by removing both occurrences of the last used letter. We note that the quantity Φ_n is the candidate for the expectation expression corresponding to the word w_0^*. Now by the induction hypothesis, there exists an $\alpha_t > 0$ such that

$$\sup_n \mathrm{E}\left[\left(\frac{I\left(L_i^T(U_{n,S}) \in \mathbb{N}_n, i \notin S, i < \max S \setminus \{i_{t+1}\}\right)}{\prod_{i \in S \setminus \{0, i_{t+1}\}} (1 - |L_{i_1}^T(U_{n,S}) - U_{n,i}|)}\right)^{1+\alpha_t}\right] < \infty.$$

Hence $\sup_n ||\Phi_n||_p < \infty$ if $(1+\alpha)p \leq (1+\alpha_t)$. Therefore

$$\alpha_{t+1} + \frac{1+\alpha_{t+1}}{1+\alpha_t} < \frac{1}{p} + \frac{1}{q} = 1 \Rightarrow \sup_n \mathrm{E}[\Phi_n \Psi_n] \leq \sup_n ||\Phi_n||_p ||\Psi_n||_q < \infty.$$

This proves the lemma for balanced Toeplitz matrices.

For balanced Hankel matrices we again use Lemma 6.1.5 and Lemma 6.1.6 and proceed in an exactly similar way to get

$$\lim_{n \to \infty} \mathrm{E}[\beta_{2k}(BH_n)] = m_{2k}^H$$

$$= \sum_{\substack{w \in \mathcal{W}_{2k}(2) \\ \text{and symmetric}}} m_{2k,w}^H$$

$$= \sum_{\substack{w \in \mathcal{W}_{2k}(2) \\ \text{and symmetric}}} \mathrm{E}\Big[\frac{\mathbb{I}(L_i^H(U_S) \in (0,1), i \notin S \cup \{2k\}}{\prod_{i \in S \setminus \{0\}} \phi_H^\infty(L_{i-1}(U_S), U_i))}\Big]. \quad (6.11)$$

It may be noted that the above sum is over *symmetric* pair-matched words. Using ideas from the proof of Theorem 2.2.3, Chapter 2, it can be shown that for any pair-matched non-symmetric word w, $m_{2k,w}^H = 0$. So the above summation is taken only over pair-matched symmetric words. $\square$

Step II (e). *Approximating matrices.* Even though the limit of the moments (6.8) and (6.11) have been established, it does not seem to be easy to show that this moment sequence determines a probability distribution uniquely (which would then be the candidate LSD). We tackle this issue by using approximating matrices whose scalings are not unbounded. We shall use the *Lévy metric* to develop this approximation. Recall that this metric metrizes weak convergence of probability measures on $\mathbb{R}$. Let μ_i, $i = 1, 2$ be two probability measures on $\mathbb{R}$. The Lévy distance between them is given by

$$\rho(\mu_1, \mu_2) = \inf\{\varepsilon > 0 : \ F_1(x - \varepsilon) - \varepsilon < F_2(x) < F_1(x + \varepsilon) + \varepsilon, \ \forall \ x \in \mathbb{R}\},$$

where F_i, $i = 1, 2$, are the distribution functions corresponding to the measures μ_i, $i = 1, 2$. $\square$

We need the following interesting lemma of Bhamidi, Evans and Sen (2009)[20].

Lemma 6.1.8. Suppose $A_{n \times n}$ is a real symmetric matrix and $B_{m \times m}$ is a principal sub-matrix of $A_{n \times n}$. Then

$$\rho(F^A, F^B) \leq \min\left(\frac{n}{m} - 1, 1\right).$$

Let $(A_k)_{k=1}^\infty$ be a sequence of real symmetric matrices. For each $\varepsilon > 0$ and each k, let $(B_k^\varepsilon)_{k=1}^\infty$ be an $n_k^\varepsilon \times n_k^\varepsilon$ principal sub-matrix of A_k. Suppose that for each $\varepsilon > 0$, $F_\infty^\varepsilon = \lim_{k \to \infty} F^{B_k^\varepsilon}$ exists and $\limsup_{k \to \infty} n_k / n_k^\varepsilon \leq 1 + \varepsilon$. Then $F_\infty = \lim_{k \to \infty} F^{A_k}$ exists and is given by $F_\infty = \lim_{\varepsilon \downarrow 0} F_\infty^\varepsilon$. $\diamond$

Going back to the proof, consider the principal submatrix BT_n^ε of BT_n obtained by *retaining* the first $\lfloor n(1-\varepsilon)\rfloor$ rows and columns of BT_n. Then for this matrix, since $|i-j| \le \lfloor n(1-\varepsilon)\rfloor$, the balancing factor becomes bounded. We shall show that the LSD of $\{F^{BT_n^\varepsilon}\}$ exists for every ε and then invoke the above result to obtain the LSD of $\{F^{BT_n}\}$. A similar argument holds for $\{BH_n\}$, by considering the principal sub-matrix obtained by *removing* the first $\lfloor n\varepsilon/2\rfloor$ and last $\lceil n\varepsilon/2\rceil$ rows and columns.

Step II (f). *Existence of the LSD of the approximating matrices.* This is done in three steps.

(i) *(M1) condition.* Clearly, for any fixed $\varepsilon > 0$, we may write

$$\frac{1}{n}\mathrm{Tr}[BT_n^\varepsilon]^h = \frac{1}{n}\sum_{\pi:\pi \text{ circuit}} \frac{X_\pi}{\prod_{1\le i\le h}\sqrt{n - |\pi(i-1)-\pi(i)|}}$$
$$\times \prod_{1\le i\le h}\mathbb{I}\big[\pi(i) \le \lfloor n(1-\varepsilon)\rfloor\big]$$

and similarly

$$\frac{1}{n}\mathrm{Tr}[BH_n^\varepsilon]^h = \frac{1}{n}\sum_{\pi:\pi \text{ circuit}} \frac{X_\pi}{\prod_{1\le i\le h}\phi_H(\pi(i-1),\pi(i))}$$
$$\times \prod_{1\le i\le h}\mathbb{I}\big[\lfloor n\varepsilon/2\rfloor \le \pi(i) \le \lfloor n(1-\varepsilon/2)\rfloor\big].$$

Since for every $\varepsilon > 0$ the scaling is bounded, the proof of the following lemma is exactly the same as the proofs of Lemma 1.4.2, Chapter 1 and Theorem 2.2.1, Chapter 2. Hence we skip its proof.

Lemma 6.1.9. (a) If h is odd, $\mathrm{E}[\beta_h(BT_n^\varepsilon)] \to 0$ and $\mathrm{E}[\beta_h(BH_n^\varepsilon)] \to 0$.

(b) If h is even $(= 2k)$, then $\lim_{n\to\infty}\mathrm{E}[\beta_h(BT_n^\varepsilon)] = \sum_w p_{BT^\varepsilon}(w) = \beta_{2k}^{T^\varepsilon}$ say, where the sum is over pair-matched words w and for every such w,

$$p_{BT^\varepsilon}(w) = \underbrace{\int_0^{1-\varepsilon}\cdots\int_0^{1-\varepsilon}}_{k+1} \frac{\prod_{i\notin S\cup\{2k\}}\mathbb{I}(0 \le L_i^T(x_S) \le 1-\varepsilon)}{\prod_{i\in S\setminus\{0\}}(1 - |L_{i-1}^T(x_S) - x_i|)}dx_S.$$

Similarly, $\lim_{n\to\infty}\mathrm{E}[\beta_h(BH_n^\varepsilon)] = \sum p_{BH^\varepsilon}(w) = \beta_k^{H^\varepsilon}$ say, where

$$p_{BH^\varepsilon}(w) = \underbrace{\int_{\frac{\varepsilon}{2}}^{1-\frac{\varepsilon}{2}}\cdots\int_{\frac{\varepsilon}{2}}^{1-\frac{\varepsilon}{2}}}_{k+1} \frac{\prod_{i\notin S\cup\{2k\}}\mathbb{I}(\frac{\varepsilon}{2} \le L_i^H(x_S) \le 1-\frac{\varepsilon}{2})}{\prod_{i\in S\setminus\{0\}}\phi_H^\infty(L_{i-1}^H(x_S),x_i)}dx_S.$$

Further, $\max\{\beta_{2k}^{T^\varepsilon}, \beta_{2k}^{H^\varepsilon}\} \le \frac{2k!}{k!2^k} \times \varepsilon^{-k}$. Hence there exists unique probability distributions F^{T^ε} and F^{H^ε} with moments $\beta_k^{H^\varepsilon}$ and $\beta_k^{H^\varepsilon}$ respectively. $\diamond$

(ii) *(M4) condition for the approximating matrices.* The almost sure convergence of $\{F^{BT_n^\varepsilon}\}$ and $\{F^{BH_n^\varepsilon}\}$ now follow from the following lemma. We omit its proof since it is essentially a repetition of arguments given in the proof of Lemma 1.4.3, Chapter 1

Lemma 6.1.10. Fix any $\varepsilon > 0$ and let $A_n = BT_n$ or BH_n. If the input sequence is uniformly bounded, independent, with mean zero and variance one then

$$\mathrm{E}\left[\frac{1}{n}\,\mathrm{Tr}(BA_n^\varepsilon)^h - \mathrm{E}\,\frac{1}{n}\,\mathrm{Tr}(BA_n^\varepsilon)^h\right]^4 = \mathrm{O}(n^{-2}). \tag{6.12}$$

As a consequence, the ESD of BA_n^ε converges to F^{A^ε} almost surely.   ◇

(iii) *Connecting the balanced matrices with the approximating matrices.* From Lemma 6.1.9 and Lemma 6.1.10, given any $\varepsilon > 0$, there exists B_ε such that $\mathbb{P}(B_\varepsilon) = 1$ and on B_ε, $F^{BT_n^\varepsilon} \Rightarrow F^{T^\varepsilon}$.

Fix any sequence $\{\varepsilon_m\}_{m=1}^\infty$ decreasing to 0. Define $B = \cap B_{\varepsilon_m}$. Using Proposition 6.1.8, on B, $F^{BT_n} \Rightarrow F^T$ for some non-random distribution function F^T where F^T is the weak limit of $\{F^{T^{\varepsilon_m}}\}_{m=1}^\infty$.

Let X^{ε_m} (resp. X) be a random variable with distribution $F^{T^{\varepsilon_m}}$ (resp. F^T) with kth moments $\beta_k^{T^{\varepsilon_m}}$ (resp. β_k^T). From Lemma 6.1.9, and (6.8) it is clear that for all $k \geq 1$,

$$\lim_{m\to\infty} \beta_{2k+1}^{T^{\varepsilon_m}} = 0 \quad \text{and} \quad \lim_{m\to\infty} \beta_{2k}^{T^{\varepsilon_m}} = m_{2k}^T = \sum_{w\in\mathcal{W}_{2k}(2)} m_{2k,w}^T.$$

From Lemma 6.1.7, m_{2k}^T is finite for every k. Hence $\{(X^{\varepsilon_m})^k\}_{m=1}^\infty$ is uniformly integrable for every k and $\lim_{m\to\infty} \beta_k^{T^{\varepsilon_m}} = \beta_k^T$. This proves that $m_k^T = \beta_k^T$ and so $\{m_k^T\}$ are the moments of F^T. The argument for BH_n is exactly the same and hence details are omitted. The proof of Theorem 6.1.1 is now complete.   □

6.2 Exercises

1. Prove Lemma 6.1.9.

2. Prove Lemma 6.1.10.

3. Look up the proof of Lemma 6.1.8.

7

Patterned band matrices

Band matrices are usually defined as matrices where the top right corner and the bottom left corner elements are zeroes. But there can be other types of banding too. With increasing dimension, the number of non-zero diagonals may be finite or may tend to ∞ at various rates. This idea of banding is crucial in the estimation of large-dimensional covariance and autocovariance matrices. Banded autocovariance matrices are discussed in Chapter 11.

Finite diagonal matrices where the number of non-zero diagonals is bounded, are not treated in this book. See Popescu (2009)[85] for some interesting limits for tri-diagonal Wigner matrices. See Bose and Sen (2011)[30] and Bose, Gangopadhyay and Saha (2013)[24] for LSD results on some other finite diagonal matrices.

Though these matrices do not fall under the unified setup of Chapter 1 directly, in this chapter we shall see how these matrices with an increasing number of non-zero diagonals may be treated by our familiar moment approach. We study the LSD of the Symmetric Circulant, Reverse Circulant, Toeplitz, Hankel matrix and Palindromic matrices under different types of banding. As we shall see, banding may change the LSD of the original matrix drastically.

Let $m_n \to \infty$ be a sequence of integers. We write m for m_n. It controls the number of non-zero elements or the number of non-zero diagonals in the matrix.

The *Type I band matrix* A_n^b of A_n is the same matrix but with input $\{x_i^*\}$ where

$$x_i^* = \begin{cases} x_i & \text{if } i \leq m, \\ 0 & \text{otherwise.} \end{cases} \tag{7.1}$$

The *Type II band matrices* H_n^B of H_n is defined with the input sequence $\{\hat{x}_i\}$ where

$$\hat{x}_i = \begin{cases} x_i & \text{if } n - m \leq i \leq n + m, \\ 0 & \text{otherwise.} \end{cases} \tag{7.2}$$

The Type II band versions RC_n^B of the symmetric Reverse Circulant RC_n and T_n^B of the symmetrix Toeplitz T_n are defined with the input sequence $\{\bar{x}_i\}$

where

$$\bar{x}_i = \begin{cases} x_i & \text{if } i \leq m \text{ or } i \geq n - m, \\ 0 & \text{otherwise.} \end{cases} \qquad (7.3)$$

Note that Type II banding does not yield any non-trivial situations for the Symmetric Circulant, the Doubly Symmetric and the Palindromic matrices. When $\alpha > 2$, all the above band matrices reduce to the full matrix with no zeroes. Further restrictions are required on α for the individual matrices to yield non-trivial situations. These are specified in the statement of the main theorem later.

If $\{x_i\}$ is square summable then, as the dimension increases, the finite-dimensional Circulant approximates the corresponding Toeplitz in various senses. Indeed this approximating property is exploited to obtain the LSD of the Toeplitz matrix in the non-random case. See Gray (2009)[60] for a recent account of this approximation idea. When the $\{x_i\}$ are i.i.d. with mean zero and variance one, then of course the square summability does not hold and approximation by the Circulant does not work. We have seen that the LSD of the Toeplitz matrix is much harder to obtain in this case.

However, when $\alpha = 0$, both the Type I and Type II band Symmetric Toeplitz matrices can still be approximated by the corresponding band Symmetric Circulant matrices in a suitable metric. As a consequence, the LSD of all these four matrices is normal. The situation changes drastically when $\alpha \neq 0$. The LSDs of the Type I Symmetric Circulant matrices continue to be the normal distribution. However, the LSD of Type I and Type II Toeplitz matrices depend on α and no explicit forms seem to be obtainable.

Likewise, after appropriate scaling, the LSD of Type I and Type II Reverse Circulants is the same as the LSD of the standard Reverse Circulant, namely, the symmetrized Rayleigh, R, irrespective of the value of α.

When $\alpha = 0$, the Type II Hankel can be approximated by the Type II Reverse Circulant and hence has the same LSD. The Type I Hankel on the other hand has a degenerate distribution when $\alpha = 0$. When $\alpha \neq 0$, both Type I and Type II Hankel matrices have complicated limit distributions which depend on the value of α.

7.1 LSD for band matrices

In this chapter we shall use the following assumptions.

Assumption I. $\{x_i\}$ are independent with mean zero and variance one which are either (i) uniformly bounded or (ii) identically distributed.

Assumption I* $\{x_i\}$ are independent with mean zero and variance one which satisfy

(i) $\sup \mathrm{E}\,|x_i|^{2+\delta} < \infty$ for some $\delta > 0$, and

(ii) for all large t, $\lim n^{-2}\sum_{i=0}^{n}\mathrm{E}[x_i^4 I(|x_i| > t)] = 0$.

Assumption II. $\{m_n\} \to \infty$ and $\lim\limits_{n\to\infty} m_n/n = \alpha$ exists.

Assumption III. $\sum\limits_{n=1}^{\infty} m_n^{-2} < \infty$. (Holds trivially when $\alpha \neq 0$ in Assumption II).

Also recall
$$k_n = \#\{L_{A_n}(i,j) : 1 \le i,j \le n\}$$
where A_n is the matrix under consideration and L is its link function. To avoid trivialities, $m_n \le k_n$. We will assume that $m_n \to \infty$ and $\frac{m_n}{n} \to \alpha < \infty$.

We now state the main theorem. Some of these results first appeared in Basak (2009)[11] and then in Basak and Bose (2011)[13]. We follow their proof here. Results similar to these have been obtained independently by Kargin (2009)[69] and Liu and Wang (2009)[71]. In particular Kargin (2009)[69] used the closeness of the Circulant and Toeplitz matrices alluded to earlier. The primary tool of Liu and Wang (2009)[71] was the representation of Toeplitz and Hankel matrices as linear combinations of backward and forward shift matrices.

Theorem 7.1.1. Suppose one of the following hold: (A) Assumption I (i) and Assumption II, (B) Assumption I (ii) and Assumption II, or (C) Assumption I *(i) (ii) and Assumption II. Then the following hold *in probability*.

(a) If $m_n \le n/2$ then ESD of $m_n^{-1/2}SC_n^b, m_n^{-1/2}DH_n^b, m_n^{-1/2}PT_n^b$ and $m_n^{-1/2}PH_n^b \Rightarrow N(0,2)$.

(b) If $m_n \le n$ then ESD of $m_n^{-1/2}RC_n^b \Rightarrow R$.

(c) If $m_n \le 2n$ then ESD of $m_n^{-1/2}H_n^b \Rightarrow H_\alpha^b$, which is symmetric and H_0^b is the degenerate distribution at zero.

(d) If $m_n \le n$ then ESD of $m_n^{-1/2}T_n^b \Rightarrow T_\alpha^b$, which is symmetric and $T_0{}^b = N(0,2)$.

(e) If $m_n \le n/2$ then ESD of $(2m_n)^{-1/2}RC_n^B \Rightarrow R$.

(f) If $m_n \le n$ then ESD of $(2m_n)^{-1/2}H_n^B \Rightarrow H_\alpha^B$, which is symmetric and $H_0^B = R$.

(g) If $m_n \le n/2$ then $m_n^{-1/2}T_n^B \Rightarrow T_\alpha^B$, which is symmetric and $T_0^B = N(0,2)$.

If Assumption III holds, then all the above convergences are almost sure in cases (A) and (B). $\diamond$

Table 7.1 summarizes the limit results. The proofs are given in detail in Section 7.2.

TABLE 7.1
Matrices and their LSD.

α	Matrix	Scaling	Limit
$\alpha = 0$	$T_n^b,\ T_n^B$	$m_n^{-1/2}$	$N(0,2)$
$0 \leq \alpha \leq 1/2$	$SC_n^b,\ DH_n^b,\ PT_n^b,\ PH_n^b$	$m_n^{-1/2}$	$N(0,2)$
$\alpha = 0$	H_n^B	$(2m_n)^{-1/2}$	R
$0 \leq \alpha \leq 1$ $0 \leq \alpha \leq 1/2$	RC_n^b RC_n^B	$m_n^{-1/2}$ $(2m_n)^{-1/2}$	R R
$0 \leq \alpha \leq 1$	$T_n^b,\ H_n^B$	$m_n^{-1/2}$	symmetric
$0 \leq \alpha \leq 1/2$	T_n^B	$m_n^{-1/2}$	symmetric
$0 < \alpha \leq 2$	H_n^b	$m_n^{-1/2}$	symmetric
$\alpha = 0$	H_n^b	$m_n^{-1/2}$	degenerate at 0

7.2   Proof

Here is an outline of the main steps in the proof. We first show that it is enough to prove the theorem under the assumption that the input sequence is uniformly bounded. Then we use the modified trace formula and show as in Chapter 1, that three or more matches are negligible. Then the (M1) condition is verified for each matrix. Finally Riesz's condition is verified in a unified way for all the matrices.

We are able to bypass verifying the (M1) condition in a few exceptional cases. When $\alpha = 0$, we do not verify (M1) for T_n^b, T_n^B and H_n^B. Instead, we establish their LSD by showing that almost surely, the first two matrices are close in W_2 metric to SC_n^b and the last one is close to RC_n^B. Further, the matrices DH_n^b, PT_n^b and PH_n^b are interrelated and they are also related to SC_n^b. Hence, no detailed arguments are necessary for them.

Figures 7.1 and 7.2 suggest that the LSD of $n^{-1/2}H_n^b$ is bimodal if $\alpha \geq 4/3$ and the density is unbounded at 0 if $\alpha \leq 5/4$.

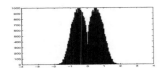
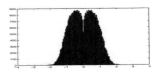

FIGURE 7.1
Histograms of the ESD of 100 realizations of H_n^b, $n = 400$, with Normal$(0, 1)$ entries for $\alpha = 2$ (left) and $\alpha = 4/3$ (right).

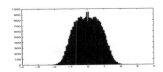
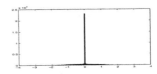

FIGURE 7.2
Histograms of the ESD of 100 realizations of H_n^b, $n = 400$, with Normal$(0, 1)$ entries for $\alpha = 5/4$ (left) and $\alpha = 1/2$ (right).

7.2.1 Reduction to uniformly bounded input

We first reduce the general case to the uniformly bounded input case. For $A_n^B, A_n = RC_n$ or T_n, let

$$
\begin{aligned}
\beta_n &= \max_{k \leq m_n} \#\{(i,j) : L_{A_n}(i,j) = k, 1 \leq i, j \leq n\} \text{ for matrices } A_n^b, \\
&= \max_{n - m_n \leq k \leq n + m_n} \#\{(i,j) : L_{H_n}(i,j) = k, 1 \leq i, j \leq n\} \text{ for } H_n^B, \\
&= \max_{|k - n/2| \geq n/2 - m_n} \#\{(i,j) : L(i,j) = k, 1 \leq i, j \leq n\}.
\end{aligned}
$$

Note that $\beta_n \leq \alpha_n = \max_k \#\{(i,j) : L_{A_n}(i,j) = k, 1 \leq i, j \leq n\}$.

The following lemma on truncation is along the lines which is by now familiar to the reader. We omit its proof.

Lemma 7.2.1. *Let A_n^b be any of the above Type I or Type II band matrices where $m_n \to \infty$ and $\beta_n = O(n)$. If for every bounded, mean 0 and variance 1 input sequence, $F^{m_n^{-1/2} A_n^b}$ converges to some fixed non-random distribution F a.s., then the same limit holds*

(a) *almost surely if $\{x_i\}$ is i.i.d. with mean zero and variance one,*

(b) *in probability if $\{x_i\}$ satisfies Assumptions I* (i) and (ii).*

In particular, for all our band matrices, $\alpha_n = O(n)$ and the conclusions are valid. ◇

7.2.2 Trace formula, circuits, words and matches

Let

$$X_\pi = \prod_{i=1}^h x_{L(\pi(i-1),\pi(i))}, \quad \mathbb{I}^h_{\pi,L} = \mathbb{I}^h_\pi = \prod_{i=1}^h \mathbb{I}(L(\pi(i-1),\pi(i)) \le m).$$

Hence

$$E[\beta_h(m^{-1/2} A_n^b)] = E\left[\frac{1}{n} \operatorname{Tr}\left(\frac{A_n^b}{\sqrt{m}}\right)^h\right] = \frac{1}{nm^{h/2}} \sum_{\pi:\pi \text{ circuit}} E X_\pi \times \mathbb{I}^h_\pi.$$

Define

$$\Pi^b(w) = \{\pi : w[i] = w[j] \Leftrightarrow L(\pi(i-1),\pi(i)) = L(\pi(j-1),\pi(j)) \text{ and } \mathbb{I}^h_\pi = 1\}.$$

Note these depend on the underlying L function but we suppress this dependence. We can rewrite $E[\beta_h(n^{-1/2} A_n)]$ and $E[\beta_h(m^{-1/2} A_n^b)]$ as

$$E[\beta_h(n^{-1/2} A_n)] = \sum_{w \text{ matched}} \frac{1}{n^{1+h/2}} \sum_{\Pi(w)} E X_\pi, \text{ and}$$

$$E[\beta_h(m^{-1/2} A_n^b)] = \sum_{w \text{ matched}} \frac{1}{nm^{h/2}} \sum_{\Pi^b(w)} E X_\pi.$$

Note that for some of the matrices, m will be replaced by $(2m)$ above. Since the outer sum is a finite sum, to check condition (M1), it is enough to show that the inner limit exists for every w. Whenever this limit exists, let

$$p(w) = \sum_{\Pi^b(w)} \lim \frac{1}{nm^{h/2}} E X_\pi.$$

$$\Pi^{b*}(w) = \{\pi : w[i] = w[j] \Rightarrow L(\pi(i-1),\pi(i)) = L(\pi(j-1),\pi(j)) \text{ and } \mathbb{I}^h_\pi = 1\}$$
$$\supseteq \Pi^b(w).$$

Let $\Pi^B(w)$ and $\Pi^{B*}(w)$ be the same as $\Pi^b(w)$ and $\Pi^{b*}(w)$, but now with additional indicators for Type II banded matrices. We shall continue to use $p(w)$ to denote the corresponding limits as above. These indicators are defined as below:

$$\mathbb{I}^h_{\pi,RC} = \prod_{i=1}^{h} \mathbb{I}\Big(L_{RC}(\pi(i-1),\pi(i)) \le m \text{ or } L_{RC}(\pi(i-1),\pi(i)) \ge n-m\Big),$$

$$\mathbb{I}^h_{\pi,H} = \prod_{i=1}^{h} \mathbb{I}(n-m \le L_H(\pi(i-1),\pi(i)) \le n+m),$$

$$\mathbb{I}^h_{\pi,T} = \mathbb{I}(|\pi(0)-\pi(1)| \le m \text{ or } |\pi(0)-\pi(1)| \ge n-m) \times$$
$$\mathbb{I}(|\pi(1)-\pi(2)| \le m \text{ or } |\pi(1)-\pi(2)| \ge n-m) \times \cdots \times$$
$$\mathbb{I}(|\pi(h-1)-\pi(h)| \le m \text{ or } |\pi(h-1)-\pi(h)| \ge n-m),$$

7.2.3 Negligibility of higher-order edges

The following lemma helps to show that the contributions from $\Pi^b(w)$ and $\Pi^B(w)$ are zero if w has at least one $e \ge 3$ and also helps to verify the (M1) condition later.

Lemma 7.2.2. (a) For SC^b, given $\pi(0),\pi(1),\ldots,\pi(i-1)$, for any non-generating vertex $\pi(i)$, $\#\pi(i) \le 6$.

(b) For any generating vertex $\pi(i)$ of SC^b, $2m \le \#\pi(i) \le 2m+1$.

(c) For any generating vertex $\pi(i)$ of RC_n^b, $m \le \#\pi(i) \le m+1$.

(d) For any generating vertex $\pi(i)$ of RC_n^B, $2m \le \#\pi(i) \le 2m+1$.   ◇

Proof. Part (a) follows easily from the relation $L_{SC}(\pi(j-1),\pi(j)) = L_{SC}(\pi(i-1),\pi(i))$ where $j < i$, which in turn implies $|n/2 - |\pi(j-1)-\pi(j)|| = |n/2 - |\pi(i-1)-\pi(i)||$.

We now prove part (b). The proofs of (c) and (d) are similar and shall be omitted.

First assume $i = 1$. Since $0 \le L_{SC}(\pi(0),\pi(1)) \le m$ and $1 \le \pi(1) \le n$, we obtain

$$0 \le \frac{n}{2} - |\frac{n}{2} - |\pi(0)-\pi(1)|| \le m \quad \Rightarrow \quad \frac{n}{2}-m \le |\frac{n}{2} - |\pi(0)-\pi(1)|| \le \frac{n}{2}.$$

It then easily follows that one of the following three holds.

(i) $\pi(0) - m \le \pi(1) \le \pi(0) + m$,
(ii) $\pi(1) \le \pi(0) + m - n$ or,
(iii) $\pi(1) \ge \pi(0) + n - m$.

If $m = n/2$ then irrespective of the value of $\pi(1)$, we have $0 \le n/2 - |n/2 - |\pi(0)-\pi(1)|| \le m$. Hence the number of choices of $\pi(1)$ is $n = 2m$.

Now assume $m < n/2$. Note that the three regions in (i), (ii) and (iii) are disjoint in this case. We consider three cases and establish (b) in each case separately.

Case A. $\pi(0) \in \{1,2,\ldots,m\}$. Note that $\pi(0) + m - n \le 2m - n < 0$ implies

$\pi(0) + m \leq 2m < n$ and the range of choices in (i) is $1 \leq \pi(1) \leq \pi(0) + m$. Hence the number of choices from (i) is $\pi(0) + m$. The number of choices from (ii) is 0. As $\pi(0) + n - m \leq n$, the range of choices from (iii) is $\pi(0) + n - m \leq \pi(1) \leq n$. Hence the number of choices from (iii) is $m - \pi(0) + 1$. In all, that is a total of $(2m + 1)$ choices for $\pi(1)$.

Case B. $\pi(0) \in \{m+1, m+2, \ldots, n-m-1\}$. As $\pi(0) - m > 0$ and $\pi(0) + m < n - m + m = n$, now the range of choices from (i) is $\pi(0) - m \leq \pi(1) \leq \pi(0) + m$. which is $2m$ many choices. As $\pi(0) + m - n < 0$ and $\pi(0) + n - m > m + n - m = n$, there are no choices from (ii) and (iii).

Case C. $\pi(0) \in \{n-m, n-m+1, \ldots, n\}$. Note that $\pi(0) + m \geq n - m + m = n$ and $\pi(0) \geq n - m > m$. Now the range of choices for $\pi(1)$ from (i) is $\pi(0) - m \leq \pi(1) \leq n$ which gives $n + m - \pi(0) + 1$ choices. Since $\pi(0) + n - m > n$, the number of choices from (iii) is 0. As $\pi(0) + m - n \geq 0$, the number of choices from (ii) is $\pi(0) + m - n$. In all, that is a total of $(2m + 1)$ choices.

Now move to $\pi(2)$. If it is a non-generating vertex then there are at most six ways to choose it and if it is a generating vertex then the above argument may be repeated. Continuing the process proves (b) completely. $\qquad\square$

The next lemma shows that the contribution from any word having edge(s) of order ≥ 3 is zero in the limit. Let $N_{h,3+}^L$ be the number of L matched circuits on $\{1, 2, \ldots, n\}$ of length h with at least one $e \geq 3$ and $N_{h,3+}^{bL}$ be the same with added restriction $\mathbb{I}_\pi^h = 1$. Define $N_{h,3+}^{BL}$ similarly for Type II banding.

Lemma 7.2.3. (a) Let $L = L_T$ or L_H. There exists a constant C_L such that $N_{h,3+}^{bL} \leq C_L n^{\lfloor (h+1)/2 \rfloor}$ and hence $n^{-(1+h/2)} N_{h,3+}^{bL} \to 0$.

(b) Let $L = L_{SC}$ or L_{RC}. Then there exists a constant C_L such that $N_{h,3+}^{bL} \leq C_L n m^{\lfloor (h-1)/2 \rfloor}$ and hence $n^{-1} m^{-h/2} N_{h,3+}^{bL} \to 0$.

(c) (a) holds for $N_{h,3+}^{BL_H}$ and $N_{h,3+}^{BL_T}$ and (b) holds for $N_{h,3+}^{BL_{RC}}$.

(d) For $\alpha = 0$, $n^{-1} m^{-h/2} N_{h,3+}^{bL_H} \to$ as $n \to \infty$. $\qquad\diamond$

Proof. (a) First observe that $N_{h,3+}^{bL} \leq N_{h,3+}^L$. Now note that if $h = 2k$ or $2k - 1$ then there are at most $k - 1$ distinct L-values, and hence, at most k generating vertices. Each generating vertex can be chosen in at most n ways and the non-generating vertices can be chosen in O(1) ways. This completes the proof of (a).

(b) Let $h = 2k$ or $2k - 1$. Since there is at least one $e \geq 3$, there must be at most $(k - 1)$ distinct L values and hence at most k generating vertices including $\pi(0)$. Note that $\#\pi(0) = n$ and by Lemma 7.2.2 $\#\pi(i) = $ O(1) or $\#\pi(i) = $ O(m) depending on whether or not $\pi(i)$ is dependent or independent. Hence $N_{h,3+}^{bL} \leq C n m^{k-1} \leq C n m^{\lfloor (h-1)/2 \rfloor}$ and (b) follows.

The proof of (c) is similar and we omit the details.

(d) Since $\alpha = 0$, for large n, $m < n$. To achieve $\mathbb{I}_\pi^h = 1$, every generating vertex $\pi(i)$, $i \geq 1$ can be chosen in at most m ways (as the link function is $L(i,j) = i + j$). Let $h = 2k$ or $h = 2k - 1$. Since there is at least one edge of order ≥ 3, we must have at most $k - 1$ generating vertices excluding $\pi(0)$. So total number of vertex choices is $\mathrm{O}(nm^{k-1})$. Hence we obtain the result. $\square$

7.2.4 (M1) condition

We now state a lemma which reduces the number of terms in the trace formula asymptotically. It also implies that the odd moments of the LSD equal zero. This is similar to Lemma 1.4.2 and Theorem 1.4.4 in Chapter 1. Its proof is omitted.

Lemma 7.2.4. (a) Suppose $\{A_n^b = RC_n^b, \ SC_n^b\}$ (for any α) or $\{A_n^b = T_n^b \text{ or } H_n^b\}$ (for $\alpha \neq 0$) with a uniformly bounded, independent mean zero and variance one input sequence $\{x_i\}$.

(i) If h is odd, $\displaystyle\lim_{n\to\infty} \mathrm{E}[\beta_h(m^{-1/2}A_n^b)] = 0$.

(ii) If $h = 2k$, then $\displaystyle\sum_{w \in \mathcal{W}_{2k}(2)} \lim_{n\to\infty} \frac{1}{nm^k} \#(\Pi^{b*}(w) - \Pi^b(w)) = 0$ and if any of the last two limits below exists

$$\lim_{n\to\infty} \mathrm{E}[\beta_{2k}(m^{-1/2}A_n^b)] = \sum_{w \in \mathcal{W}_{2k}(2)} \lim_{n\to\infty} \frac{1}{nm^k} \#\Pi^b(w)$$

$$= \sum_{w \in \mathcal{W}_{2k}(2)} \lim_{n\to\infty} \frac{1}{nm^k} \#\Pi^{b*}(w). \qquad (7.4)$$

(b) All conclusions of (a) above hold for $\{m^{-1/2}RC_n^B\}$ (for any α) and for $\{n^{-1/2}H_n^B\}$ and $\{n^{-1/2}T_n^B\}$ (when in either case $\alpha \neq 0$) with Π^b and Π^{b*} replaced by Π^B and Π^{B*} respectively.

(c) Similar conclusions hold for $\{m^{-1/2}H_n^b\}$ when $\alpha = 0$. $\qquad \diamond$

Hence to establish (M1), we restrict consideration to only pair-matched words. We show that the inner limit in (7.4) (and hence of $p(w)$) exists in each case, for every pair-matched w.

For the cases when $\alpha \neq 0$, it will be convenient to obtain expressions for $\displaystyle\lim_{n\to\infty} \mathrm{E}[\beta_h(n^{-1/2}A_n^b)]$ where $A_n^b = T_n^b, H_n^b, T_n^B$ or H_n^B. Observe that then the expression for $\displaystyle\lim_{n\to\infty} \mathrm{E}[\beta_h(m^{-1/2}A_n^b)]$ is given by the relation

$$\lim_{n\to\infty} \mathrm{E}[\beta_h(m^{-1/2}A_n^b)] = \alpha^{-h/2} \times \lim_{n\to\infty} \mathrm{E}[\beta_h(n^{-1/2}A_n^b)].$$

Similar comment holds for Type II banding.

(M1) condition for SC_n^b. Define the slopes $u_i = \pi(i-1) - \pi(i)$. Note that $w[i] = w[j]$ implies that $u_i \pm u_j = 0, \pm n$. We first show that only those matchings with $u_i + u_j = 0, \pm n$ contribute in the limit.

Lemma 7.2.5. Fix an L_{SC} matched word w of length $2k$, with $|w| = k$. Let $\mathcal{N}^b$ be the number of matched circuits π on $\{1, 2, \ldots, n\}$ of length $2k$ such that $\mathbb{I}^h_\pi = 1$ and it has at least one pair $i < j$ with $w[i] = w[j]$ such that $u_i - u_j = 0, \pm n$. Then $\mathcal{N}^b = O(nm^{k-1})$ and hence $n^{-1}m^{-k}\mathcal{N}^b \to 0$. ◇

Proof. Let $(i_1, j_1), (i_2, j_2), \ldots (i_k, j_k)$ denote the pair-partition corresponding to the word w, i.e., $w[i_l] = w[j_l], 1 \le l \le k$. Suppose without loss of generality, $u_{i_k} - u_{j_k} = 0, \pm n$. Clearly a circuit π becomes completely specified if we know $\pi(0)$ and all the "slopes" u_i's.

By Lemma 7.2.2(a), if we fix some value for u_{i_l}, each u_{j_l} has six possible values. Now $L_{SC}(\pi(i-1), \pi(i)) \le m$ implies either $|u_i| \le m$ or $|u_i| \ge n - m$. Hence for every i, each u_i has $O(m)$ possible values. Further, $\pi(0)$ has at most n possible values. For any such valid choice, from the sum restriction $\sum_{i=1}^{2k} u_i = \pi(2k) - \pi(0) = 0$ we know $u_{i_k} + u_{j_k}$ and on the other hand by hypothesis, $u_{i_k} - u_{j_k} = 0, +n, -n$. Thus, the pair (u_{i_k}, u_{j_k}) has 3 possibilities. Thus, there are at most $O(nm^{k-1})$ circuits with the given restrictions and the proof of the lemma is complete. □

As a consequence of the above lemma we have

$$\sum_{w \in \mathcal{W}_{2k}(2)} \lim_{n\to\infty} \frac{1}{nm^k} \#\Pi^{b*}(w) = \sum_{w \in \mathcal{W}_{2k}(2)} \lim_{n\to\infty} \frac{1}{nm^k} \#\Pi^{b'}(w), \qquad (7.5)$$

where

$$\Pi^{b'}(w) = \{\pi : \pi \text{ is a circuit}, w[i] = w[j] \Rightarrow u_i + u_j = 0, \pm n \text{ and } I^h_\pi = 1\}.$$

Lemma 7.2.6. For any pair-matched word w with $|w| = k$, we have $\frac{1}{nm^k} \#\Pi^{b'}(w) \to 2^k$. Hence

$$\lim \mathrm{E}[\beta_{2k}(m^{-1/2}SC^b_n)] = 2^k \frac{(2k)!}{2^k k!} = \mathrm{E}[N(0,2)^{2k}].$$

◇

Proof. Suppose for some $i < j, u_i + u_j = 0, \pm n$. If we know the circuit up to position $(j-1)$ then $\pi(j)$ has to take one of the values $A - n, A, A + n$ where $A = \pi(j-1) - \pi(i) + \pi(i-1)$. Noting that $-(n-2) \le A \le (2n-1)$, exactly one of the three values will fall within 1 and n and be a valid choice for $\pi(j)$. From Lemma 7.2.2 (b) if i is a generating vertex and the circuit is known up to position $i-1$ then there are $2m + c_i$ choices for $\pi(i)$ to be between 1 and n, where $c_i = 0$ or 1, depending on the value of the previous vertices. Assume for the moment that such a choice of all $\pi(i)$ satisfies the circuit condition. Then $n \times (2m)^k \le \#\Pi^{b'}(w) \le n \times (2m+1)^k$. So we have, $\lim_{n\to\infty} \frac{1}{nm^k} \#\Pi^{b'}(w) = 2^k$.

Now we show why the circuit condition $\pi(0) = \pi(2k)$ is satisfied. Observe that

$$\pi(2k) - \pi(0) = \sum_{i=1}^{2k} u_i = dn$$

for some $d \in \mathbb{Z}$. But since $|\pi(2k) - \pi(0)| \le n - 1$, we must have $d = 0$. Thus, the circuit condition is satisfied. Since the total number of pair-matched words is $\frac{(2k)!}{2^k k!}$, the proof is complete. $\qquad\square$

Closeness of T_n^b and SC_n^b when $\alpha = 0$. We now show that for $\alpha = 0$, T_n^b and SC_n^b have the same limit behavior.

Lemma 7.2.7. (a) Suppose Assumption I (i) or (ii) hold and $\alpha = 0$. Then

$$W_2^2(F^{m^{-1/2}T_n^b}, F^{m^{-1/2}SC_n^b}) \to 0 \quad \text{almost surely.}$$

(b) Suppose Assumption I* (i) and (ii) hold and $\alpha = 0$. Then

$$W_2^2(F^{m^{-1/2}T_n^b}, F^{m^{-1/2}SC_n^b}) \to 0 \quad \text{in probability.}$$

$\diamond$

Proof. Define the upper (lower) kth diagonal as those indices (i, j) such that $i - j = k$ (respectively, $i - j = -k$). Note that both the matrices SC_n^b and T_n^b have all upper and lower kth diagonal entries equal to x_k, for $k = 1, 2, \ldots, m$ and moreover SC_n^b has all the upper and lower kth diagonal entries equal to x_{n-k}, for $k = n - 1, n - 2, \ldots, n - m$. Hence

$$
\begin{aligned}
W_2^2(F^{m^{-1/2}T_n^b}, F^{m^{-1/2}SC_n^b}) &\le \frac{1}{n} \operatorname{Tr}\Big[\frac{(T_n^b - SC_n^b)^2}{m}\Big] \\
&= \frac{2}{mn}(mx_m^2 + (m-1)x_{m-1}^2 + \cdots + x_1^2) \\
&\le \frac{2}{n}(x_m^2 + x_{m-1}^2 + \cdots + x_1^2) \\
&= 2 \times \Big(\frac{m}{n}\Big) \times \frac{x_1^2 + x_2^2 + \cdots + x_m^2}{m}.
\end{aligned}
$$

(a) First let $\{x_i\}$ be i.i.d. Then $\frac{x_1^2 + x_2^2 + \cdots + x_m^2}{m} \overset{a.s.}{\to} \mathrm{E}[x_0^2] = 1$ since $\frac{m}{n} \to 0$ and the result follows. If instead, the $\{x_i\}$ are uniformly bounded, a simple modification in the above proof does the job.

(b) Now suppose Assumption I* (i) and (ii) hold. Then by SLLN, $\mathrm{E}[\frac{x_1^2 + x_2^2 + \cdots + x_m^2}{n}] = \frac{m}{n} \to 0$ as $n \to \infty$ and

$$
\begin{aligned}
\operatorname{Var}\Big[\frac{x_1^2 + x_2^2 + \cdots + x_m^2}{n}\Big] &\le \frac{\sum\limits_{i=1}^m \mathrm{E}[x_i]^4}{n^2} \\
&\le \frac{1}{n^2} \times \Big[mt^4 + \sum_{i=1}^m \mathrm{E}[x_i^4 \mathbb{I}(|x_i| > t)]\Big] \\
&= \frac{mt^4}{n^2} + \Big(\frac{m^2}{n^2}\Big) \times \frac{\sum\limits_{i=1}^m \mathrm{E}[x_i^4 \mathbb{I}(|x_i| > t)]}{m^2} \to 0.
\end{aligned}
$$

Thus, $\frac{x_1^2+x_2^2+\cdots+x_m^2}{n} \xrightarrow{P} 0$. This completes the proof. $\qquad\square$

(M1) condition for RC_n^b. We shall show that for any non-symmetric word (7.4) equals zero and for any symmetric word (7.4) equals 1.

Lemma 7.2.8. (a) For L_H, if w is non-symmetric, pair-matched, then there exists a linear function $\Lambda(\pi(i),\ i \in S)$ such that all its coefficients are not zero and $\Lambda(\pi(i),\ i \in S) = 0$.

(b) For L_{RC}, (a) holds with the modification $\Lambda(\pi(i),\ i \in S) = dn$ for some integer d. $\qquad\diamond$

Proof. Let $t_i = \pi(i-1) + \pi(i)$ and for any word let S be the set of generating vertices. As we have done so many times before, write $\pi(i)$ as a linear combination $L_i^H(\pi_S)$ of the generating vertices. Observe that if $i \in S$ then $L_i^H(\pi_S) = \pi(i)$. Since the RC_n link function is $L(i,j) = i + j$ mod n, one can check that for every non-generating vertex i one must have $\pi(i) = L_i^H(\pi_S) + a_i n$ for some integer a_i.

(a) First note that $\pi(0) - \pi(2k) = (t_1 + t_3 + t_5 + \cdots + t_{2k-1}) - (t_2 + t_4 + t_6 + \cdots + t_{2k}) = 0$. Since w is pair-matched let $\{(i_s, j_s), s \geq 1\}$ be such that $w[i_s] = w[j_s],\ i_s < j_s$ and $i_s,\ s = 1, 2, \ldots, k$ are arranged in ascending order of i_s. Since w is not a symmetric word we have at least one pair (i_*, j_*) such that both of them are either at odd places or at even places. Now

$$(t_1+t_3+\cdots+t_{2k-1})-(t_2+t_4+\cdots+t_{2k}) = 2\Big(\sum_{\substack{i_s:(i_s,j_s) \\ \text{are both odd}}} t_{i_s} - \sum_{\substack{i_s:(i_s,j_s) \\ \text{are both even}}} t_{i_s} \Big). \quad (7.6)$$

If there is any non-generating vertex $\pi(i)$ in the above expression we replace it by the linear combination $L_i^H(\pi_S)$. Let i_* be the largest index that appears in the above equation. Note that i^* is a generating vertex. Since any non-generating vertex can be expressed as a linear combination of generating vertices to its left, the coefficient of $\pi(i_*)$ is non-zero. Hence we obtain a linear function $\Lambda(\pi(i),\ i \in S) = 0$.

(b) For the RC_n matrix

$$(t_1+t_3+\cdots+t_{2k-1})-(t_2+t_4+\cdots+t_{2k}) = 2\Big(\sum_{\substack{i_s:(i_s,j_s) \\ \text{are both odd}}} t_{i_s} - \sum_{\substack{i_s:(i_s,j_s) \\ \text{are both even}}} t_{i_s} \Big)+cn \quad (7.7)$$

for some integer c. The proof can be completed by repeating the arguments of part (a). $\qquad\square$

Lemma 7.2.9. $\lim_{n\to\infty} \frac{1}{nm^k}\#\Pi^{b*}(w) = 0$ whenever w is non-symmetric pair-matched. $\qquad\diamond$

Proof. By Lemma 7.2.8 there exists a non-zero linear function $\Lambda = \Lambda(\pi(i),\ i \in S)$ such that $\Lambda = dn$ for some integer d. Let i_* be the largest index having non-zero coefficient in Λ. Now choose vertices of π from left to right. $\pi(0)$ has n choices. By Lemma 7.2.2(c), for any $1 \leq i \leq i^* - 1$, $\pi(i)$ has at most $m + 1$ choices or at most one choice, depending on whether or not the vertex is generating. Having chosen all these vertices, the number of choices of $\pi(i_*)$ is O(1). Next we move on to $\pi(i_* + 1)$ and proceed as before. Clearly, the total number of possible choices is $O(nm^{k-1})$. This completes the proof. $\square$

Lemma 7.2.10. For any pair-matched word, given the generating vertices, each non-generating vertex has exactly one value. $\diamond$

The proof is easy and is skipped.

Lemma 7.2.11. $\lim\limits_{n \to \infty} \frac{1}{nm^k} \#\Pi^{b*}(w) = 1$ whenever w is pair-matched symmetric. Hence

$$\lim E[\beta_{2k}(m^{-1/2} RC_n^b)] = k!.$$

$\diamond$

Proof. The initial vertex $\pi(0)$ has n choices. From Lemma 7.2.2(c), for any other generating vertex $\pi(i)$, $m \leq \#\pi(i) \leq m + 1$. From Lemma 7.2.10, given the generating vertices, every non-generating vertex has only one value. We show that such a choice automatically satisfies the circuit condition. Let the partition generated by the word w be $\{(i_s, j_s), s \geq 1\}$, $i_s < j_s$. Let $t_i = \pi(i-1) + \pi(i)$. We will have $t_{i_s} = t_{j_s} + d_s n$, for some integer d_s. Since w is symmetric, one of (i_s, j_s) will occur in an odd place and another at an even place. Now for any choice of the generating vertices, there exists some integer d so that

$$\pi(0) - \pi(2k) = (t_1 + t_3 + t_5 + \cdots + t_{2k-1}) - (t_2 + t_4 + t_6 + \cdots + t_{2k}) = dn.$$

As $|\pi(0) - \pi(2k)| \leq n - 1$, d must be 0, proving that the circuit condition is satisfied. Hence $n \times m^k \leq \#\Pi^{b*}(w) \leq n \times (m+1)^k$. So we have, $\lim\limits_{n \to \infty} \frac{1}{nm^k} \#\Pi^{b*}(w) = 1$. Since there are exactly $k!$ symmetric words, the lemma is proved completely. $\square$

(M1) condition for T_n^b, $\alpha \neq 0$. The following result is similar to Lemma 7.2.11 and is essentially proved in Lemma 2.2.2, Chapter 2.

Lemma 7.2.12. Fix $k \in \mathbb{N}$. Let $\mathcal{M}$ be the number of L_T matched circuits π on $\{1, 2, \ldots, n\}$ of length $2k$ with at least one pair of L_T matched edges $(\pi(i-1), \pi(i))$ and $(\pi(j-1), \pi(j))$ such that $\pi(i) - \pi(i-1) + \pi(j) - \pi(j-1) \neq 0$. Let $\mathcal{M}^b$ be the same count with the extra condition $\mathbb{I}_\pi^{2k} = 1$. Then $n^{-(k+1)}\mathcal{M}^b \leq n^{-(k+1)}\mathcal{M} \to 0$. $\diamond$

Earlier we have shown that, $E[\beta_{2k+1}(n^{-1/2}T_n^b)] \to 0$. By Lemma 7.2.4 and 7.2.12, *if* the second limit below exists

$$\lim E[\beta_{2k}(n^{-1/2}T_n^b)] \quad = \quad \lim \frac{1}{n^{1+k}} \sum_{\substack{w \text{ matched,} \\ |w|=k}} \#\Pi^{b*}(w) \qquad (7.8)$$

$$= \quad \lim \frac{1}{n^{1+k}} \sum_{\substack{w \text{ matched,} \\ |w|=k}} \#\Pi^{b**}(w) \qquad (7.9)$$

$$= \quad \lim \left(\frac{m}{n}\right)^k \times \frac{1}{nm^k} \sum_{\substack{w \text{ matched,} \\ |w|=k}} \#\Pi^{b**}(w) \quad (7.10)$$

$$= \quad \alpha^k \times \beta_{2k}(T_\alpha^b) \quad \text{say,} \qquad (7.11)$$

where

$$\Pi^{b**}(w) = \left\{\pi : w[i] = w[j] \Rightarrow \pi(i-1) - \pi(i) + \pi(j-1) - \pi(j) = 0 \text{ and } \mathbb{I}_\pi^{2k} = 1\right\}.$$

Taking $x_i = \pi(i)/n$, $\#\Pi^{b**}(w)$ can be expressed as

$$\#\{(x_0, x_1, \ldots, x_{2k}) : x_0 = x_{2k}, x_i \in \{j/n, \ 1 \le j \le n,$$
$$|x_{i-1} - x_i| \le m \text{ and } x_{i-1} - x_i + x_{j-1} - x_j = 0 \text{ if } w[i] = w[j]\}.$$

Let $S = \{0\} \cup \{\min(i,j) : w[i] = w[j], i \ne j\}$ be the set of all indices of the generating vertices of word w and let $x_S = \{x_i : i \in S\}$. Each x_i is a unique linear combination $L_i^T(x_S)$. L_i^T depends on the word w, but for notational ease we suppress this dependence.

Clearly $L_i^T(x_S) = x_i$ if $i \in S$ and also summing the k equations would imply $L_{2k}^T(x_S) = x_0$. So

$$\#\Pi^{b**}(w) \quad = \quad \#\{x_S : L_i^T(x_S) \in \mathbb{N}_n \text{ and } |L_{i-1}^T(x_S) - L_i^T(x_S)| \le m/n$$
$$\text{for all } i = 0, 1, \ldots, 2k\}$$

where $\mathbb{N}_n = \{1/n, 2/n, \ldots, 1\}$. If $w[i] = w[j]$ then, $|L_{i-1}^T(x_S) - L_i^T(x_S)| = |L_{j-1}^T(x_S) - L_j^T(x_S)|$. So we can replace the restriction $|L_{i-1}^T(x_S) - L_i^T(x_S)| \le m/n$, $1 \le i \le 2k$ by $|L_{i-1}^T(x_S) - L_i^T(x_S)| \le m/n$ for $i \in S$. Hence as in Chapter 2,

$$\frac{\#\Pi^{b**}(w)}{n^{1+k}} \quad = \quad E\left[\mathbb{I}\big(0 \le L_i^T(U_{n,S}) \le 1, \ \forall \ i \notin S \cup \{2k\}\big)\right.$$
$$\left. \times \mathbb{I}\big(|L_{i-1}^T(U_{n,S}) - U_{n,i}| \le \alpha_n \ \forall \ i \in S\big)\right],$$

where $\alpha_n = m/n$ and for each $i \in S$, $U_{n,i}$ is a discrete uniform on $\mathbb{N}_n$ and $U_{n,S}$ is the random vector on $\mathbb{R}^{k+1}$ whose co-ordinates are $U_{n,i}$ and $U_{n,i}$'s are independent of each other.

Taking limits

$$\lim_{n\to\infty} \frac{\#\Pi^{b**}(w)}{n^{1+k}} = \underbrace{\int_0^1 \int_0^1 \cdots \int_0^1}_{k+1} \prod_{i\notin S\cup\{2k\}} \times \mathbb{I}(0 \le L_i^T(x_S) \le 1)$$

$$\times \prod_{i\in S} \mathbb{I}(|L_{i-1}^T(x_S) - x_i| \le \alpha) dx_S$$

$$= p_{T_\alpha^b}(w) \quad \text{(say)},$$

and

$$\lim E[\beta_{2k}(m^{-1/2}T_n^b)] = \beta_{2k}(T_\alpha^b) = \alpha^{-k} \sum_{w\in\mathcal{W}_{2k}(2)} p_{T_\alpha^b}(w)$$

where $p_{T_\alpha^b}(\cdot)$ is as above.

(M1) condition for H_n^b, $\alpha \ne 0$. Similar arguments as above lead to

$$\beta_{2k}(H_\alpha^b) = \alpha^{-k} \lim_{n\to\infty} \frac{1}{n^{1+k}} \sum_{w:w \text{ pair-matched},|w|=k} \#\Pi^{b*}(w)$$

where

$$\Pi^{b*}(w) = \{\pi : w[i] = w[j] \Rightarrow \pi(i-1) + \pi(i) = \pi(j-1) + \pi(j) \text{ and } \mathbb{I}_\pi^{2k} = 1\}.$$

As before, each x_i is a unique linear combination $L_i^H(x_S)$ of the generating vertices and $L_i^H(x_S) = x_i$ for all $i \in S$ and $n^{-(k+1)}\Pi^{b*}(w)$ may be written as an expectation with independent discrete uniform on $\mathbb{R}^{k+1}$ and then we take the limit.

Unlike the previous case, we do not automatically have $L_{2k}^H(x_S) = x_{2k}$ for every word w and that explains the extra indicator in the integrand below.

$$p_{H_\alpha^b}(w) = \underbrace{\int_0^1 \cdots \int_0^1}_{k+1} \prod_{i\notin S\cup\{2k\}} \mathbb{I}(0 \le L_i^H(x_S) \le 1) \times$$

$$\prod_{i\in S} \mathbb{I}(L_{i-1}^H(x_S) + x_i \le \alpha) \times \mathbb{I}(x_0 = L_{2k}^H(x_S)) dx_S, \quad (7.12)$$

and $\beta_{2k}(H_\alpha^b) = \alpha^{-k} \sum_{w\in\mathcal{W}_{2k}(2)} p_{H_\alpha^b}(w)$.

From the proof of Theorem 2.2.3, Chapter 2, $\mathbb{I}(x_0 = x_{2k}) = 1$ iff w is a symmetric word. Since for all other words the restriction $\mathbb{I}(x_0 = x_{2k}) = 1$ is one extra linear restriction, the integral above is zero and hence $p_{H_\alpha^b}(w) = 0$ for all non-symmetric words.

Thus, we can write

$$\beta_{2k}(H_\alpha^b) = \alpha^{-k} \sum_{w \text{ symmetric},w\in\mathcal{W}_{2k}(2)} p_{H_\alpha^b}(w).$$

(M1) condition for H_n^b, $\alpha = 0$.

Lemma 7.2.13. For any $w \in \mathcal{W}_{2k}(2)$, $\lim\limits_{n \to \infty} \frac{1}{nm^k} \#\Pi^b(w) = 0$. ◇

Proof. Since the link function is $L(i,j) = i+j$, to have $\mathbb{I}_\pi^h = 1$, every generating vertex, including $\pi(0)$, has at most m choices and every non-generating verticex has at most one choice. This proves that $\#\Pi^b(w) = O(m^{k+1})$ and hence we get the result. $\qquad\square$

Hence we get $\lim E[\beta_h(m^{-1/2}H_n^b)] = 0$ for all h and the limit is degenerate.

(M1) condition for RC_n^B and H_n^B. The first lemma shows that non-symmetric words do not contribute to the limit of RC_n^B. The second lemma yields the LSD for RC_n^B. Their proofs are omitted.

Lemma 7.2.14. For RC_n^B $\lim\limits_{n \to \infty} \frac{1}{nm^k} \#\Pi^{B*}(w) = 0$ whenever $w \in \mathcal{W}_{2k}(2)$ is non-symmetric. ◇

Lemma 7.2.15. For RC_n^B, $\lim\limits_{n \to \infty} \frac{1}{nm^k} \#\Pi^{B*}(w) = 2^k$ if $w \in \mathcal{W}_{2k}(2)$ is symmetric. Hence

$$\lim E[\beta_{2k}((2m_n)^{-1/2}RC_n^B)] = k! = E[R^{2k}].$$

◇

For H_n^B, when $\alpha \neq 0$, proceeding as in Section 7.2.4

$$\beta_{2k}(H_\alpha^B) = \lim_{n \to \infty} \frac{1}{n(2m_n)^k} \sum_{w \in \mathcal{W}_{2k}(2)} \#\Pi^{B**}(w), \text{ where,}$$

$$\begin{aligned}
\#\Pi^{B**}(w) = \#\{&(x_0, x_1, \ldots, x_{2k}) : x_0 = x_{2k}, x_i \in \{1/n, 2/n, \ldots, (n-1)/n, 1\} \\
&-m \leq x_{i-1} + x_i - n \leq m, \ x_{i-1} + x_i = x_{j-1} + x_j \text{ if } w[i] = w[j]\}.
\end{aligned}$$

Hence

$$\begin{aligned}
\frac{\#\Pi^{B*}(w)}{n^{1+k}} =\ & E\Big[\mathbb{I}(0 \leq L_i^H(U_{n,S}) \leq 1, \ \forall \ i \notin S \cup \{2k\}) \times \mathbb{I}(x_0 = x_{2k}) \\
&\times \mathbb{I}\big(|L_{i-1}^H(U_{n,S}) + U_{n,i} - 1| \leq \tfrac{m}{n} \ \forall \ i \in S\big)\Big], \qquad (7.13)
\end{aligned}$$

$$\begin{aligned}
\frac{1}{n^{1+k}} \#\Pi^{B*}(w) \ \to\ & \underbrace{\int_0^1 \cdots \int_0^1}_{k+1} \prod_{i \notin S \cup \{2k\}} \mathbb{I}(0 \leq L_i^H(x_S) \leq 1) \\
&\times \prod_{i \in S} \mathbb{I}(|L_{i-1}^H(x_S) + x_i - 1| \leq \alpha) \times \mathbb{I}(x_0 = x_{2k}) dx_S \\
=\ & p_{H_\alpha^B}(w) \text{ say.}
\end{aligned}$$

Hence

$$\beta_{2k}(H_\alpha^B) = (2\alpha)^{-k} \sum_{w\in\mathcal{W}_{2k}(2)} p_{H_\alpha^B}(w)$$

Now we will deal with the case $\alpha = 0$.

Lemma 7.2.16. (a) Suppose Assumption I (i) or (ii) holds and $\alpha = 0$. Then

$$W_2^2(F^{m^{-1/2}H_n^B}, F^{m^{-1/2}RC_n^B}) \overset{a.s.}{\to} 0.$$

(b) Suppose Assumption I* (i) and (ii) holds and $\alpha = 0$. Then

$$W_2^2(F^{m^{-1/2}H_n^B}, F^{m^{-1/2}RC_n^B}) \overset{P}{\to} 0.$$

◇

Proof. We construct another pair of RC_n^B and H_n^B matrices as follows: Let

$$y_{i+j,n} = \begin{cases} x_{i+j} & \text{if } i+j \le n-1 \\[2mm] x_{i+j-n} & \text{if } i+j \ge n. \end{cases} \tag{7.14}$$

$$\widetilde{H}_n^B(i,j) = \begin{cases} y_{i+j,n} & \text{if } |i+j-n| \le m \\[2mm] 0 & \text{otherwise .} \end{cases} \tag{7.15}$$

$$RC_n^B(i,j) = \begin{cases} y_{i+j,n} & \text{if } i+j \mod n \le m \text{ or } i+j \mod n \ge n-m \\[2mm] 0 & \text{otherwise.} \end{cases} \tag{7.16}$$

As $m/n \to 0$ for sufficiently large n, $m < n/2$. Hence $y_{i_1+j_1,n} \ne y_{i_2+j_2,n}$ for any (i_1, j_1) and (i_2, j_2) such that $i_1 + j_1 \le n-1 < i_2 + j_2$.

It is not hard to see that for sufficiently large n

$$E[\beta_h(m^{-1/2}H_n^B)] = E[\beta_h(m^{-1/2}\widetilde{H}_n^B)]$$

and

$$E[\beta_h(m^{-1/2}H_n^B) - E[\beta_h(m^{-1/2}H_n^B)]]^4 = E[\beta_h(m^{-1/2}\widetilde{H}_n^B) - E[\beta_h(m^{-1/2}\widetilde{H}_n^B)]]^4$$

for each h. This proves that if the LSD of one of these sequences exists, then so does the other. Moreover, if almost sure convergence (via (M4)) holds or, in probability convergence (via (M2)) holds for one sequence, then the same holds for the other. Hence it is enough to show that

$$W_2^2(F^{m^{-1/2}\widetilde{H}_n^B}, F^{m^{-1/2}RC_n^B}) \overset{a.s.}{\to} 0.$$

On the other hand, the LSD of the above RC matrix is the same as obtained before.

Now

$$W_2^2(F^{m^{-1/2}\widetilde{H}_n^B}, F^{m^{-1/2}RC_n^B}) \leq \frac{1}{n}\operatorname{Tr}\Big[\frac{(\widetilde{H}_n^B - RC_n^B)^2}{m}\Big]$$

$$= \frac{1}{mn}[(m-1)x_m^2 + \cdots + x_2^2]$$

$$+ \frac{1}{mn}[x_0^2 + x_{n-1}^2 + \cdots + (m+1)x_{n-m}^2]$$

$$\leq \frac{1}{n}(x_0^2 + x_2^2 + \cdots + x_m^2)$$

$$+ \frac{2}{n}(x_{n-m}^2 + \cdots + x_{n-1}^2)$$

$$= \left(\frac{m}{n}\right) \times \frac{x_0^2 + x_2^2 + \cdots + x_m^2}{m}$$

$$+ \frac{2}{n}(x_{n-m}^2 + \cdots + x_{n-1}^2).$$

(a) First assume $\{x_i\}$ are i.i.d. As $m/n \to 0$, by SLLN, the first term in the above expression $\to 0$, and the second term is

$$\frac{x_{n-m}^2 + \cdots + x_{n-1}^2}{n} = \frac{x_0^2 + \cdots + x_{n-1}^2}{n} - \frac{x_0^2 \cdots + x_{n-m-1}^2}{n-m}\frac{n-m}{n}$$

$$\overset{a.s.}{\to} \operatorname{E}[x_0^2] - \operatorname{E}[x_0^2](1) = 0.$$

If the input sequence is uniformly bounded, a slight modification of the above argument yields the result.

(b) Now assume that the input sequence $\{x_i\}$ satisfies Assumption I* (i) and (ii). We then have

$$W_2^2(F^{m^{-1/2}\widetilde{H}_n^B}, F^{m^{-1/2}RC_n^B}) \leq \frac{x_0^2 + x_1^2 + \cdots + x_m^2}{n} + \frac{2}{n}(x_{n-m}^2 + \cdots + x_{n-1}^2).$$

The proof that the first term goes to 0 is similar to the proof of Lemma 7.2.7.

It remains to show that $\frac{1}{n}(x_{n-m}^2 + \cdots + x_{n-1}^2) \overset{P}{\to} 0$. First note that $\frac{1}{n}\operatorname{E}(x_{n-m}^2 + \cdots + x_{n-1}^2) \to 0$. On the other hand, using Assumption I* (ii),

$$\frac{1}{n^2}\operatorname{Var}(x_{n-m}^2 + \cdots + x_{n-1}^2) \leq \frac{\sum_{i=n-m}^{n-1}\operatorname{E}[x_i^4]}{n^2}$$

$$\leq \frac{\sum_{i=1}^{n-1}\operatorname{E}[x_i^4\mathbb{I}(|x_i| > t)]}{n^2} + \frac{1}{n}t^4 \to 0.$$

So the proof is complete.   □

(M1) Condition for T_n^B ($\alpha \neq 0$). The following lemma on slopes is similar to Lemma 7.2.12 proved earlier for T_n^b. We omit the proof.

Lemma 7.2.17. Let $\mathcal{M}$ be the number of L_T matched circuits π on $\{1, 2, \ldots, n\}$ of length $2k$ with at least one pair of L^T matched edges $(\pi(i-1), \pi(i))$ and $(\pi(j-1), \pi(j))$ such that $\pi(i) - \pi(i-1) + \pi(j) - \pi(j-1) \neq 0$. Let $\mathcal{M}^B$ be the same count with the extra condition $\mathbb{I}^{2k}_{\pi, T} = 1$. Then $n^{-(k+1)} \mathcal{M}^B \leq n^{-(k+1)} \mathcal{M} \to 0$. ◇

Now we proceed to verify the (M1) condition and derive the limit. Earlier we have shown that, $\mathrm{E}[\beta_{2k+1}(n^{-1/2} T_n^B)] \to 0$. Further, *if* the limit below exists

$$\beta_{2k}(T_\alpha^B) = \lim_{n \to \infty} \mathrm{E}[\beta_{2k}(m^{-1/2} T_n^B)] = \alpha^{-k} \lim_{n \to \infty} \frac{1}{n^{1+k}} \sum_{\substack{w \in \mathcal{W}_{2k}(2) \\ |w| = k}} \#\Pi^B(w).$$

Using Lemma 7.2.4 and Lemma 7.2.17,

$$\lim_{n \to \infty} \frac{1}{n^{1+k}} \#\Pi^B(w) = \lim_{n \to \infty} \frac{1}{n^{1+k}} \#\Pi^{B*}(w) = \lim_{n \to \infty} \frac{1}{n^{1+k}} \#\Pi^{B**}(w).$$

Taking $x_i = \pi(i)/n$,

$$\#\Pi^{b**}(w) = \#\{(x_0, x_1, \ldots, x_{2k}) : x_0 = x_{2k}, x_i \in \{1/n, 2/n, \ldots, (n-1)/n, 1\},$$
$$|x_{i-1} - x_i| \leq m \text{ or } |x_{i-1} - x_i| \geq n - m, \text{ and}$$
$$x_{i-1} - x_i + x_{j-1} - x_j = 0 \text{ if } w[i] = w[j]\}.$$

Let $S = \{0\} \cup \{\min(i, j) : w[i] = w[j], i \neq j\}$ be the set of all indices corresponding to the generating vertices. Let $x_S = \{x_i : i \in S\}$. Since x_i's satisfy k equations, each is a unique linear combination $L_i^T(x_S)$ of x_j's, $j \in S$, $j \leq i$. These L_i^T's depend on the word w but we suppress this dependence. Clearly $L_i^T(x_S) = x_i$ if $i \in S$ and also summing the k equations would imply $L_{2k}^T(x_S) = x_0$. So

$$\#\Pi^{B**}(w) = \#\{x_S : L_i^T(x_S) \in \mathbb{N}_n \text{ and } |L_{i-1}^T(x_S) - L_i^T(x_S)| \leq m/n$$
$$\text{or } |L_{i-1}^T(x_S) - L_i^T(x_S)| \geq 1 - m/n \ \forall \ i = 0, \ldots, 2k\}$$

where $\mathbb{N}_n = \{1/n, 2/n, \ldots, 1\}$. Since

$$m \leq n/2 \times |L_{i-1}^T(x_S) - L_i^T(x_S)| \leq m/n \tag{7.17}$$

$$\text{or} \qquad |L_{i-1}^T(x_S) - L_i^T(x_S)| \geq 1 - m/n \tag{7.18}$$

can be written as $||L_{i-1}^T(x_S) - L_i^T(x_S)| - 1/2| \geq 1/2 - m/n$, we can write $\#\Pi^{B**}(w)$ as

$$\#\{x_S : L_i^T(x_S) \in \mathbb{N}_n, ||L_{i-1}^T(x_S) - L_i^T(x_S)| - 1/2| \geq 1/2 - m/n, \forall 0 \leq i \leq 2k\}.$$

Note that $|L_{i-1}^T(x_S) - L_i^T(x_S)| = |L_{j-1}^T(x_S) - L_j^T(x_S)|$ if $w[i] = w[j]$. So we can replace the restrictions $||L_{i-1}^T(x_S) - L_i^T(x_S)| - 1/2| \geq 1/2 - m/n$, for all $i = 1, 2, \ldots, 2k$ by $||L_{i-1}^T(x_S) - L_i^T(x_S)| - 1/2| \geq 1/2 - m/n$ for $i \in S$. Hence

$$\frac{\#\Pi^{B**}(w)}{n^{1+k}} = \mathrm{E}\left[\mathbb{I}\left(0 \leq L_i^T(U_{n,S}) \leq 1, \, \forall \, i \notin S \cup \{2k\}\right)\right.$$
$$\left. \times \mathbb{I}\left(||L_{i-1}^T(U_{n,S}) - U_{n,i}| - 1/2| \geq 1/2 - \alpha_n \, \forall \, i \in S\right)\right],$$

where $\alpha_n = m/n$ and taking limits

$$\frac{1}{n^{1+k}}\#\Pi^{B**}(w) \rightarrow \underbrace{\int_0^1 \int_0^1 \cdots \int_0^1}_{k+1} \prod_{i \notin S \cup \{2k\}} \mathbb{I}(0 \leq L_i^T(x_S) \leq 1)$$
$$\times \prod_{i \in S} \mathbb{I}(||L_{i-1}^T(x_S) - x_i| - 1/2| \geq 1/2 - \alpha)dx_S$$
$$= p_{T_\alpha^B}(w), \quad \text{say.}$$

Hence $\beta_{2k}(T_\alpha^B) = \alpha^{-k} \sum_{w:w\in\mathcal{W}_{2k}(2)} p_{T_\alpha^B}(w)$.

Closeness of T_n^B and SC_n^b when $\alpha = 0$. The next lemma is similar to Lemma 7.2.7 and its proof is omitted.

Lemma 7.2.18. (a) Suppose Assumption I (i) or (ii) holds and $\alpha = 0$. Then

$$\Delta_n \equiv W_2^2(F^{m^{-1/2}T_n^B}, F^{m^{-1/2}SC_n^b}) \xrightarrow{a.s.} 0.$$

(b) Suppose Assumption I* (i) and (ii) holds and $\alpha = 0$. Then

$$\Delta_n \equiv W_2^2(F^{m^{-1/2}T_n^B}, F^{m^{-1/2}SC_n^b}) \xrightarrow{P} 0.$$

$\diamond$

DH_n^b, PH_n^b and PT_n^b matrices. Note that $PH_n^b J_n = J_n PH_n^b = PT_n^b$ where J_n is the matrix with entries 1 in the main anti-diagonal and zero elsewhere. Since $J_n^2 = I_n$ $(PT_n^b)^{2k} = (PH_n^b)^{2k}$, LSDs of PT_n^b and PH_n^b, if they exist, are the same.

Moreover, one can easily check that (i) the $n \times n$ principal minor of PH_{n+1}^b is DH_n^b and (ii) the $n \times n$ principal minor of PT_{n+1}^b is SC_n^b. Hence all these four Type I band matrices have the same LSD.

Since the ESD of $m_n^{-1/2}SC_n^b \Rightarrow N(0,2)$ (provided (M2) or (M4) condition, which we shall verify, is satisfied), the same is true for the other three matrices.

Carleman's/Riesz's condition.

Lemma 7.2.19. Let L_{2k} be the limit obtained in any of the (M1) conditions. Then for some constants C, $L_{2k} \leq C^k \frac{(2k)!}{k!}$ and hence Carleman's/Riesz's condition (R) is satisfied. $\diamond$

Proof. Proof for all the matrices are similar. Here is an outline. $\pi(0)$ has at most n possible values and each remaining generating vertex has at most $C_1 m$ possible values where C_1 is a constant. Each non-generating vertex has at most C_2 possible values for some constant C_2. Hence we get

$$\sum_{w \in \mathcal{W}_{2k}(2)} \lim_{n \to \infty} \frac{1}{nm^k} \# \Pi^b(w) \le \sum_{w \in \mathcal{W}_{2k}(2)} \lim_{n \to \infty} \frac{(C_1 C_2)^k nm^k}{nm^k} = \frac{2k!}{k!2^k} \times (C_1 C_2)^k.$$

(7.19)

The expression on the right side agrees with the moments of a $N(0, C_1 C_2)$ variable. This completes the proof. $\qquad\square$

(M2) and (M4) conditions. Let $Q_{h,4}$ be the number of circuits $(\pi_1, \pi_2, \pi_3, \pi_4)$ of length h which are jointly and cross matched with respect to link function L and let $Q_{h,4}^b$ be the same count with the added restriction $\mathbb{I}_{\pi_i}^h = 1$, $i = 1, 2, 3, 4$.

Lemma 7.2.20. (a) For L_{SC} or L_{RC}, there exists a constant K such that $Q_{h,4}^b \le K n^4 m^{2h-2}$.

(b) If the input sequence is uniformly bounded, independent, with mean zero and variance one, and $\{A_n^b = SC_n^b$ or $RC_n^b\}$ then

$$\mathrm{E} \left[\frac{1}{n} \mathrm{Tr} \left(\frac{A_n^b}{\sqrt{m}} \right)^h - \mathrm{E} \frac{1}{n} \mathrm{Tr} \left(\frac{A_n^b}{\sqrt{m}} \right)^h \right]^4 = \mathrm{O} \left(\frac{1}{m^2} \right).$$

(7.20)

Hence if $m_n \to \infty$, (M2) holds. If $\sum_{n=1}^{\infty} \frac{1}{m_n^2} < \infty$ then (M4) holds. The same conclusions hold for $\{RC_n^B\}$.

(c) If the input sequence is uniformly bounded, independent, with mean zero and variance one, and $\{A_n^b = SC_n^b$ or $RC_n^b\}$ then

$$\mathrm{Var}[\beta_h(A_n^b)] = \mathrm{E} \left[\frac{1}{n} \mathrm{Tr} \left(\frac{A_n^b}{\sqrt{m}} \right)^h - \mathrm{E} \frac{1}{n} \mathrm{Tr} \left(\frac{A_n^b}{\sqrt{m}} \right)^h \right]^2 = \mathrm{O} \left(\frac{1}{m} \right).$$

(7.21)

Hence (M2) holds. The same conclusions hold for $\{RC_n^B\}$.

(d) For L_T or L_H, there exists a constant K depending on L such that $Q_{h,4}^b \le K n^{2h+2}$.

(e) If the input sequence is uniformly bounded, independent, with mean zero and variance one, and $\{A_n^b = T_n^b$ or $H_n^b\}$ then,

$$\mathrm{E} \left[\frac{1}{n} \mathrm{Tr} \left(\frac{A_n^b}{\sqrt{n}} \right)^h - \mathrm{E} \frac{1}{n} \mathrm{Tr} \left(\frac{A_n^b}{\sqrt{n}} \right)^h \right]^4 = \mathrm{O} \left(\frac{1}{n^2} \right).$$

(7.22)

Hence (M4) holds. The same conclusion holds for $\{T_n^B\}$ and $\{H_n^B\}$.

(f) For $\alpha = 0$ and the Hankel link function, there exists a constant K

depending on the link function such that for sufficiently large n $Q_{h,4}^b \leq K_m^{2h+2}$. Hence for sufficiently large n

$$\mathrm{E}\left[\frac{1}{n}\,\mathrm{Tr}\left(\frac{A_n^b}{\sqrt{m}}\right)^h - \mathrm{E}\,\frac{1}{n}\,\mathrm{Tr}\left(\frac{A_n^b}{\sqrt{m}}\right)^h\right]^4 = \mathrm{O}\left(\frac{1}{n^2}\right). \tag{7.23}$$

◇

Proof. (a) This proof is exactly the same as the proof of Lemma 1.4.3, Chapter 1 the only difference being that, excluding $\pi(0)$, all remaining generating vertices can be chosen in at most $m+1$ (respectively $2m+1$) ways for SC_n^b (respectively RC_n^b). We omit the details.

(b) We write the fourth moment as

$$\frac{1}{n^4 m^{2h}}\,\mathrm{E}[\mathrm{Tr}(A_n^b)^h - \mathrm{E}(\mathrm{Tr}(A_n^b)^h)]^4 = \frac{1}{n^4 m^{2h}}\sum_{\pi_1,\pi_2,\pi_3,\pi_4}\mathrm{E}[\prod_{i=1}^4 (\mathrm{X}_{\pi_i} - \mathrm{E}\,\mathrm{X}_{\pi_i})]\times\prod_{i=1}^4 \mathbb{I}_{\pi_i}^h.$$

Using independence, it follows easily that whenever $(\pi_1, \pi_2, \pi_3, \pi_4)$ are (i) not jointly matched, or (ii) jointly matched but are not cross-matched, the corresponding expectation in the above expression is zero. Since the entries are bounded, so is $\mathrm{E}[\prod_{i=1}^4 (\mathrm{X}_{\pi_i} - \mathrm{E}\,\mathrm{X}_{\pi_i})]$. Therefore by part (a)

$$\mathrm{E}\left[\frac{1}{n}\,\mathrm{Tr}\left(\frac{A_n^b}{\sqrt{m}}\right)^h - \mathrm{E}\,\frac{1}{n}\,\mathrm{Tr}\left(\frac{A_n^b}{\sqrt{m}}\right)^h\right]^4 \leq K\frac{n^4 m^{2h-2}}{n^4 (m^{h/2})^4} = \mathrm{O}(m^{-2}). \tag{7.24}$$

(c) This is an easy consequence of the Cauchy-Schwartz inequality and part (b). Observe

$$\mathrm{E}\left[\frac{1}{n}\,\mathrm{Tr}\left(\frac{A_n^b}{\sqrt{m}}\right)^h - \mathrm{E}\,\frac{1}{n}\,\mathrm{Tr}\left(\frac{A_n^b}{\sqrt{m}}\right)^h\right]^2 \leq \left(\mathrm{E}\left[\frac{1}{n}\,\mathrm{Tr}\left(\frac{A_n^b}{\sqrt{m}}\right)^h - \mathrm{E}\,\frac{1}{n}\,\mathrm{Tr}\left(\frac{A_n^b}{\sqrt{m}}\right)^h\right]^4\right)^{\frac{1}{2}}. \tag{7.25}$$

Now using part (b) completes the proof of (c).

(d) Since $Q_{h,4}^b \leq Q_{h,4}$, the claim follows.

Proof of (e) and (f) are along similar lines and are omitted. This completes all the steps in the proof of Theorem 7.1.1. □

Tables 7.2–7.5 on the following pages consider the situations where the LSDs are not known explicitly. Using the results obtained in Section 7.2.4, and performing the necessary integrations, we provide the contribution from the different words for moments of order two and four, and adding these contributions, provide the moments of the LSD. It may also be noted that if we follow the threads of proofs in Section 7.2.4, then it is clear that the even moments of the LSD of $n^{-1/2}T_n^b$ and $n^{-1/2}H_n^b$ increase with α.

TABLE 7.2
Type I Toeplitz limit ($\alpha \neq 0$). Word and moments of order 2 and 4.

α	w	Indicator set	$p_{T_\alpha^b}(w)$	Moments
$0 < \alpha < 1$	aa	$|x_0 - x_1| \leq \alpha$	$\alpha(2 - \alpha)$	$\beta_2(T_\alpha^b) = (2 - \alpha)$
$0 < \alpha \leq \frac{1}{2}$	$abba$ $aabb$ $abab$	$|x_i - x_{i+1}| \leq \alpha, i = 0, 1$ $|x_i - x_{i+1}| \leq \alpha, i = 0, 1,$ $0 \leq x_0 - x_1 + x_2 \leq 1$	$\frac{2}{3}\alpha^2(6 - 5\alpha)$ as above $4\alpha^2(1 - \alpha)$	$\beta_4(T_\alpha^b) = \frac{4}{3}(9 - 8\alpha)$
$\frac{1}{2} < \alpha \leq 1$	$abba$ $aabb$ $abab$		$\frac{-1+6\alpha-2\alpha^3}{3}$ as above $\frac{2}{3}(2\alpha^3 - 6\alpha^2 + 6\alpha - 1)$	$\beta_4(T_\alpha^b) = \frac{4(-3\alpha^2+6\alpha-1)}{3\alpha^2}$

7.3 Exercises

1. Prove Lemma 7.2.1.

2. Prove Lemma 7.2.4.

3. Look up Kargin (2009)[69] to learn how he uses the closeness of the Circulant and Toeplitz matrices. In particular see the proof of the result that under suitable conditions the Type I banded Toeplitz EESD converges to a non-normal LSD when $\alpha \neq \frac{4}{7}$.

4. Prove Lemma 7.2.10. Hint: Look at the relevant portion of the proof of Theorem 2.3.2 in Chapter 2.

5. Look up Liu and Wang (2009)[71] to learn how they prove the existence of the LSD of Toeplitz and Hankel band matrices by using representations of these matrices as linear combinations of backward and forward shift matrices.

6. Look up Popescu (2009)[85] for some interesting limits for tridiagonal Wigner matrices.

7. Look up Bose and Sen (2011)[30] and Bose, Gangopadhyay and Saha (2013)[24] for LSD results on some other finite diagonal matrices.

TABLE 7.3
Type I Hankel limit ($\alpha \neq 0$). Word and moments of order 2 and 4.

α	w	Indicator set	$p_{H_\alpha^b}(w)$	Moments
$0 < \alpha \leq 1$	aa	$x_0 + x_1 \leq \alpha$	$\frac{1}{2}\alpha^2$	$\beta_2(H_\alpha^b) = \frac{1}{2}\alpha$
$1 \leq \alpha \leq 2$	aa		$\alpha - 1 + \frac{\alpha(2-\alpha)}{2}$	$\beta_2(H_\alpha^b) =$ $1 - \frac{1}{\alpha} + \frac{(2-\alpha)}{2}$
$0 < \alpha \leq 1$	$abab$ $abba$ $aabb$	$x_i + x_{i+1} \leq \alpha, i = 0,1$	0 $\frac{\alpha^3}{3}$ as above	$\beta_4(H_\alpha^b) = \frac{2\alpha}{3}$
$1 < \alpha \leq 2$	$abab$ $abba$ $aabb$		0 $\alpha - 1 + \frac{1-(\alpha-1)^3}{3}$ as above	$\beta_4(H_\alpha^b) =$ $2(\frac{\alpha-1}{\alpha^2} + \frac{1-(\alpha-1)^3}{3\alpha^2})$

TABLE 7.4
Type II Toeplitz limit ($\alpha \neq 0$). Word and moments of order 2 and 4.

α	w	Indicator set	$p_{T_\alpha^B}(w)$	Moments
$\alpha \in (0, 1/2]$	aa	$|x_0 - x_1| \leq \alpha$ or $|x_0 - x_1| \geq 1 - \alpha$	2α	$\beta_2(T_\alpha^B) = 2$
$\alpha \in (0, 1/2]$	$abba$ $aabb$ $abab$	$|x_0 - x_1| \leq \alpha$ or $|x_0 - x_1| \geq 1 - \alpha,$ $|x_1 - x_2| \leq \alpha$ or $|x_1 - x_2| \geq 1 - \alpha$ $|x_0 - x_1| \leq \alpha$ or $|x_0 - x_1| \geq 1 - \alpha,$ $|x_1 - x_2| \leq \alpha$ or $|x_1 - x_2| \geq 1 - \alpha,$ $0 \leq x_0 - x_1 + x_2 \leq 1$	$4\alpha^2$ as above $4\alpha^2(1 - \frac{2}{3}\alpha)$	$\beta_4(T_\alpha^B) =$ $4(3 - \frac{2}{3}\alpha)$

TABLE 7.5
Type II Hankel limit ($\alpha \neq 0$). Word and moments of order 2 and 4.

α	w	Indicator set	$p_{H_\alpha^B}(w)$	Moments
$\alpha \in (0, 1]$	aa	$\|x_0 + x_1 - 1\| \leq \alpha$	$\alpha(2 - \alpha)$	$\beta_2(H_\alpha^B) = 1 - \frac{\alpha}{2}$
$\alpha \in (0, 1/2]$	$abab$ $abba$ $aabb$	$\|x_0 + x_1 - 1\| \leq \alpha,$ $\|x_1 + x_2 - 1\| \leq \alpha$	0 $\frac{2}{3}\alpha^2(6 - 5\alpha)$ as above	$\beta_4(H_\alpha^B) = \frac{1}{3}(6 - 5\alpha)$
$\alpha \in [1/2, 1]$	$abab$ $abba$ $aabb$		0 $\frac{-1+6\alpha-2\alpha^3}{3}$ as above	$\beta_4(H_\alpha^B) = \frac{(-1+6\alpha-2\alpha^3)}{6\alpha^2}$

8

Triangular matrices

Triangular random matrices have gained importance recently. For example, Dykema and Haagerup (2004)[52] were led to the consideration of the following *asymmetric* triangular version of the Wigner matrix

$$
T_n = \begin{bmatrix}
t_{1,1} & t_{1,2} & t_{1,3} & \cdots & t_{1,n-1} & t_{1,n} \\
0 & t_{2,2} & t_{2,3} & \cdots & t_{2,n-1} & t_{2,n} \\
0 & 0 & t_{3,3} & \cdots & t_{3,n-1} & t_{3,n} \\
& & & \vdots & & \\
0 & 0 & 0 & \cdots & 0 & t_{n,n}
\end{bmatrix} \tag{8.1}
$$

where $(t_{i,j})_{1 \leq i \leq j \leq n}$ are i.i.d. complex Gaussian random variables having mean 0 and variance 1. It does not seem to be easy to obtain the LSD of this non-symmetric matrix. They obtained the LSD of $T_n^* T_n$.

In this chapter our goal is to study the spectral properties of $n^{-1/2} X_n^u$ where X_n^u is a *real symmetric* triangular patterned random matrix. We use the moment method to show that the limiting spectral distribution (LSD) of triangular Wigner, Toeplitz, Hankel and Symmetric Circulant matrices exist when the entries satisfy Assumption I, II or III of Chapter 1. All the limit results are universal. In the light of developments of Chapter 9 and the next Chapter 10, these LSD results for one patterned matrix at a time can be extended to joint convergence of any combination of these when we assume that the sequences are independent and the entries are independent with mean zero and variance one and have uniformly bounded moments of all orders.

The identification of the LSD is a non-trivial issue. By exploiting the Catalan recursions inbuilt in the Wigner matrix, it follows that the $(2k)$th moment of the LSD of the symmetric triangular Wigner matrix is given by $\frac{k^k}{(k+1)!}$ and the odd moments are zero. However, moment formulae for other matrices remain elusive and appear rather difficult to obtain.

Identification of any structure in the joint limit moments appears much harder, even for the triangular Wigner matrix. Unlike the full Wigner, the triangular Wigner matrices are not asymptotically free.

In Section 8.1 we describe our setup, notation and state and prove the main existence results for the LSD. In Section 8.2 we concentrate on the triangular Wigner matrix and derive the moments of the LSD.

8.1   General pattern

Consider patterned matrices of the form

$$X_n = ((x_{L(i,j)})). \tag{8.2}$$

Let X_n^u be the upper triangular version of the X_n matrix where the (i,j)-th entry of X_n^u is given by $x_{L(i,j)}$ if $(i+j) \le n+1$ and 0 otherwise. We shall often drop the subscript n, and simply denote the matrix by X^u.

For instance, triangular symmetric Wigner, the Hankel, Toeplitz and Symmetric Circulant matrices are:

$$W_n^u = \begin{bmatrix} x_{11} & x_{12} & x_{13} & \cdots & x_{1(n-1)} & x_{1n} \\ x_{12} & x_{22} & x_{23} & \cdots & x_{2(n-1)} & 0 \\ & & & \vdots & & \\ x_{1(n-1)} & x_{2(n-1)} & 0 & \cdots & 0 & 0 \\ x_{1n} & 0 & 0 & \cdots & 0 & 0 \end{bmatrix}, \tag{8.3}$$

$$H_n^u = \begin{bmatrix} x_2 & x_3 & x_4 & \cdots & x_n & x_{n+1} \\ x_3 & x_4 & x_5 & \cdots & x_{n+1} & 0 \\ & & & \vdots & & \\ x_n & x_{n+1} & 0 & \cdots & 0 & 0 \\ x_{n+1} & 0 & 0 & \cdots & 0 & 0 \end{bmatrix}, \tag{8.4}$$

$$T_n^u = \begin{bmatrix} x_0 & x_1 & x_2 & \cdots & x_{n-2} & x_{n-1} \\ x_1 & x_0 & x_1 & \cdots & x_{n-3} & 0 \\ x_2 & x_1 & x_0 & \cdots & 0 & 0 \\ & & & \vdots & & \\ x_{n-1} & 0 & 0 & \cdots & 0 & 0 \end{bmatrix}. \tag{8.5}$$

and

$$SC_n^u = \begin{bmatrix} x_0 & x_1 & x_2 & \cdots & x_2 & x_1 \\ x_1 & x_0 & x_1 & \cdots & x_3 & 0 \\ x_2 & x_1 & x_0 & \cdots & 0 & 0 \\ & & & \vdots & & \\ x_1 & 0 & 0 & \cdots & 0 & 0 \end{bmatrix}. \tag{8.6}$$

It is to be noted that a triangular Reverse Circulant matrix is the same as a triangular Hankel matrix.

Recall the key assumptions from Chapters 1 and 2.

Assumption I $\{x_i, x_{ij}\}$ are i.i.d. and uniformly bounded with mean 0 and variance 1.

Assumption II $\{x_i, x_{ij}\}$ are i.i.d. with mean zero and variance 1.

Assumption III $\{x_i, x_{ij}\}$ are independent with mean zero and variance 1 and with uniformly bounded moments of all order.

The restriction of Property B to the triangular matrices is now:

Property B' The link function L satisfies

$$\Delta = \sup_n \sup_t \sup_{1 \leq k \leq n} \#\{l : 1 \leq l \leq n, k+l \leq n+1, \ L(k,l) = t\} < \infty.$$

For a pair-matched word w of length $2k$, let

$$\Pi_X^*(w) = \{\pi : w[i] = w[j] \Rightarrow L(\pi(i-1), \pi(i)) = L(\pi(j-1), \pi(j))\}$$
$$\Pi_X(w) = \{\pi : w[i] = w[j] \Leftrightarrow L(\pi(i-1), \pi(i)) = L(\pi(j-1), \pi(j))\}.$$

We shall write now

$$\frac{1}{n} \text{Tr}((\frac{1}{\sqrt{n}} X^u)^k) = \frac{1}{n^{1+k/2}} \sum_w \sum_{\pi \in \Pi(w)} \prod_{i=1}^k x_{L(\pi(i-1), \pi(i))} 1_{\{\pi(i-1)+\pi(i) \leq n+1\}}.$$

Let us denote

$$X_\pi = \prod_{i=1}^k x_{L(\pi(i-1), \pi(i))} 1_{\{\pi(i-1)+\pi(i) \leq n+1\}}.$$

For a fixed k, let us define the class Π_k^u as follows.

$$\Pi_k^u = \{\pi : \pi \text{ is circuit of length } k, \ \pi(i-1) + \pi(i) \leq n+1, 1 \leq i \leq k\}.$$

Let us denote

$$\Pi_{1,X}(w) = \Pi_X(w) \bigcap \Pi_{2k}^u \text{ and } \Pi_{1,X}^*(w) = \Pi_X^*(w) \bigcap \Pi_{2k}^u.$$

For every pair-matched word w of length $2k$, define, if the limit exists,

$$p_{u,X}(w) = \lim_{n \to \infty} \frac{\#\Pi_{1,X}^*(w)}{n^{1+k}}.$$

When there is no chance for confusion, we simply write $p_u(w)$ for $p_{u,X}(w)$ and $\Pi_1(w)$ and $\Pi_1^*(w)$ for $\Pi_{1,X}(w)$ and $\Pi_{1,X}^*(w)$ respectively.

By using the arguments of Chapter 1, we immediately get the following theorem. We omit its proof. Recall the definitions of α_n and k_n given in 1.4.2 of Chapter 1. We continue to use that definition.

Theorem 8.1.1. Suppose $\{X_n^u\}$ satisfies Assumptions II or III and the L function satisfies Property B' and $\alpha_n k_n^d = O(n^2)$. Suppose for every pair-matched word w, $p_u(w)$ exists. Then the LSD of $\frac{X^u}{\sqrt{n}}$ exists a.s. The LSD is universal, symmetric about 0, and is determined by the even moments

$$\beta_{2k} = \sum_{w \in W_{2k}(2)} p_u(w), \quad k \geq 1. \tag{8.7}$$

$\diamond$

Figures 8.1 and 8.2 show the plots of the ESD of triangular matrices listed earlier. We can state the following theorem on their LSD.

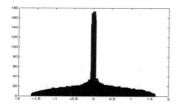
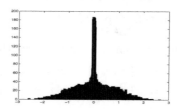

FIGURE 8.1
Histogram plots of the ESD of triangular matrices ($n = 1000$) with entries $N(0,1)$ of (i)
Wigner (left) (ii) Toeplitz (right).

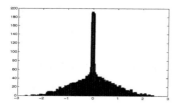
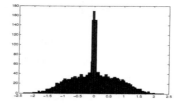

FIGURE 8.2
Histogram plots of empirical distribution of triangular matrices ($n = 1000$) with entries
$N(0,1)$ of (i) Symmetric Circulant (left) (ii) Hankel (right).

Theorem 8.1.2. Let $\{X_n^u\}$ be any of the following triangular matrices with
an input sequence that satisfies Assumption I, II or III: triangular Wigner, triangular Hankel, triangular Toeplitz or triangular Symmetric Circulant. Then
the LSD for $\frac{X_n^u}{\sqrt{n}}$ exists almost surely. The LSD are universal and are symmetric
with even moments given by (8.7).

Proof. As before, without loss, we shall work under Assumption I. Notice that
the link functions of all these matrices satisfy Property B'. Thus, we need
to only establish Condition (M1). The combinatorics to show the existence
of $p_u(w)$ is done case-by-case. However, all the proofs are similar and use
the volume method. We provide a sketch only for the triangular Wigner and
the triangular Hankel matrices. Recall that for the full matrix version of the
matrices, $p(w)$ was evaluated as an integral of an indicator function. Here
also, $p_u(w)$ turns out to be an integral, but of a different indicator function
corresponding to $\Pi_1^*(w)$ instead of $\Pi^*(w)$.

Triangular Wigner Matrix: Let L_W denote the Wigner link function. We already know that (see (2.4)) if $w \in \mathcal{W}_{2k}(2)$ is not Catalan then,

$$p(w) = \lim_{n \to \infty} \frac{\#\Pi^*(w)}{n^{1+k}} = 0.$$

As $\Pi^*(w) \supseteq \Pi_1^*(w)$, it follows that $p_u(w) = 0$ if w is non-Catalan. Thus, we now focus on Catalan words of length $2k$.

Lemma 8.1.3. Let w be a Catalan word of length $2k$. Let S denote the set of all generating vertices of w. Then for all $j \notin S$, there exists a unique $i \in S$ such that $i < j$ and $\pi(j) = \pi(i)$ for all $\pi \in \Pi^*(w)$. ◇

Proof. Let $\pi \in \Pi^*(w)$. Let j be the minimum index of a non-generating vertex of w. Clearly then, $w[j-1] = w[j]$ and hence $\pi(j-2) = \pi(j)$. Since $j > j-2 \in S$, the result is true in this case. Now let j be any non-generating vertex. Let us assume that for every non-generating vertex with index less than j, the result holds. Let $i < j$ be the index of the first occurrence of $w[j]$. Let w_1 be the sub-word formed by letters between $w[i]$ and $w[j]$. Since w is Catalan, w_1 is also Catalan, and it can be easily shown that, $\pi(i) = \pi(j-1)$ and hence $\pi(j) = \pi(i-1)$. If $(i-1) \in S$, then we are already done. If $(i-1) \notin S$, then also by induction hypothesis the result holds.

If the i corresponding to a fixed j is not unique then we have a non-trivial relation between two generating vertices which implies $\#\Pi^*(w) = O(n^k)$, hence contradicting the fact that $\lim_{n \to \infty} \frac{\#\Pi^*(w)}{n^{1+k}} = 1$. This proves the uniqueness. □

Definition 8.1.1. For any j (not necessarily in S), let us denote by $\phi(j)$ the *unique* vertex such that

$$\phi(j) \in S, \ \phi(j) \leq j \ \text{ and } \ \pi(j) = \pi(\phi(j)) \ \text{ for all } \ \pi \in \Pi^*(w).$$

◇

Note that if $j \in S$, then $j = \phi(j)$. Next we note that among the $2k$ inequalities, $\pi(i-1) + \pi(i) \leq n+1$, $1 \leq i \leq 2k$, each inequality is repeated twice, as $w[i] = w[j] \Rightarrow \pi(i-1) + \pi(i) = \pi(j-1) + \pi(j)$. So we can write

$$\begin{aligned}
\Pi_1^*(w) &= \{\pi : w[i] = w[j] \Rightarrow L_W(\pi(i-1), \pi(i)) = L_W(\pi(j-1), \pi(j)), \\
&\quad \pi(i-1) + \pi(i) \leq n+1 \ \forall i \in S - \{0\}\} \\
&= \{\pi : \pi(j) = \pi(\phi(j)) \ \forall j \notin S, \\
&\quad \pi(\phi(i-1)) + \pi(\phi(i)) \leq n+1 \ \forall i \in S - \{0\}\}.
\end{aligned}$$

Now we use the standard volume method arguments. Let us define

$$v_i = \frac{\pi(i)}{n}, \ U_n = \{\frac{1}{n}, ..., \frac{n-1}{n}, 1\} \ \text{ and } \ v_S = \{v_i : i \in S\}. \tag{8.8}$$

Then,

$$\#\Pi_1^*(w) = \#\{(v_0, ..., v_{2k}) : v_i \in U_n \; \forall 0 \le i \le 2k, v_i = v_{\phi(i)} \; \forall i \notin S,$$
$$v_{\phi(i-1)} + v_{\phi(i)} \le 1 + 1/n \; \forall i \in S - \{0\}, v_0 = v_{2k}\}$$
$$= \#\{v_S : v_i \in U_n \; \forall i \in S, v_{\phi(i-1)} + v_{\phi(i)} \le 1 + 1/n \; \forall i \in S - \{0\}\}.$$

From the above equation it follows that $\frac{\#\Pi_1^*(w)}{n^{1+k}}$ is nothing but the Riemann sum for the function

$$I_W(v_S) = I(v_{\phi(i-1)} + v_{\phi(i)} \le 1, i \in S - \{0\})$$

over $[0,1]^{k+1}$. Since the function is clearly Riemann integrable, the following convergence holds:

$$\lim_{n \to \infty} \frac{1}{n^{1+k}} \#\Pi_1^*(w) = \int_{[0,1]^{k+1}} I_W(v_S) dv_S. \tag{8.9}$$

It follows that $p_u(w) = \lim_{n \to \infty} \frac{1}{n^{1+k}} \#\Pi_1^*(w)$ exists.

Triangular Hankel Matrix: The Hankel link function is $L(i,j) = i + j$. Here

$$\Pi_1^*(w) = \{\pi : w[i] = w[j] \Rightarrow \pi(i-1) + \pi(i) = \pi(j-1) + \pi(j),$$
$$\pi(i-1) + \pi(i) \le n + 1\}.$$

Let S denote the set of all generating vertices of w. For every $i \in S - \{0\}$, let j_i denote the index such that $w[i] = w[j_i]$. Let us define v_i, U_n, v_S as in (8.8). Then

$$\#\Pi_1^*(w) = \#\{(v_0, ..., v_{2k}) : v_i \in U_n, \forall i; v_{(i-1)} + v_i = v_{(j_i-1)} + v_{j_i}, i \in S;$$
$$v_{(i-1)} + v_i \le 1 + 1/n, i \in S - \{0\}; v_0 = v_{2k}\}.$$

It can easily be seen from the above equations (other than $v_0 = v_{2k}$) that each of the $\{v_i : i \notin S\}$ can be written uniquely as an integer linear combination $L_i(v_S)$. Moreover, $L_i(v_S)$ only contains generating vertices of index less than i with non-zero coefficients. For all $i \in S$, let us define $L_i(v_s) = v_i$.

Clearly,

$$\#\Pi_1^*(w) = \#\{(v_0, ..., v_{2k}) : v_i \in U_n \; \forall \, i, v_0 = v_{2k}, v_i = L_i(v_S) \; \forall i \notin S,$$
$$v_{i-1} + v_i \le 1 + 1/n \; \forall i \in S - \{0\}\}.$$

Integer linear combinations of elements of U_n are again in U_n iff they are between 0 and 1. Hence,

$$\#\Pi_1^*(w) = \#\{v_S : v_i \in U_n \; \forall i \in S, v_0 = L_{2k}(v_S), 0 < L_i(v_S) \le 1 \; \forall i \notin S,$$
$$L_{i-1}(v_S) + L_i(v_s) \le 1 + 1/n \; \forall i \in S - \{0\}\}. \tag{8.10}$$

From (8.10) it follows that $\frac{\#\Pi_1^*(w)}{n^{1+k}}$ is nothing but the Riemann sum for the function $I_H(v_S)$ defined by

$$I(0 \le L_i(v_S) < 1, i \notin S, v_0 = L_{2k}^l(v_S), L_{i-1}(v_S) + L_i(v_S) \le 1 \ \forall i \in S - \{0\})$$

over $[0,1]^{k+1}$. Hence, as before,

$$\lim_{n\to\infty} \frac{1}{n^{1+k}} \#\Pi_1^*(w) = \int_{[0,1]^{k+1}} I_H(v_S)dv_S. \tag{8.11}$$

Hence $p_u(w)$ exists for every pair-matched word w.

Incidentally, for a Hankel matrix, $p(w) = 0$ if $w \notin \mathcal{S}_{2k}$. Hence it also follows that $p_u(w) = 0$ for every such word w for the triangular Hankel matrix. We know that the LSD of the Reverse Circulant, the Symmetric Circulant, the Toeplitz and the Hankel matrices are unbounded. The same is true for their triangular versions.

Theorem 8.1.4. The LSD of triangular Hankel, Toeplitz and Symmetric Circulant matrices have unbounded support. ◇

Proof. We shall prove the theorem only for the triangular Hankel matrices. The proof for other matrices is similar. Recall that the $2k$-th moment of the LSD for the full and triangular versions are given by

$$\beta_{2k} = \sum_{w\in\mathcal{W}_{2k}(2)} p(w), \quad \beta_{2k}' = \sum_{w\in\mathcal{W}_{2k}(2)} p_u(w).$$

We claim that for every w,

$$p_u(w) \ge \frac{1}{2^k}p(w).$$

To see this, first note that

$$p_u(w) = \int_{[0,1]^{k+1}} I_H(v_S)dv_S. \tag{8.12}$$

Also, recall that (see (2.24)), for the full Hankel matrix,

$$p(w) = \int_{[0,1]^{k+1}} I(0 \le L_i(v_S) < 1, i \notin S, v_0 = L_{2k}^l(v_S))dv_S.$$

Making a change of variable $y_i = v_i/2$ for all $i \in S$, since L_i are affine linear functions, we see that

$$\frac{p(w)}{2^k} = \int_{[0,1/2]^{k+1}} I(0 \le L_i(y_S) < 1/2, i \notin S, y_0 = L_{2k}^l(y_S))dv_S$$

$$= \int_{[0,1/2]^{k+1}} K(v_S)dy_S \text{ say.} \tag{8.13}$$

Now it is trivial to note that

$$K(v_S) \leq I_H(v_S) \quad \text{for all} \quad v_S.$$

The moment inequality claim now follows from the above observation in conjunction with (8.12) and (8.13). Since the Hankel LSD has unbounded support, so does the triangular Hankel LSD.   □

8.2   Triangular Wigner matrix

Recall that for the full Wigner matrix, $p(w) = 1$ for each Catalan word, and as a consequence the LSD is the semi-circle law. In Table 8.1 we list the values of $p_u(w)$ for Catalan words of small lengths for the triangular Wigner matrix. The contributions are only equal within certain isomorphic classes. It does not seem easy to obtain the individual $p_u(w)$'s for different Catalan words. Nevertheless their total contribution can be calculated in a relatively simple way and hence identify the LSD.

TABLE 8.1
$p_u(w)$ for Catalan words for W_n^u.

Word	$p_u(w)$
aa	1/2
aabb	1/3
abba	1/3
aabbcc	1/4
abbcca	1/4
abbacc	5/24
aabccb	5/24
abccba	5/24

8.2.1   LSD

Theorem 8.2.1. Let W_n^u be a triangular Wigner matrix with an input sequence that satisfies Assumption I. Then almost surely

$$\lim_{n \to \infty} \frac{1}{n} \mathrm{Tr} \left(\frac{W_n^u}{\sqrt{n}} \right)^{2k} = \frac{k^k}{(k+1)!}.$$

The sequence $\{ \frac{k^k}{(k+1)!} \}$ are moments of the unique probability measure ν supported on $[0, \ e]$ and given by

$$dv(x) = \psi(x)dx \quad \text{where } \psi : (0, \ e) \to \mathbb{R}^+ \tag{8.14}$$

is the unique solution of

$$\psi\left(\frac{\sin v}{v} \exp(v \cot v)\right) = \frac{1}{\pi} \sin v \exp(-v \cot v). \tag{8.15}$$

Hence the density of the LSD is given by $|x|\psi(x^2)$ where ψ satisfies (8.15) and its support is contained in $[-\sqrt{e}, \ \sqrt{e}]$. Since the LSD is universal, the above limit moments are also the $(2k)$th moments of the LSD under any of the Assumptions I, II or III. ◇

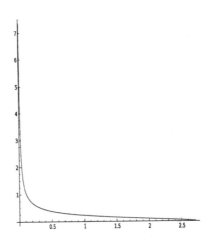

FIGURE 8.3
Plot of function ψ.

Remark 8.2.1. The plot of the function ψ is given in Figure 8.3. The above moments are related to the *Lambert W function*. This function is defined through the equation

$$W(x) \exp(W(x)) = x.$$

If x is real then the above equation has two possible branches and the one satisfying $W(x) \geq -1$ is called the principal branch and is denoted by W_0. The principal branch W_0 is analytic at zero and one can use Lagrange's inversion theorem to see that,

$$1 + \frac{1}{xW_0(1/x)} = \sum_{k=0}^{\infty} \frac{k^k}{(k+1)!} x^{-(k+1)}$$

for x lying outside $(-e, \ e)$. For details we refer the reader to the exciting article on the Lambert W function by Corless et al. (1996)[44]. ◇

Proof. To begin with, we need the following simple lemma whose proof is omitted.

Lemma 8.2.2. (a) Let w be a Catalan word of length $2k$. Let π be a function $\pi : \{0, 1, ..., k\} \rightarrow \{1, 2, ..., n\}$ such that $w[i] = w[j] \Rightarrow (\pi(i-1), \pi(i)) = (\pi(j), \pi(j-1))$. Then $\pi(0) = \pi(2k)$ and hence $\pi \in \Pi^*(w)$.

(b) Let $\phi(j)$ be as in Definition 8.1.1. If $w = w_1 w_2$ where w_1 and w_2 are Catalan words of length $2k_1$ and $2k_2$ respectively. Then for every vertex $i > 2k_1$, $\phi(i) \geq 2k_1$. ◇

Let w be a Catalan word of length $2k$. Let S be the set of generating vertices for w. It has already been shown that (see (8.9))

$$p_u(w) = \int \cdots \int_{[0,1]^{k+1}} I(v_{\phi(i-1)} + v_{\phi(i)} \leq 1, i \in S - \{0\}) dv_S. \qquad (8.16)$$

This integral can be evaluated as an iterated integral. It is clear that integrating out all the variables other than v_0 leaves a polynomial in v_0, say Q_w. It follows that

$$p_u(w) = \int_0^1 Q_w(v_0) dv_0 \qquad (8.17)$$

for some *polynomial* $Q_w(\cdot)$. Let us illustrate with a few examples. Let $w = aa$. Then

$$p_u(w) = \int_{v_0+v_1 \leq 1} dv_0 dv_1 = \int_0^1 (1 - v_0) dv_0$$

and hence

$$Q_{aa}(x) = 1 - x.$$

Now let $w = abba$. Then

$$\begin{aligned}
p_u(w) &= \int \cdots \int_{[0,1]^3} I(v_0 + v_1 \leq 1, v_1 + v_2 \leq 1) dv_0 dv_1 dv_2 \\
&= \int_0^1 \int_0^{1-v_0} \int_0^{1-v_1} dv_2 dv_1 dv_0 \\
&= \int_0^1 \int_0^{1-v_0} (1 - v_1) dv_1 dv_0 = \int_0^1 \frac{(1 - v_0^2)}{2} dv_0.
\end{aligned}$$

Hence

$$Q_{abba}(x) = \frac{1 - x^2}{2}.$$

The following two lemmata collect the required properties of $Q_w(\cdot)$.

Lemma 8.2.3. (a) Let $w = w_1 w_2$ be a Catalan word of length $2k$ where w_1 and w_2 are both Catalan. Then $Q_w(x) = Q_{w_1}(x) Q_{w_2}(x)$.

(b) Let $w = a w_1 a$ be a Catalan word with w_1 Catalan. Then

$$Q_w(x) = \int_0^{1-x} Q_{w_1}(y)dy.$$

◇

Proof. (a) Let w_1 and w_2 be Catalan words of length $2k_1$ and $2k_2$ respectively. We divide the set of inequalities in the indicator function in (8.16) above into two classes:

1. $v_{\phi(i-1)} + v_{\phi(i)} \leq 1, i \in S - \{0\}, i \leq 2k_1$; i.e., the inequalities corresponding to generating vertices in w_1.

2. $v_{\phi(i-1)} + v_{\phi(i)} \leq 1, i \in S - \{0\}, i > 2k_1$; i.e., the inequalities corresponding to generating vertices in w_2.

Lemma 8.2.2 (a) implies that $\phi(2k_1) = 0$ and Lemma 8.2.2 (b) implies that the inequalities in items (1) and (2) above do not have any common variable except v_0. Hence in (8.16) integrating with respect to the variables corresponding to the generating vertices in w_1 we get $Q_{w_1}(v_0)$ and integrating with respect to the variables corresponding to the generating vertices in w_2, we get $Q_{w_2}(v_0)$. The result now follows from the definition of Q_w.

(b) The proof of this is similar. Note that the variable v_0 does not occur anywhere in the equations corresponding to the generating vertices in w_1. Integrating with respect to all these variables (other than v_1), we get $Q_{w_1}(v_1)$. Hence the last step of evaluating the iterated integral is by (8.16)

$$p_u(w) = \int_{v_0+v_1\leq 1} Q_{w_1}(v_1)dv_0 dv_1 = \int_0^1 \left[\int_0^{1-v_0} Q_{w_1}(v_1)dv_1\right] dv_0,$$

and the result follows. □

Let

$$G_0(x) = 1 \text{ and } G_{2n}(x) = \sum_{w\in C_{2n}} Q_w(x). \tag{8.18}$$

Lemma 8.2.4. With $\{G_{2n}, n \geq 0\}$ as in (8.18),

(a) for $n \geq 1$ we have

$$G_{2n}(x) = \sum_{k=1}^n G_{2(n-k)}(x) \int_0^{1-x} G_{2(k-1)}(y)dy;$$

(b) for all $n \geq 0$,

$$G_{2n}(x) = \frac{(1-x)(n+1-x)^{n-1}}{n!}.$$

◇

Proof. (a) For $n = 1$, the only Catalan word of length 2 is aa. Hence $G_2(x) = Q_{aa}(x) = 1 - x$ and the result is true for $n = 1$. Let $n \geq 2$ and let $G_{2n,k}(x)$ be the sum of $Q_w(x)$ over all Catalan words w such that the first letter is repeated at the $2k$-th place. Clearly, for such a w, $w = aw_1aw_2$ where w_1 is a Catalan word of length $2(k-1)$ and w_2 is a Catalan word of length $2(n-k)$. Using the previous lemma, it follows that

$$
\begin{aligned}
G_{2n,k}(x) &= \sum_{|w_1|=(k-1),|w_2|=(n-k)} Q_{aw_1aw_2}(x) \\
&= \sum_{|w_1|=(k-1),|w_2|=(n-k)} Q_{aw_1a}(x)Q_{w_2}(x) \\
&= \sum_{|w_1|=(k-1)} Q_{aw_1a}(x) \sum_{|w_2|=(n-k)} Q_{w_2}(x) \\
&= \sum_{|w_1|=(k-1)} \int_0^{1-x} Q_{w_1}(y)dy \sum_{|w_2|=(n-k)} Q_{w_2}(x) \\
&= \sum_{|w_2|=(n-k)} Q_{w_2}(x) \int_0^{1-x} \Big(\sum_{|w_1|=(k-1)} Q_{w_1}(y) \Big) dy \\
&= G_{2(n-k)}(x) \int_0^{1-x} G_{2(k-1)}(y)dy.
\end{aligned}
$$

As $G_{2n}(x) = \sum_{k=1}^n G_{2n,k}(x)$, part (a) follows.

(b) We prove this by induction. The cases $n = 0, 1$ are clear. Now suppose that the result is true for all $j < n$. Then by (a) and the induction hypothesis we have

$$
\begin{aligned}
G_{2n}(x) &= \sum_{k=1}^n G_{2(n-k)}(x) \int_0^{1-x} G_{2(k-1)}(y)dy \\
&= \sum_{k=1}^n \frac{(1-x)(n-k+1-x)^{n-k-1}}{(n-k)!} \int_0^{1-x} \frac{(1-y)(k-y)^{k-2}}{(k-1)!}dy \\
&= \sum_{k=1}^n \frac{(1-x)(n-k+1-x)^{n-k-1}}{(n-k)!} \frac{(1-x)(k-1+x)^{k-1}}{k!} \\
&= \frac{(-1)^{n-1}}{n!} \sum_{k=1}^n \binom{n}{k} p_k(z)p_{n-k}(-z),
\end{aligned}
$$

where $z = 1 - x$ and

$$
p_n(x) = x(x - n)^{n-1}
$$

is an Abel polynomial of degree n. It is a well-known fact (see Riordan (1979)[89]) that Abel polynomials satisfy the following combinatorial identity:

$$p_n(x+y) = \sum_{k=0}^{n} \binom{n}{k} p_k(x) p_{n-k}(y).$$

It follows that

$$
\begin{aligned}
G_{2n}(x) &= \frac{(-1)^{n-1}}{n!} \sum_{k=0}^{n} \binom{n}{k} p_k(z) p_{n-k}(-z) + \frac{(-1)^n}{n!} p_0(z) p_n(-z) \\
&= \frac{(-1)^{n-1}}{n!} p_n(0) + \frac{(-1)^n}{n!} p_n(-z) \\
&= \frac{(-1)^n (-z)(-z-n)^{n-1}}{n!} \\
&= \frac{z(z+n)^{n-1}}{n!} \\
&= \frac{(1-x)(n+1-x)^{n-1}}{n!}
\end{aligned}
$$

and the proof of Lemma 8.2.4 is complete. $\square$

To conclude the proof of Theorem 8.2.1, by the previous lemmata,

$$
\begin{aligned}
\lim_{n\to\infty} \frac{1}{n} \mathrm{Tr}\Big(\frac{W_n^u}{\sqrt{n}}\Big)^{2k} &= \sum_{w\in\mathcal{C}_{2k}} p_u(w) \\
&= \int_0^1 G_{2k}(x)\,dx \qquad \text{(see equation(8.18))} \\
&= \int_0^1 \frac{(1-x)(k+1-x)^{k-1}}{k!}\,dx \quad \text{(Lemma 8.2.4(b))} \\
&= \frac{k^k}{(k+1)!}.
\end{aligned}
$$

$\square$

8.2.2   Contribution of Catalan words

It appears difficult, even for the triangular Wigner matrix, to determine $p_u(w)$ for each Catalan word w. As we have already seen, unlike what happens for a full Wigner matrix, $p_u(w)$ is not the same for all Catalan words. Here we record some observations about contributions of different Catalan words for triangular Wigner matrices.

Theorem 8.2.5. (a) For w of the form $w = aabbcc\ldots$, $p_u(w) = 1/(k+1)$.

(b) If w and w' are Catalan words of the form $w = aw_1aw_2$ and $w' = aaw_1w_2$ then $p_u(w) \le p_u(w')$.

(c) If w and w' are Catalan words of the form $w = abbaw_1w_2$ and $w' = abw_1baw_2$ are Catalan words then $p_u(w) \ge p_u(w')$.

(d) For any word w of length $2k$, $p_u(w) \leq 1/(k+1)$.   ◇

Proof. (a) Clearly, $\pi \in \Pi^*(w)$ implies

$$(\min(\pi(0),\pi(1)),\max(\pi(0),\pi(1))) = (\min(\pi(1),\pi(2)),\max(\pi(1),\pi(2)))$$

and hence $\pi(0) = \pi(2)$. Arguing similarly, it follows that,

$$\Pi^*(w) = \{\pi : \pi(0) = \pi(2) = \pi(4) = \ldots = \pi(2k)\}.$$

Hence,

$$\#\Pi_1^*(w) = \#\{\pi : \pi(0)+\pi(1) \leq n, \pi(0)+\pi(2) \leq n, \ldots, , \pi(0)+\pi(k) \leq n\}.$$

Now

$$
\begin{aligned}
p_u(w) &= \lim_{n\to\infty} \frac{\#\Pi_1^*(w)}{n^{1+k}} \\
&= \int \cdots \int_{[0,1]^{k+1}} I(v_0 + v_1 \leq 1, \ldots, v_0 + v_k \leq 1) dv_0 dv_1 ... dv_k \\
&= \int_0^1 \left[\int_0^{1-v_0} \cdots \int_0^{1-v_0} dv_1 \cdots dv_k \right] dv_0 \\
&= \int_0^1 (1-v_0)^k dv_0 = \frac{1}{k+1}.
\end{aligned}
$$

□

(b) We need the following lemma. The proof is easy by induction and we omit the details.

Lemma 8.2.6. Let w be a Catalan word of length $2k$. Let π be a function $\pi : \{0,1,...,k\} \to \{1,2,...,n\}$ such that $w[i] = w[j] \Rightarrow (\pi(i-1),\pi(i)) = (\pi(j),\pi(j-1))$. Then $\pi(0) = \pi(2k)$ and hence $\pi \in \Pi^*(w)$.   ◇

Let us denote, for any set $U \in \mathbb{R}^{k+1}$, by $Vol(U)$, the Lebesgue measure of the set $U \bigcap [0,1]^{k+1}$. From what we have already proved, it follows that for every Catalan word w_0 of length $2k$, $p_u(w_0) = Vol(U_0)$ where U_0 is the region in $\mathbb{R}^{k+1}$ given by the set of all v's determined by the set of inequalities

$$N = \{v_{\phi(i-1)} + v_{\phi(i)} \leq 1, i \in S - \{0\}\}. \tag{8.19}$$

Now let $w = aw_1 aw_2$ and let w_1 and w_2 be Catalan words of lengths $2k_1$ and $2k_2$. Now we partition the set N into three classes.

1. $N_1 = \{v_0 + v_1 \leq 1\}$: This is the first inequality since $\phi(0) = 0, \phi(1) = 1$ and $1 \in S$.

2. $A_{w_1}(v_1)$: This is the set of all inequalities $\{v_{\phi(i-1)} + v_{\phi(i)} \leq 1 : 2 \leq i \leq 2k_1 + 1, i \in S - \{0\}\}$; i.e., the inequalities corresponding to the generating vertices in w_1. By Lemma 8.2.2 it follows that these inequalities do not involve the variable v_0. Also, the inequalities only depend on w_1 and v_1.

3. $A_{w_2}(v_0)$: This is the set of all the inequalities $\{v_{\phi(i-1)} + v_{\phi(i)} \leq 1 : 2k_1 + 3 \leq i \leq 2k, i \in S - \{0\}\}$; i.e., the class of inequalities corresponding to the generating vertices in w_2. From Lemma 8.2.6, it follows that $\pi(2k + 2) = \pi(0)$ for every $\pi \in \Pi^*(w)$ and hence $\phi(2k + 2) = 0$. Now Lemma 8.2.2 implies that all the variables occurring in these inequalities are independent of the variables that occurred previously, apart from v_0. Also, the inequalities only depend on w_1 and v_0.

It follows that $p_u(w) = Vol(B_w)$ where

$$B_w = N_1 \cup A_{w_1}(v_1) \cup A_{w_2}(v_0).$$

Similarly, for the word $w' = aaw_1w_2$ the three classes of inequalities are:

1. N_1 (as earlier).

2. $A_{w_1}(v_0)$: In this case $\phi(2) = 0$. By renaming the variables other than v_0, if necessary, the class of inequalities corresponding to the generating vertices in w_1 in this case is the same as that in the previous case with v_0 replaced by v_1. By Lemma 8.2.2 it follows that these inequalities do not involve the variable v_1.

3. $A_{w_2}(v_0)$: The class of inequalities corresponding to the generating vertices in w_2 in this case is the same as that in the previous case after renaming the variables other than v_0. Lemma 8.2.2 implies that all the variables occurring in these inequalities are independent of the variables that occurred previously, apart from v_0.

It follows that $p_u(w) = Vol(B_{w'})$ where

$$B_{w'} = N_1 \cup A_{w_1}(v_0) \cup A_{w_2}(v_0).$$

Now

$$
\begin{aligned}
Vol(B_w) &= Vol(B_w^1) + Vol(B_w^2) \quad \text{and} \\
Vol(B_{w'}) &= Vol(B_{w'}^1) + Vol(B_{w'}^2)
\end{aligned}
$$

where

$$
\begin{aligned}
B_w^1 &= N_1 \cup A_{w_1}(v_1) \cup A_{w_2}(v_0) \cup \{v_0 \leq v_1\}, \\
B_w^2 &= N_1 \cup A_{w_1}(v_1) \cup A_{w_2}(v_0) \cup \{v_1 \leq v_0\}
\end{aligned}
$$

and

$$
\begin{aligned}
B_{w'}^1 &= N_1 \cup A_{w_1}(v_0) \cup A_{w_2}(v_0) \cup \{v_0 \leq v_1\}, \\
B_{w'}^2 &= N_1 \cup A_{w_1}(v_0) \cup A_{w_2}(v_0) \cup \{v_1 \leq v_0\}.
\end{aligned}
$$

It is easy to see that $B_{w'}^1 \supseteq B_w^1$ and $B_w^2 \supseteq B_{w'}^2$.

Now,

$$B^1_{w'} - B^1_w = N_1 \cup A_{w_1}(v_0) \cup A_{w_2}(v_0) \cup \{v_0 \leq v_1\} \cup (A_{w_1}(v_1))';$$
$$B^2_w - B^2_{w'} = n_1 \cup A_{w_1}(v_1) \cup A_{w_2}(v_0) \cup \{v_1 \leq v_0\} \cup (A_{w_1}(v_0)'.$$

Now by interchange of variables v_0 and v_1, it follows that $Vol(B^1_{w'} - B^1_w) = Vol(C_{w,w'})$ where

$$C_{w,w'} = N_1 \cup A_{w_1}(v_1) \cup A_{w_2}(v_1) \cup \{v_1 \leq v_0\} \cup (A_{w_1}(v_0))'.$$

Again, $v_1 \leq v_0$ and $A_{w_2}(v_0)$ together imply $A_{w_2}(v_1) \leq 1$ and hence $C_{w,w'} \supseteq B^2_w - B^2_{w'}$. It follows that,

$$
\begin{aligned}
Vol(C_{w,w'}) &\geq Vol(B^2_w - B^2_{w'}) \\
\Rightarrow Vol(B^1_{w'} - B^1_w) &\geq Vol(B^2_w - B^2_{w'}) \\
\Rightarrow Vol(B^1_{w'}) + Vol(B^1_{w'}) &\geq Vol(B^1_w) + Vol B^2_w)
\end{aligned}
$$

and this completes the proof of (b).

(c) As in part (b), now $p_u(w) = Vol(B_w)$ and $p_u(w') = Vol(B_{w'})$ where

$$B_w = N_1 \cup \{v_1 + v_2 \leq 1\} \cup A_{w_1}(v_0) \cup A_{w_2}(v_0) \text{ and}$$
$$B_{w'} = N_1 \cup \{v_1 + v_2\} \cup A_{w_1}(v_2) \cup A_{w_2}(v_0).$$

Now

$$
\begin{aligned}
Vol(B_w) &= Vol(B^1_w) + Vol(B^2_w) \text{ and} \\
Vol(B_{w'}) &= Vol(B^1_{w'}) + Vol(B^2_{w'})
\end{aligned}
$$

where

$$B^1_w = N_1 \cup \{v_1 + v_2 \leq 1\} \cup A_{w_1}(v_0) \cup A_{w_2}(v_0) \cup \{v_0 \leq v_2\},$$
$$B^2_w = N_1 \cup \{v_1 + v_2 \leq 1\} \cup A_{w_1}(v_0) \cup A_{w_2}(v_0) \cup \{v_2 \leq v_0\},$$

and

$$B^1_{w'} = N_1 \cup \{v_1 + v_2 \leq 1\} \cup A_{w_1}(v_2) \cup A_{w_2}(v_0) \cup \{v_0 \leq v_2\},$$
$$B^2_{w'} = N_1 \cup \{v_1 + v_2 \leq 1\} \cup A_{w_1}(v_2) \cup A_{w_2}(v_0) \cup \{v_2 \leq v_0\}.$$

As $v_0 \leq v_2$ and $A_{w_1}(v_2)$ together imply $A_{w_1}(v_0)$, $B^1_w \supseteq B^1_{w'}$. Similarly it follows that $B^2_{w'} \supseteq B^2_w$.

Now,

$$B^1_w - B^1_{w'} = N_1 \cup \{v_1 + v_2 \leq 1\} \cup A_{w_1}(v_0) \cup A_{w_2}(v_0)\{v_0 \leq v_2\} \cup (A_{w_1}(v_2))';$$
$$B^2_{w'} - B^2_w = N_1 \cup \{v_1 + v_2 \leq 1\} \cup A_{w_1}(v_2) \cup A_{w_2}(v_0)\{v_2 \leq v_0\} \cup (A_{w_1}(v_0))'.$$

Once again by interchange of two variables v_0 and v_2 we see that $Vol(B^2_{w'} - B^2_w) = Vol(C_w, w')$ where

$$C_{w,w'} = N_1 \cup \{v_1 + v_2 \leq 1\} \cup A_{w_1}(v_0) \cup A_{w_2}(v_2)\{v_0 \leq v_2\} \cup (A_{w_1}(v_2)'.$$

Again, $v_0 \leq v_2$ and $A_{w_2}(v_2)$ together imply $A_{w_2}(v_0)$ and hence $Cw, w' \subseteq B_w^1 - B_{w'}^1$. It follows that,

$$
\begin{aligned}
Vol(Cw, w') &\leq & Vol(B_w^1 - B_{w'}^1) \\
\Rightarrow Vol(B_{w'}^1 - B_w^1) &\geq & Vol(B_w^2 - B_{w'}^2)
\end{aligned}
$$

and the proof of (c) is complete.

(d) If $w = a_1 a_1 a_2 a_2 ... a_k a_k$, then the result follows from part (a). If not, by left rotation, from w we can obtain a word w' such that w' does not start with a double letter and w' is not of the form $aw_1 a$ either. Since $p_u(w)$ is invariant under rotation, $p_u(w) = p_u(w')$. By hypothesis, w' must be of the form $w' = aw_1 aw_2$ where w_1 and w_2 are Catalan. By part (b), $p_u(w) = p_u(w') \leq p_u(aaw_1w_2) = p_u(w_1w_2aa)$ and the number of consecutive double letters at the end of w_1w_2aa is strictly greater than that of w. The proof can now be completed by induction on the number of consecutive double letters at the end of w.   □

8.3   Exercises

1. Prove Theorem 8.1.1.

2. By using the moment formula, show that the support of the LSD of the triangular Wigner is contained in $[-\sqrt{e}, \sqrt{e}]$.

3. Show that the density of the LSD of the triangular Wigner is unbounded at zero.

4. In the spirit of Chapter 5, consider triangular Wigner type matrices and show that they have the same LSD as the triangular Wigner matrix.

5. Consider the lower triangular Wigner matrix

$$
W_n^l = \begin{bmatrix}
0 & 0 & 0 & \cdots & 0 & x_{1n} \\
0 & 0 & 0 & \cdots & x_{2(n-1)} & x_{2n} \\
 & & & \vdots & & \\
0 & x_{2(n-1)} & x_{3(n-1)} & \cdots & x_{(n-1)(n-1)} & x_{(n-1)n} \\
x_{1n} & x_{2n} & x_{3n} & \cdots & x_{(n-1)n} & x_{nn}
\end{bmatrix}.
$$

Show that the LSD of this matrix coincides with the LSD of the triangular Wigner matrix.

9

Joint convergence of i.i.d. patterned matrices

So far we have studied the convergence of a single sequence of random matrices via the convergence of their ESDs. A natural question is how can we give meaning to joint convergence of several sequences of random matrices? Since matrices are in general non-commutative objects, this needs to be done with care. We need to first introduce the notion of a *non-commutative probability space* which consists of an algebra $\mathcal{A}$ and a state ϕ on it (which is a linear functional with certain properties). Square matrices of a given order in particular form a natural non-commutative probability space with the state being the average expected trace or simply the average trace. The convergence of a collection of variables from a sequence of non-commutative probability spaces is defined in terms of the convergence of the state of all monomials formed out of this collection.

In this chapter we study the joint convergence of p independent symmetric matrices with an identical pattern. The case when the different sequences are governed by possibly different patterns will be considered in Chapter 10. We extend the concept of words to *indexed words* and continue to use the trace-moment approach. Under a mild condition, if the marginal limit exists then we show that the joint tracial limit exists and can be expressed in terms of the marginals with the help of words and indexed words.

By this approach, we are also able to show that independent copies of any one of the Wigner, Symmetric Circulant, Reverse Circulant Toeplitz or Hankel matrices converge in average expected trace and also average trace almost surely.

The first three limits have special structures which we shall introduce later. The limits are, respectively, free semi-circle, classical independent normal and half-independent symmetrized Rayleigh. The Toeplitz and Hankel joint limits are not free, independent or half-independent. No characterization of the dependence in the joint limit is known for these matrices.

9.1 Non-commutative probability space

A *non-commutative probability space* is a pair $(\mathcal{A}, \phi)$ where $\mathcal{A}$ is a unital complex algebra (with unity 1) and $\phi : \mathcal{A} \to \mathbb{C}$ is a linear functional which satisfies

$\phi(1) = 1$. Such a ϕ is called a *state*. The state ϕ is called *tracial* if $\phi(ab) = \phi(ba)$ for all a, b. Elements of $\mathcal{A}$ will be called *variables*.

Example 9.1.1. Let $(X, \mathcal{B}, \mu)$ be a probability space. Let $L(\mu) = \bigcap_{1 \le p < \infty} L^p(\mu)$ be the algebra of random variables with finite moments of all orders. Then $(L(\mu), \phi)$ becomes a (commutative) probability space where the state ϕ is the expectation functional, that is, integration with respect to μ and it is clearly a tracial state.   ◇

Example 9.1.2. Let $(X, \mathcal{B}, \mu)$ be a probability space. Let $\mathcal{A}_n = Mat_n(L(\mu))$ be the space of $n \times n$ complex random matrices with elements from $L(\mu)$. Then ϕ_n equal to $\frac{1}{n} \mathrm{E}_\mu[\mathrm{Tr}(\cdot)]$ or $\frac{1}{n} \mathrm{Tr}(\cdot)$ both yield non-commutative probabilities. These ϕ_n are both tracial.   ◇

In the previous chapters we have tied distributional convergence of random variables to convergence of moments. The place of expectation is now taken by the state ϕ and we define convergence of non-commutative variables as follows.

For any (non-commuting) variables $x_1, \ldots, x_n$, let $\mathbb{C}\langle x_1, x_2, ..., x_n \rangle$ be the unital algebra of all complex polynomials in these variables. If $a_1, a_2, \ldots, a_n \in \mathcal{A}$ then their *joint distribution* $\mu_{\{a_i\}}$ is defined canonically by their *mixed moments*

$$\mu_{\{a_i\}}(x_{i_1} \cdots x_{i_m}) = \phi(a_{i_1} \cdots a_{i_m}).$$

That is,

$$\mu_{\{a_i\}}(P) = \phi(P(\{a_i\})) \quad \text{for} \quad P \in \mathbb{C}\langle x_1, x_2, ..., x_n \rangle.$$

Definition 9.1.1. Let $(\mathcal{A}_n, \phi_n)$, $n \ge 1$ and $(\mathcal{A}, \phi)$ be non-commutative probability spaces and let $\{a_i^n\}_{i \in J}$ be a sequence of subsets of $\mathcal{A}_n$ where J is any finite subset of $\mathbb{N}$. Then we say that $\{a_i^n\}_{i \in J}$ *converges in law* to $\{a_i\}_{i \in J} \subset \mathcal{A}$ if for all complex polynomials P,

$$\lim_{n \to \infty} \mu_{\{a_i^n\}_{i \in J}}(P) = \mu_{\{a_i\}_{i \in J}}(P).$$

◇

Due to the linearity of states, to verify convergence in the above definition, it is enough to verify the convergence for all monomials $q = x_{i_1} \cdots x_{i_k}$, $k \ge 1$ where each x_i is any of the $\{a_i^n\}, i \in J$.

We note here in passing that often the algebra is a $*$-algebra and in that case it is natural for moments to include $*$ variables. In a similar way, it is then natural for convergence in law to also include these variables. For simplicity, we shall stick to the weaker non-starred version of the notions and that shall suffice for the purposes of this book.

9.2 Joint convergence

Let $(\mathcal{A}_n = Mat_n(L(\mu)), \phi_n)$ be the non-commutative probability space of $n \times n$ real symmetric matrices with the state $\phi_n = n^{-1} \mathrm{E}[\mathrm{Tr}]$. Let $(\Omega, \mathcal{B}, \mu)$ be a probability space and let $X_{i,n} : \Omega \to \mathcal{A}_n$ for $1 \le i \le p$ be symmetric patterned random matrices of order n with a common link function L. We shall refer to $1 \le i \le p$ as p distinct *indices*. The (j, k)-th entry of the matrix $X_{i,n}$ will be denoted by $X_{i,n}(L(j, k))$. We now extend Assumption III of Chapter I in a natural way. Recall Property B from Definition 1.4.1 in Chapter 1.

Assumption III For the collection of the $n \times n$ matrices $\{X_{i,n}\}_{1 \le i \le p}$, the input sequence of each matrix is independent with mean zero and variance 1. These input sequences are also independent across i. The matrices have a common link function L which satisfies Property B and

$$\sup_{n \in \mathbb{N}} \sup_{1 \le i \le p} \sup_{1 \le m \le l \le n} \mathrm{E}[|X_{i,n}(L(m, l))|^k] \le c_k < \infty. \text{ for all } k \ge 1.$$

Suppressing the dependence on n, we shall simply write X_i for $X_{i,n}$. We view $\{\frac{1}{\sqrt{n}} X_i\}_{1 \le i \le p}$ as elements of $(\mathcal{A}_n = Mat_n(L(\mu)), \phi_n)$ where $\phi_n = n^{-1} \mathrm{E}[\mathrm{Tr}]$. Denote the joint distribution of $\{\frac{1}{\sqrt{n}} X_i\}_{1 \le i \le p}$ by $\widehat{\mu_n}$. Then $\{\frac{1}{\sqrt{n}} X_i\}_{1 \le i \le p}$ converges in law if

$$\widehat{\mu_n}(q) = \phi_n(q) \tag{9.1}$$

$$= \frac{1}{n^{1+k/2}} \mathrm{E}[\mathrm{Tr}(X_{t_1} \cdots X_{t_k})] \tag{9.2}$$

$$= \frac{1}{n^{1+k/2}} \sum_{j_1, \ldots, j_k} \mathrm{E}[X_{t_1}(L(j_1, j_2)) X_{t_2}(L(j_2, j_3)) \cdots X_{t_k}(L(j_k, j_1))] \tag{9.3}$$

converges for all monomials q of the form $q(\{X_i\}_{1 \le i \le p}) = X_{t_1} \cdots, X_{t_k}$. Here $\{t_1, t_2 \ldots t_k\}$ is any subset of $\{1, 2 \ldots, p\}$.

To upgrade to almost sure convergence, we also define

$$\widetilde{\mu_n}(q) = \frac{1}{n^{1+k/2}} \mathrm{Tr}[X_{t_1} \cdots X_{t_k}]$$

$$= \frac{1}{n^{1+k/2}} \sum_{j_1, \ldots, j_k} X_{t_1}(L(j_1, j_2)) X_{t_2}(L(j_2, j_3)) \cdots X_{t_k}(L(j_k, j_1)).$$

Note that $\tilde{\phi}_n = \frac{1}{n} \mathrm{Tr}$ is also a tracial state on $\mathcal{A}_n$. Convergence in law of $\{\frac{1}{\sqrt{n}} X_i\}_{1 \le i \le p}$ as elements of $(\mathcal{A}_n, \tilde{\phi}_n = \frac{1}{n} \mathrm{Tr})$ is the same as the almost sure convergence of $\widetilde{\mu_n}(q)$ for every q. *All developments below are with respect to one fixed monomial q at a time.*

Recall the notions of circuit, L-values, matching and pair-matching from

Chapter 1. Leaving aside the scaling factors $\sqrt{n}$, a typical element in (9.3) can be now written as

$$\mathrm{E}\Big[\prod_{j=1}^{k} X_{t_j}(L(\pi(j-1),\pi(j)))\Big]. \tag{9.4}$$

If for a circuit π, every L-value is repeated within the same index, then it is *index-matched*. If a circuit is not index-matched then the expectation in (9.4) is zero. Let

$$H = \{\pi : \pi \text{ is an index-matched circuit}\}.$$

Define an equivalence relation on H by defining $\pi_1 \sim_I \pi_2$ if and only if,

$$t_i = t_j \text{ and } X_{t_i}(L(\pi_1(i-1),\pi_1(i))) = X_{t_j}(L(\pi_1(j-1),\pi_1(j)))$$

$$\Longleftrightarrow$$

$$X_{t_i}(L(\pi_2(i-1),\pi_2(i))) = X_{t_j}(L(\pi_2(j-1),\pi_2(j))).$$

Any equivalence class induces a partition of $\{1,2,\ldots,k\}$ and each block of the partition is associated with an index. Any such class can be expressed as an *indexed word* w where letters appear in alphabetic order of their first occurrence and with a subscript to distinguish the index. For example the partition $(\{\{1,3\},1\},\{\{2,4\},2\},\{\{5,7\},1\},\{\{6,8\},3\})$ is identified with the word $a_1b_2a_1b_2c_1d_3c_1d_3$. A typical position in an indexed word would be referred to as $w_{t_i}[i]$.

The class of indexed circuits corresponding to an indexed matched word w and the class of indexed pair-matched words are denoted, respectively, by

$$\Pi_{I_q}(w) = \{\pi : w_{t_i}[i] = w_{t_j}[j] \text{ iff}$$
$$X_{t_i}(L(\pi(i-1),\pi(i))) = X_{t_j}(L(\pi(j-1),\pi(j)))\}$$
$$\mathcal{IW}_k(2) = \{\text{all indexed paired-matched words } w \text{ of length } k\}(\text{even}). \tag{9.5}$$

Note that we already have the corresponding *non-indexed* versions of the above notions from Chapter 1. If we drop all the indices from an indexed word, then we obtain a **non-indexed** word. For any monomial q, dropping the index amounts to dealing with only one matrix, that is, with the marginal distribution.

Recall that $w[i]$ denotes the i-th entry of a non-indexed word w. Also recall

$$\Pi(w) = \{\pi : w[i] = w[j] \Leftrightarrow L(\pi(i-1),\pi(i)) = L(\pi(j-1),\pi(j))\} \tag{9.6}$$
$$\mathcal{W}_k(2) = \{\text{all paired-matched words } w \text{ of length } k\} \ (k \text{ is even}). \tag{9.7}$$

For any word $w \in \mathcal{IW}_k(2)$, consider the non-indexed word w' obtained by dropping the index. Then $w' \in \mathcal{W}_k(2)$. Since we are dealing with one fixed monomial at a time, this yields a bijective mapping, say

$$\psi_q : \mathcal{IW}_k(2) \to \mathcal{W}_k(2). \tag{9.8}$$

For any $w \in \mathcal{IW}_k(2)$, define $p_{I_q}(w)$ (in short $p(w)$) by

$$p_{I_q}(w) = p(w) = \lim_{n \to \infty} \frac{1}{n^{k/2+1}} \# \ \Pi_{I_q}(w) \quad \text{if the limit exists.}$$

The following result is due to Bose, Hazra and Saha (2010)[26].

Theorem 9.2.1. Let $\{X_i\}_{1 \le i \le p}$ be patterned matrices which satisfy Assumption III. Fix any monomial $q = X_{t_1} X_{t_2} \cdots X_{t_k}$. Assume that, whenever k is even,

$$p(w) = \lim_{n \to \infty} \frac{1}{n^{k/2+1}} \# \ \Pi(w) \text{ exists for all } w \in \mathcal{W}_k(2). \tag{9.9}$$

(a) Then $p_{I_q}(w) = p(\psi_q(w))$ and for any k, the limit below exists.

$$\lim_{n \to \infty} \widehat{\mu_n}(q) = \sum_{w \in \mathcal{IW}_k(2)} p_{I_q}(w) = \alpha(x_{t_1} \cdots x_{t_k}) \quad \text{(say)}$$

with

$$|\alpha(x_{t_1} \cdots x_{t_k})| \begin{cases} \le & \frac{k! \Delta(L)^{k/2}}{(k/2)! 2^{k/2}} \text{ if } k \text{ is even and each index appears even times} \\ = & 0 \qquad\qquad \text{if } k \text{ is odd or some index appears odd times.} \end{cases}$$

(b) $\mathrm{E}\left[|\widehat{\mu_n}(q) - \widehat{\mu_n}(q)|^4\right] = \mathrm{O}(n^{-2})$ and hence $\lim_{n \to \infty} \widehat{\mu_n}(q) = \alpha(x_{t_1} \cdots x_{t_k})$ almost surely. ◇

Remark 9.2.1. As we have seen in Chapter 1, (9.9) is true for Wigner, Toeplitz, Hankel, Reverse Circulant and Symmetric Circulant matrices. The quantity $\frac{k!}{(k/2)! 2^{k/2}}$ above is the total number of non-indexed pair-matched words of length k (k even). If all pair-matched words do not contribute to the limit, then this bound can be improved.

Consider the polynomial algebra $\mathbb{C}\langle a_1, a_2, \ldots, a_p \rangle$ in non-commutative indeterminates $\{a_i\}_{1 \le i \le p}$ and define a linear functional ϕ on it by

$$\phi(a_{t_1} \cdots a_{t_k}) = \lim_{n \to \infty} \widehat{\mu_n}(X_{t_1} \cdots X_{t_k}).$$

Then Theorem 9.2.1 implies we have convergence in law of $(\{\frac{1}{\sqrt{n}} X_i\}_{1 \le i \le p}, \ \phi_n)$ to $(\{a_i\}_{1 \le i \le p}, \ \phi)$ in the sense of Definition 9.1.1 where $\phi_n = \frac{1}{n} \mathrm{E}[\mathrm{Tr}(\cdot)]$ or $\frac{1}{n} \mathrm{Tr}(\cdot)$. In the latter case, the convergence is almost sure. ◇

Remark 9.2.2. From Theorem 9.2.1, if $q = X_{t_1} \cdots X_{t_k}$ is a symmetric matrix then the EESD of $n^{-1/2} q$ converges under Assumption III. ◇

Proof of Theorem 9.2.1. (a) Fix a monomial $q = q(\{X_i\}_{1 \le i \le p}) = X_{t_1} \cdots X_{t_k}$. Since ψ_q is a bijection,

$$\Pi_{I_q}(w) = \Pi(\psi_q(w)) \quad \text{for} \ w \in \mathcal{IW}_k(2).$$

Hence using (9.9),

$$\lim_{n\to\infty} \frac{1}{n^{k/2+1}} \#\Pi_{I_q}(w) = \lim_{n\to\infty} \frac{1}{n^{k/2+1}} \#\Pi(\psi_q(w)) = p(\psi_q(w)) = p_{I_q}(w).$$

For simplicity denote

$$\mathbb{T}_{\boldsymbol{j}} = \mathrm{E}[X_{t_1}(L(j_1,j_2))X_{t_2}(L(j_2,j_3))\cdots X_{t_k}(L(j_k,j_1))] \text{ for } \boldsymbol{j} = (j_1,\ldots,j_k).$$

Then

$$\widehat{\mu_n}(q) = \frac{1}{n^{k/2+1}} \sum_{j_1,\ldots,j_k} \mathbb{T}_{\boldsymbol{j}}. \tag{9.10}$$

In the monomial, if any index appears once, then by independence and mean zero condition, $\mathbb{T}_{\boldsymbol{j}} = 0$ for every $\boldsymbol{j}$. Hence $\widehat{\mu_n}(q) = 0$.

So henceforth assume that each index appearing in the monomial, appears at least twice. Now again, if $\boldsymbol{j}$ belongs to a circuit which is not index-matched, then $\mathbb{T}_{\boldsymbol{j}} = 0$.

Now form the following matrix M whose (i,j)th element is given by:

$$M(L(i,j)) = |X_{t_1}(L(i,j))| + |X_{t_2}(L(i,j))| + \cdots + |X_{t_k}(L(i,j))|.$$

Observe that,

$$|\mathbb{T}_{\boldsymbol{j}}| \leq \mathrm{E}[M(L(j_1,j_2))\cdots M(L(j_k,j_1))].$$

From Lemma 1.4.2 of Chapter 1 it is known that the total contribution of all circuits which have at least one three or more match, is zero in the limit.

As a consequence of the above discussion, if k is odd, then $\widehat{\mu_n}(q) \to 0$.

So assume k is even. In that case, we need to consider only circuits which are pair-matched. Further, this pair-matching must occur within the same index. If $\boldsymbol{j}$ belongs to any such circuit, then by independence, mean zero and variance one condition, $\mathbb{T}_{\boldsymbol{j}} = 1$.

Then using all the facts established so far,

$$\lim_{n\to\infty} \widehat{\mu_n}(q) = \lim_{n\to\infty} \frac{1}{n^{k/2+1}} \sum_{\substack{\pi \text{ pair}-\text{matched} \\ \text{within indices}}} \mathrm{E}[X_{t_1}(L(\pi(0),\pi(1)))\cdots$$
$$\times X_{t_k}(L(\pi(k-1),\pi(k)))]$$
$$= \lim_{n\to\infty} \frac{1}{n^{k/2+1}} \sum_{w\in\mathcal{IW}_k(2),\ \pi\in\Pi_{I_q}(w)} \mathrm{E}[X_{t_1}(L(\pi(0),\pi(1)))\cdots$$
$$\times X_{t_k}(L(\pi(k-1),\pi(k)))]$$
$$= \sum_{w\in\mathcal{IW}_k(2)} p_{I_q}(w).$$

The last claim in part (a) follows since

$$\sum_{w\in\mathcal{IW}_{2k}(2)} p_{I_q}(w) = \sum_{w\in\mathcal{IW}_{2k}(2)} p(\psi_q(w)) \leq \sum_{w\in\mathcal{W}_{2k}(2)} p(w) \leq \frac{(2k)!\Delta(L)^k}{k!2^k}.$$

The last inequality above was established in Theorem 1.4.4 of Chapter 1.

(b) We can write

$$E\left[|\widetilde{\mu_n}(q) - \widehat{\mu_n}(q)|^4\right] = \frac{1}{n^{2k+4}} \sum_{\pi_1,\pi_2,\pi_3,\pi_4} E\left[\prod_{l=1}^{4}\left(\mathbb{X}_{\pi_l} - E\,\mathbb{X}_{\pi_l}\right)\right]. \qquad (9.11)$$

where

$$\mathbb{X}_\pi = X_{t_1}(L(\pi(0),\pi(1)))\cdots X_{t_k}(L(\pi(k-1),\pi(k))).$$

If $(\pi_1, \pi_2, \pi_3, \pi_4)$ are not jointly-matched, then one of the circuits, say π_j, has an L-value which does not occur anywhere else. Also note that $E\,\mathbb{X}_{\pi_j} = 0$. Hence, using independence,

$$E\left[\prod_{l=1}^{4}\left(\mathbb{X}_{\pi_l} - E\,\mathbb{X}_{\pi_l}\right)\right] = E\left[\mathbb{X}_{\pi_j}\prod_{l=1,l\neq j}^{4}\left(\mathbb{X}_{\pi_l} - E\,\mathbb{X}_{\pi_l}\right)\right] = 0. \qquad (9.12)$$

If $(\pi_1, \pi_2, \pi_3, \pi_4)$ are jointly-matched but are not cross-matched, then one of the circuits, say π_j, is only self-matched, that is, none of the L-values is shared by the other circuits. Then by independence,

$$E\left[\prod_{l=1}^{4}\left(\mathbb{X}_{\pi_l} - E\,\mathbb{X}_{\pi_l}\right)\right] = E\left[\left(\mathbb{X}_{\pi_j} - E\,\mathbb{X}_{\pi_j}\right)\prod_{l=1,l\neq j}^{4}\left(\mathbb{X}_{\pi_l} - E\,\mathbb{X}_{\pi_l}\right)\right] = 0. \quad (9.13)$$

Since $\{X_{i,n}\}_{1\leq i\leq n}$ satisfy Assumption III, $E\left[\prod_{l=1}^{4}\left(\mathbb{X}_{\pi_l} - E\,\mathbb{X}_{\pi_l}\right)\right]$ is uniformly bounded over all $(\pi_1, \pi_2, \pi_3, \pi_4)$.

Using Lemma 1.4.3 of Chapter 1, and (9.11)–(9.13),

$$E\left[|\widetilde{\mu_n}(q) - \widehat{\mu_n}(q)|^4\right] \leq K\frac{n^{2k+2}}{n^{2k+4}} = O(n^{-2}).$$

Now by an easy application of the Borel-Cantelli lemma, $\widetilde{\mu_n}(q)$ converges almost surely. □

9.3 Nature of the limit

From the above result, $\lim \widetilde{\mu_n}(q) = 0$ when k is odd or when there is an index which appears an odd number of times in the monomial q. Henceforth we thus assume that the order of the monomial is even and each index appears an even number of times.

(A) Symmetric Circulants

Definition 9.3.1. Suppose $\{\mathcal{A}_i\}_{i\in J} \subset \mathcal{A}$ are unital sub-algebras. Then they are said to be *independent* if they commute and $\phi(a_1 \cdots a_n) = \phi(a_1)\cdots\phi(a_n)$ for all $a_i \in \mathcal{A}_{k(i)}$ whenever $i \neq j \implies k(i) \neq k(j)$. ◇

It can be easily checked that Symmetric Circulant matrices with different input sequences are commutative and so the limit is also commutative. We have the following theorem.

Theorem 9.3.1. Let $\{SC_i\}_{1\leq i\leq p}$ be p independent sequences of $n \times n$ Symmetric Circulant matrices which satisfy Assumption III. Then $\{n^{-1/2}SC_i\}_{1\leq i\leq p}$ converges in law to $\{a_i\}_{1\leq i\leq p}$ which are independent and standard Gaussian. ◇

This really means that the variables $\{a_i\}$ are independent in accordance with Definition 9.3.1.

Proof. First recall that the total number of non-indexed pair-matched words of length $2k$ equals $\frac{(2k)!}{k!2^k} = C_k$, (say). Further, for *any* pair-matched word $w \in \mathcal{W}_{2k}(2)$,

$$p(w) = \lim_n \frac{1}{n^{1+k}}\#\Pi(w) = 1.$$

Now consider an order $2k$ monomial where each index appears an even number of times. Hence, from Theorem 9.2.1, for any fixed monomial q,

$$\lim_{n\to\infty} \widehat{\mu_n}(q) = \sum_{w\in \mathcal{IW}_{2k}(2)} p_{I_q}(w) = \#\mathcal{IW}_{2k}(2).$$

Let l be the total number of distinct indices (distinct matrices) in the monomial $q = x_{t_1}x_{t_2}\cdots x_{t_{2k}}$. Let $2n_i$ be the number of matrices of the ith index. Then the set of all indexed pair-matched words of length $2k$ is obtained by forming pair-matched sub-words of index i of lengths $2n_i$, $1 \leq i \leq l$. Hence

$$\phi(a_{i_1}\cdots a_{i_{2k}}) = \prod_{i=1}^{l} C_{n_i}. \tag{9.14}$$

Now if $\{a_1,\ldots,a_p\}$ denotes i.i.d. standard normal random variables then the above is the mixed moment $\mathrm{E}[\prod_{i=1}^{l} a_i^{2n_i}]$. This proves the theorem. □

(B) Reverse Circulants

Definition 9.3.2. Let $\{a_i\}_{i\in J} \subset \mathcal{A}$. We say that they *half-commute* if $a_i a_j a_k = a_k a_j a_i$, for all $i, j, k \in J$. Observe that if $\{a_i\}_{i\in J}$ half-commute then a_i^2 commutes with a_j and a_j^2 for all $i, j \in J$. ◇

It is easily checked that if A, B, C are arbitrary Reverse Circulant matrices of the same order, then $ABC = CBA$.

Definition 9.3.3. Suppose $\{a_i\}_{i \in J} \subset \mathcal{A}$. For any $k \geq 1$, and any $\{i_j\} \subset J$, let $a = a_{i_1} a_{i_2} \cdots a_{i_k}$ be an element of $\mathcal{A}$. For any $i \in J$, let $E_i(a)$ and $O_i(a)$ be respectively, the number of times a_i has occurred in the even positions and in the odd positions in a. The monomial a is said to be *symmetric* (with respect to $\{a_i\}_{i \in J}$) if $E_i(a) = O_i(a)$ for all $i \in J$, else it is called *non-symmetric*. $\diamond$

Definition 9.3.4. Let $\{a_i\}_{i \in J}$ in $(\mathcal{A}, \phi)$ be half-commuting. They are said to be *half-independent* if (i) $\{a_i^2\}_{i \in J}$ are independent and (ii) whenever a is non-symmetric with respect to $\{a_i\}_{i \in J}$, we have $\phi(a) = 0$. $\diamond$

Remark 9.3.1. Half-independence arises in classification results for easy quantum groups and some quantum analogue of de Finetti's theorem. The above definition is equivalent to that given in Banica, Curran, and Speicher (2012)[10], although there is no notion of symmetric monomials there. As pointed out in Speicher (1997)[97], the concept of half-independence does not extend to sub-algebras. $\diamond$

Example 9.3.1. This is Example 2.4 of Banica, Curran, and Speicher (2012)[10]. Let $(\Omega, \mathcal{B}, \mu)$ be a probability space and let $\{\eta_i\}$ be a family of independent complex Gaussian random variables. Define $a_i \in (M_2(L(\mu)), \mathrm{E}[\mathrm{Tr}(\cdot)])$ by

$$a_i = \begin{bmatrix} 0 & \eta_i \\ \bar{\eta}_i & 0 \end{bmatrix}.$$

Then $\{a_i\}$ are half-independent. $\diamond$

Theorem 9.3.2. Let $\{RC_i\}_{1 \leq i \leq p}$ be an independent sequence of symmetric $n \times n$ Reverse Circulant matrices which satisfy Assumption III. Then $\{n^{-1/2} RC_i\}_{1 \leq i \leq p}$ converges in law to half-independent $\{a_i\}_{1 \leq i \leq p} \in (M_2(L(\mu)), \mathrm{E}[\mathrm{Tr}(\cdot)])$ where $a_i = \begin{bmatrix} 0 & \eta_i \\ \bar{\eta}_i & 0 \end{bmatrix}$ and η_i are i.i.d. complex standard Gaussian. Each a_i has the symmetrized Rayleigh distribution. $\diamond$

To prove the result, we need the following notion:

Definition 9.3.5. Fix $k \geq 2$. A word $w \in \mathcal{IW}_k(2)$ is called *indexed symmetric* if each letter occurs once each in an odd and an even position *within the same index*. $\diamond$

Clearly, every indexed Catalan word is an indexed symmetric word but the converse is not true.

Proof of Theorem 9.3.2. Consider a monomial q of length $2k$ where each index appears an even number of times. From the single matrix case, it follows that $p(w) = 0$ if w is not a symmetric word. If w is not an indexed symmetric word then $\psi_q(w)$ is not a symmetric word and hence for such w, $p_{I_q}(w) = p(\psi_q(w)) = 0$. Hence we may restrict to indexed symmetric words and then we have by Theorem 9.2.1 (a) that

$$\lim_{n \to \infty} \widehat{\mu_n}(q) = \#(IS_q(w)),$$

where $IS_q(w)$ is the collection of all indexed symmetric words of length $2k$.

The number of symmetric words of length $2k$ is $k!$. Let l be the number of distinct indices in the monomial and $2n_i$ be the number of matrices of the ith index in the monomial q. All symmetric words are obtained by arranging the $2n_i$ letters of the ith index in a symmetric way for $i = 1, 2, \ldots l$.

It is then easy to see that these arguments imply that

$$\#(IS_q(w)) = n_1! \times n_2! \times \cdots \times n_l!. \tag{9.15}$$

Observe that if the monomial $a_{i_1} a_{i_2} \cdots a_{i_k} \in (M_2(L(\mu)), \mathrm{E}[\mathrm{Tr}(\cdot)])$ is non-symmetric, then

$$E(\mathrm{Tr}[a_{i_1} a_{i_2} \cdots a_{i_k}]) = 0.$$

If instead $q(\{a_i\}) = a_{i_1} a_{i_2} \cdots a_{i_{2k}}$ is symmetric then using half-independence,

$$E(\mathrm{Tr}[a_{i_1} \cdots a_{i_{2k}}]) = n_1! \times n_2! \times \cdots \times n_l! = \lim_{n \to \infty} \widehat{\mu_n}(q).$$

Hence from (9.15) it follows that the joint limit is asymptotically half-independent. We already know that $\{k!, \ k \geq 1\}$ are the $(2k)$th moments of the symmetrized Rayleigh distribution. This completes the proof. $\square$

(C) **Wigner matrices**

Joint convergence of the Wigner matrices was first studied in Voiculescu (1991)[102]. For details and further extensions, we refer the readers to Anderson, Guionnet and Zeitouni (2010)[1]. One of the fundamental results in joint convergence in RMT is that, generally speaking, Wigner matrices converge jointly and the limits are free. Here we give a quick proof of the asymptotic freeness, essentially translating the concept of non-crossing partitions that is used in the standard proof, into the language of words.

Definition 9.3.6. Fix a monomial of length $k \geq 2$. If for a $w \in \mathcal{IW}_k(2)$, sequentially deleting all double letters of the same index leads to the empty word then we call w an *indexed Catalan word*. ◇

For example, the monomial $X_1 X_2 X_2 X_1 X_1 X_1$ has exactly two indexed Catalan words $a_1 b_2 b_2 a_1 c_1 c_1$ and $a_1 b_2 b_2 c_1 c_1 a_1$. An indexed Catalan word associated with $X_1 X_2 X_2 X_1 X_1 X_1 X_2 X_2$ is $a_1 b_2 b_2 a_1 c_1 c_1 d_2 d_2$ which is not even a valid indexed word for the monomial $X_1 X_1 X_1 X_1 X_2 X_2 X_2 X_2$.

Definition 9.3.7. Suppose $\{\mathcal{A}_i\}_{i \in J} \subset \mathcal{A}$ are unital sub-algebras. These sub-algebras are called *freely independent* or simply *free* if $\phi(a_j) = 0$, $a_j \in \mathcal{A}_{i_j}$ and $i_j \neq i_{j+1}$ for all j implies $\phi(a_1 \cdots a_n) = 0$. The random variables (or elements of an algebra) $(a_1, a_2, \ldots, a_n)$ will be called free if the sub-algebras generated by them are free. ◇

Theorem 9.3.3. Let $\{W_i\}_{1 \leq i \leq p}$ be an independent sequence of $n \times n$ Wigner matrices which satisfy Assumption III. Then $\{n^{-1/2} W_i\}_{1 \leq i \leq p}$ converges in law to $\{a_i\}_{1 \leq i \leq p}$ which are free and semi-circular. ◇

Proof. The convergence follows from Theorem 9.2.1 and Remark 9.2.1. From the proof of Theorem 2.1.3 in Chapter 2, we already know that for non-indexed words, $p(w) = 1$ if w is a Catalan word and 0 otherwise. As a consequence, the marginals are semi-circular. It remains to show that $\{a_i\}_{1 \le i \le p}$ are free.

Now fix any monomial $q = x_{t_1} x_{t_2} \cdots x_{t_{2k}}$ where each index appears an even number of times in the monomial. Let w be an indexed Catalan word. It remains Catalan when we ignore the indices. Hence from above, $p_{I_q}(w) = p(\psi_q(w)) = 1$. Likewise, if w is not indexed Catalan then the word $\psi_q(w)$ cannot be Catalan and hence $p_{I_q}(w) = p(\psi_q(w)) = 0$. Hence if $ICAT_q$ denotes the set of indexed Catalan words corresponding to a monomial q then from the above discussion,

$$\lim_{n \to \infty} \widehat{\mu_n}(q) = \#(ICAT_q).$$

Any double letter corresponds to a pair-partition (within the same index) by the equivalence relation $\sim_I$. From Lemma 2.1.2 of Chapter 2, it is known that the number of Catalan words of length $2k$ is the same as the number of non-crossing pair-partitions (denoted by $\mathcal{NC}_{2k}(2)$) of length $2k$.

Since the elements of the same pair-partition must belong to the same index, we have

$$\#(ICAT_q) = \sum_{\pi \in \mathcal{NC}_{2k}(2)} \prod_{(j,j') \in \pi} \mathbb{I}_{t_j = t_{j'}}.$$

This holds for every monomial q. However, this is nothing but the freeness and joint semi-circularity (see Theorem 5.4.2 of Anderson, Guionnet and Zeitouni (2010)[1]). $\qquad\square$

(D) Toeplitz and Hankel.

Consider first the Toeplitz matrix. Since $p(w)$ exists, from Theorem 9.2.1, we have the joint convergence for Toeplitz matrices. For any fixed monomial q let SNC_q be the indexed symmetric words which are not Catalan. Then we obtain the following:

$$
\begin{aligned}
\phi(a_{i_1} \cdots a_{i_k}) &= \sum_{w \in \mathcal{IW}_k(2)} p_{I_q}(w) \\
&= \sum_{w \in ICAT_q} p_{I_q}(w) + \sum_{w \in SNC_q} p_{I_q}(w) + \sum_{\text{other } w} p_{I_q}(w) \\
&= \#(ICAT_q) + \sum_{w \in SCN_q} p_{I_q}(w) + \sum_{\text{other } w} p_{I_q}(w).
\end{aligned}
$$

Consider $q(X_1, X_2, X_3) = X_1 X_2 X_3 X_1 X_2 X_3$ where X_1, X_2 and X_3 are scaled independent Toeplitz matrices. From Table 2.2 of Chapter 2,

$$p_{C_q}(a_1 b_2 c_3 a_1 b_2 c_3) = p(\psi_q(a_1 b_2 c_3 a_1 b_2 c_3)) = p(abcabc) = \frac{1}{2}.$$

For this monomial, the only indexed pair-matched word possible is $a_1 b_2 c_3 a_1 b_2 c_3$ and hence

$$\phi(a_1 a_2 a_3 a_1 a_2 a_3) = \frac{1}{2} \neq 0.$$

Thus, there is no asymptotic freeness.

Now let $q(X_1, X_2, X_3) = X_1 X_2 X_3 X_2 X_3 X_1$. Then the only indexed pair-matched word is $a_1 b_2 c_3 b_2 c_3 a_1$ and

$$\phi(a_1 a_2 a_3 a_2 a_3 a_1) = p_{C_q}(a_1 b_2 c_3 b_2 c_3 a_1) = p(abcbca) = \frac{2}{3}.$$

On the other hand we have already seen that $\phi(a_1 a_2 a_3 a_1 a_2 a_3) = \frac{1}{2}$. Since the two contributions are not equal, the Toeplitz limit is not independent.

They are not half-independent since if they were, then

$$\phi(a_1 a_2 a_3 a_1 a_2 a_3) = \phi(a_1^2)\phi(a_2^2)\phi(a_3^2) = 1,$$

but that is not the case. Thus, the Toeplitz limit is not free, independent or half-independent.

For Hankel matrices, the indexed non-symmetric words *do not* contribute to the limit. So for any fixed monomial q we have

$$\phi(a_{i_1} \cdots a_{i_k}) = \#(ICAT_q) + \sum_{w \in SNC_q} p_{I_q}(w).$$

The Hankel limit is also not free, half-independent or independent. This can be checked along the above lines by considering appropriate monomials and their contributions. It is interesting to note that Hankel matrices do not half-commute and that is why even though the limits vanish on non-symmetric words they cannot be half-independent.

9.4   Exercises

1. Verify the claim made in Remark 9.2.2.

2. Show that the joint limit of Hankel matrices is not free, independent and half-independent.

3. Show that Symmetric Circulant matrices of the same order commute.

4. Show that for any Reverse Circulant matrices A, B, C of the same order, $ABC = CAB$.

5. Check that $a = \begin{bmatrix} 0 & \eta \\ \bar{\eta} & 0 \end{bmatrix}$, where η is a complex standard Gaussian, has $(2k)$th moments as $k!$ under $\phi(\cdot) = \frac{1}{2} \operatorname{E} \operatorname{Tr}(\cdot)$.

6. Show that Hankel matrices do not half-commute in general.

7. Show that pyramidal multiplicativity discussed in Chapter 2 holds on the class of indexed Catalan words w for any of the matrices considered in this chapter.

8. Show that independent Wigner-type matrices considered in Chapter 5 converge jointly in law to free semi-circle variables under Assumption III.

9. Show that i.i.d. triangular matrices considered in Chapter 8 converge jointly with respect to $E \frac{1}{n} \operatorname{Tr}$ when the entries satisfy Assumption III. Show that they are not asymptotically free.

10

Joint convergence of independent patterned matrices

In Chapter 9 we showed the joint convergence of independent copies of a *single* patterned matrix. In this chapter our goal is to show joint convergence when the sequences involved are not necessarily of the same pattern.

There are many reasons for doing this. For instance, trace of non-commutative polynomials of random matrices has found applications in operator algebras and telecommunications. The use of random matrix theory and free probability in CDMA (Code Division Multiple Access) and MIMO (multiple input and multiple output) systems has been shown in many articles. See Rashidi Far et al. (2008)[87], Couillet, Debbah, and Silverstein (2011)[46] and Tulino and Verdu (2004)[101]. For a MIMO system with n_1 and n_2 transmitter receiver antennae, the received signal is represented as $\mathbb{Y}_n = \mathbb{H}\mathbb{A}_n + \mathbb{B}_n$ where $\mathbb{A}_n$ is an n_1-dimensional vector depending on n and $\mathbb{B}_n$ is a noise signal and $\mathbb{H}$ is the channel matrix which generally has a block structure as given below and $\mathbb{Y}_n$ is an n_2-dimensional vector.

$$\mathbb{H} = \begin{bmatrix} C_1 & C_2 & \dots & C_L & 0 & \dots & & \dots & 0 \\ 0 & C_1 & C_2 & \dots & C_L & 0 & & & \vdots \\ \vdots & 0 & C_1 & C_2 & \dots & C_L & 0 & & \\ & \ddots & \ddots & \ddots & & \ddots & \ddots & & \vdots \\ \vdots & & & \ddots & \ddots & \ddots & & \ddots & 0 \\ 0 & \dots & & \dots & 0 & C_1 & C_2 & \dots & C_L \end{bmatrix}.$$

One of the main issues in the study of a MIMO system is the eigenvalue distribution of $\mathbb{H}\mathbb{H}^*$ since this is linked to the capacity of the channel. Here $\{C_i\}$ can be Wigner matrices or more general matrices. It may also happen that some of the blocks are Toeplitz or Hankel or any other patterned matrices. Studying the spectral properties of such matrices boils down to studying the joint convergence of different patterned matrices. The results of this chapter can be used for studying such systems. We refer the readers to Male (2012)[72] for related results.

In Theorem 10.2.1, we provide sufficient conditions for the joint convergence to hold. Under Property B (see Definition 1.4.1) and Assumption III (see Chapter 1) on the input sequence, we show that if a criterion (Condition 10.9) holds for one copy each of any sub-collection of matrices, then the joint convergence holds for multiple copies. This Condition 10.9 is satisfied by the five matrices: Wigner, Toeplitz, Hankel, Reverse Circulant and Symmetric Circulant. As an application of Theorem 10.2.1, the following holds: If $\mathbb{P}$ is a symmetric polynomial in any two of the following scaled matrices: Wigner, Toeplitz, Hankel, Reverse Circulant and Symmetric Circulant with uniformly bounded entries, then the LSD of the matrix $\mathbb{P}$ converges to a non-random measure μ on $\mathbb{R}$ weakly almost surely. Moreover, any collection of Wigner matrices is free of any collection of the other four matrices (see Theorem 10.3.1).

10.1   Definitions and notation

Let $(X, \mathcal{B}, \mu)$ be a probability space. Let $L(\mu) := \bigcap_{p \geq 1} L^p(X, \mu)$ be the algebra of random variables with finite moments of all orders. Let

$$A_n := Mat_n(L(\mu)) \tag{10.1}$$

be the space of $n \times n$ complex random matrices with entries coming from $L(\mu)$. Then $(\mathcal{A}_n, \phi_j)$, $j = 1, 2$ are two non-commutative probability spaces where

$$\phi_1(A) = \frac{1}{n} \operatorname{Tr}(A) \text{ and } \phi_2(A) = \frac{1}{n} \operatorname{E}[\operatorname{Tr}(A)]. \tag{10.2}$$

Consider h different types of patterned matrices where type j has p_j independent copies, $1 \leq j \leq h$. The different link functions shall be referred to as *colors* and different independent copies of the matrices of any given color shall be referred to as *indices*. Let $\{X_{i,n}^j, 1 \leq i \leq p_j\}$ be $n \times n$ symmetric patterned matrices with link functions L_j, $j = 1, \ldots, h$. Let $X_i^j(L_j(p, q))$ denote the (p, q)-th entry of $X_{i,n}^j$. We suppress the dependence on n to simplify notation. Two natural assumptions on the link function and the input sequence are:

Assumption (A1). All link functions $\{L_j, j = 1, \ldots, h\}$ satisfy *Property B*, that is,

$$\max_{1 \leq j \leq h} \sup_{n \geq 1} \sup_t \sup_{1 \leq p \leq n} \#\{q : 1 \leq q \leq n, L_j(p, q) = t\} \leq \Delta < \infty.$$

Assumption (A2). Input sequences $\{X_i^j(k) : k \in \mathbb{Z} \text{ or } \mathbb{Z}^2\}$ are real random variables independent across i, j and k with mean zero and variance 1 and the moments are uniformly bounded, that is, for every k,

$$\sup_{1 \leq j \leq h} \sup_{1 \leq i \leq p_j} \sup_{n \geq 1} \sup_{1 \leq p,q \leq n} \operatorname{E}\left[|X_i^j(L_j(p, q))|^k\right] \leq c_k < \infty.$$

We consider $\{\frac{1}{\sqrt{n}} X_{i,n}^j, 1 \leq i \leq p_j\}_{1 \leq j \leq h}$ as elements of $\mathcal{A}_n$ given in (10.1) and investigate the joint convergence with respect to the normalized tracial states ϕ_1 or ϕ_2 (as in (10.2)). The sequence of matrices jointly converge if and only if for every monomial q,

$$\phi_d\big(q\big(\frac{1}{\sqrt{n}}\{X_{i,n}^j, 1 \leq i \leq p_j\}_{1 \leq j \leq h}\big)\big)$$

converges to a limit as $n \to \infty$ for either $d = 1$ or $d = 2$. For $d = 1$, the convergence is in the almost sure sense. The case of $h = 1$ and $p_1 = 1$ (a single patterned matrix) was dealt in Chapter 2 and $h = 1$ and $p_1 > 1$ (i.i.d. copies of a single patterned matrix) was dealt in Chapter 9. In particular, we know convergence holds for i.i.d. copies of any one of the five patterned matrices. The starting point in showing this was the trace formula. The related concepts of circuits, matches and words will now be extended below to multiple copies of several matrices.

Since our primary aim is to show convergence for every monomial, we shall from now on, *fix an arbitrary monomial q of length k*. We generally denote the colors and indices present in q by $(c_1, \ldots, c_k)$ and $(t_1, \ldots, t_k)$ respectively. Then we may write,

$$q\big(\frac{1}{\sqrt{n}}\{X_{i,n}^j, 1 \leq i \leq p_j\}_{1 \leq j \leq h}\big) = \frac{1}{n^{k/2}} Z_{c_1,t_1} \cdots Z_{c_k,t_k}, \qquad (10.3)$$

where $Z_{c_m,t_m} = X_{t_m}^{c_m}$ for $1 \leq m \leq k$.

From (10.3) we get,

$$\widetilde{\mu_n}(q) := \frac{1}{n} \operatorname{Tr} \Big[\frac{1}{n^{k/2}} Z_{c_1,t_1} Z_{c_2,t_2} \cdots Z_{c_k,t_k} \Big]$$

$$= \frac{1}{n^{1+k/2}} \sum_{j_1,\cdots,j_k} Z_{c_1,t_1}(L_{c_1}(j_1,j_2)) Z_{c_2,t_2}(L_{c_2}(j_2,j_3)) \cdots Z_{c_k,t_k}(L_{c_k}(j_k,j_1))$$

$$= \frac{1}{n^{1+k/2}} \sum_{\pi:\pi \text{ is a circuit}} \prod_{i=1}^{k} Z_{c_i,t_i}(L_{c_i}(\pi(i-1),\pi(i)))$$

$$= \frac{1}{n^{1+k/2}} \sum_{\pi:\pi \text{ is a circuit}} \mathbf{Z}_\pi \quad \text{say.} \qquad (10.4)$$

Also define,

$$\overline{\mu_n} = \operatorname{E}[\widetilde{\mu_n}]. \qquad (10.5)$$

Due to independence and mean zero of the input sequences,

$$\operatorname{E}[\mathbf{Z}_\pi] = 0 \quad \text{if } \pi \text{ has at least one edge of order one.} \qquad (10.6)$$

If all L-values appear more than once then we say the circuit is *matched* and only these circuits are relevant due to the above.

A circuit is said to be *color-matched* if all the L-values are repeated within the same color. A circuit is said to be *color- and index-matched* if in addition, all the L-values are also repeated within the same index.

We can define an equivalence relation on the set of color- and index-matched circuits, extending the ideas of Chapters 1 and 9. We say $\pi_1 \sim \pi_2$ if and only if their matches take place at the same colors and at the same indices. In other words,

$$c_i = c_j, t_i = t_j \ \text{ and } \ L_{c_i}(\pi_1(i-1), \pi_1(i)) = L_{c_j}(\pi_1(j-1), \pi_1(j))$$

$$\Updownarrow$$

$$c_i = c_j, t_i = t_j \ \text{ and } \ L_{c_i}(\pi_2(i-1), \pi_2(i))) = L_{c_j}(\pi_2(j-1), \pi_2(j)).$$

An equivalence class can be expressed as a colored and indexed word w: each word is a string of letters in alphabetic order of their first occurrence with a subscript and a superscript to distinguish the index and the color respectively. The i-th position of w is denoted by $w[i]$.

For example, if

$$q = X_1^1 X_2^1 X_1^2 X_1^2 X_2^2 X_2^2 X_2^1 X_1^1 = Z_{1,1} Z_{1,2} Z_{2,1} Z_{2,1} Z_{2,2} Z_{2,2} Z_{1,2} Z_{1,1},$$

then $a_1^1 b_2^1 c_1^2 c_1^2 d_2^2 d_2^2 b_2^1 a_1^1$ is *one of the* colored and indexed words corresponding to q. Any colored and indexed word uniquely determines the monomial it corresponds to. A colored and indexed word is *pair-matched* if all its letters appear exactly twice. We shall see later that under *Property B*, only such circuits and words survive in the limits of (10.4) and (10.5).

Now we define some useful subsets of circuits. For a colored and indexed word w, let

$$\Pi_{CI}(w) = \{\pi : w[i] = w[j] \Leftrightarrow$$
$$(c_i, t_i, L_{c_i}(\pi(i-1), \pi(i))) = (c_j, t_j, L_{c_j}(\pi(j-1), \pi(j)))\}.(10.7)$$
$$\Pi_{CI}^*(w) = \{\pi : w[i] = w[j] \Rightarrow$$
$$(c_i, t_i, L_{c_i}(\pi(i-1), \pi(i)) = (c_j, t_j, L_{c_j}(\pi(j-1), \pi(j)))\}. (10.8)$$

Every colored and indexed word has a corresponding non-indexed version which is obtained by dropping the indices from the letters (i.e. the subscripts). For example, $a_1^1 b_2^1 c_1^2 c_1^2 d_2^2 d_2^2 b_2^1 a_1^1$ yields $a^1 b^1 c^2 c^2 d^2 d^2 b^1 a^1$. For any monomial q, dropping the indices amounts to replacing, for every j, the independent copies X_i^j by a single X^j with link function L_j. In other words it corresponds to the case where $p_j = 1$ for $1 \leq j \leq h$.

Let $\psi(q)$ be the monomial obtained by dropping the indices from q. For example,

if $q = Z_{1,1} Z_{1,2} Z_{2,1} Z_{2,1} Z_{2,2} Z_{2,2} Z_{1,2} Z_{1,1}$ then $\psi(q) = Z_1 Z_1 Z_2 Z_2 Z_2 Z_2 Z_1 Z_1$.

(10.7) and (10.8) get mapped to the following subsets of non-indexed colored words w' via ψ:

$$\Pi_C(w) = \{\pi : w[i] = w[j] \Leftrightarrow c_i = c_j \text{ and } L_{c_i}(\pi(i-1), \pi(i)) = L_{c_j}(\pi(j-1), \pi(j))\},$$

$$\Pi_C^*(w) = \{\pi : w[i] = w[j] \Rightarrow c_i = c_j \text{ and } L_{c_i}(\pi(i-1), \pi(i)) = L_{c_j}(\pi(j-1), \pi(j))\}.$$

Since pair-matched words are going to be crucial, let us define:

$CIW(2) = \{w : w \text{ is indexed colored pair-matched corresponding to } q\}$

$CW(2) = \{w : w \text{ is non-indexed colored pair-matched corresponding to } \psi(q)\}.$

For $w \in CIW(2)$, let us consider the word obtained by dropping the indices of w. This defines an injective mapping into $CW(2)$ and we continue to denote this mapping by ψ.

For any $w \in CW(2)$ and $w' \in CIW(2)$, we define (*whenever the limits exist*),

$$p_C(w) = \lim_{n \to \infty} \frac{1}{n^{1+k/2}} \# \Pi_C^*(w) \quad \text{and} \quad p_{CI}(w') = \lim_{n \to \infty} \frac{1}{n^{1+k/2}} \# \Pi_{CI}^*(w').$$

10.2 Joint convergence

We now give a criterion for joint convergence from Basu et al. (2012)[15].

Theorem 10.2.1. Let $\{\frac{1}{\sqrt{n}} X_{i,n}^j, 1 \le i \le p_j\}_{1 \le j \le h}$ be a sequence of real symmetric patterned random matrices which satisfy Assumptions A1 and A2. Fix a monomial q of length k and assume that, for all $w \in CW(2)$,

$$p_C(w) = \lim_{n \to \infty} \frac{1}{n^{1+k/2}} \#(\Pi_C^*(w)) \quad \text{exists.} \tag{10.9}$$

Then,

(a) for all $w \in CIW(2)$, $p_{CI}(w)$ exists and $p_{CI}(w) = p_C(\psi(w))$;

(b) we have

$$\lim_{n \to \infty} \overline{\mu_n}(q) = \sum_{w \in CIW(2)} p_{CI}(w) = \alpha(q) \quad (\text{say})$$

with

$$|\alpha(q)| \le \begin{cases} \frac{k! \Delta^{k/2}}{(k/2)! 2^{k/2}} & \text{if } k \text{ is even and each index appears} \\ & \text{an even number of times} \\ 0 & \text{otherwise; and} \end{cases}$$

(c) $\lim_{n \to \infty} \widetilde{\mu_n}(q) = \alpha(q)$ almost surely.

As a consequence, if (10.9) holds for every q then $\{\frac{1}{\sqrt{n}} X_{i,n}^j, 1 \le i \le p_j\}_{1 \le j \le h}$ converges jointly in both the states ϕ_1 and φ_2 and the limit is universal. ◇

Remark 10.2.1. (a) Theorem 10.2.1 asserts that if the joint convergence holds for $p_j = 1, j = 1, \ldots, h$ (that is, if condition (10.9) holds), then the joint convergence continues to hold for $p_j \geq 1$. There is no general way of checking (10.9). However, see the next theorem.

(b) Under the conditions of Theorem 10.2.1, if the monomial q yields a symmetric matrix, then the corresponding LSD exists almost surely and is symmetric.

(c) The moment conditions may be reduced to some extent depending on the monomial when the input sequences are i.i.d.   ⋄

The next theorem identifies situations where $p_C(w)$ exist.

Theorem 10.2.2. Suppose Assumption (A2) holds. Then $p_C(w)$ exists for all monomials q and $w \in CW(2)$, for any of the two of the matrices Wigner, Toeplitz, Hankel, Symmetric Circulant and Reverse Circulant at a time.   ⋄

In general, the value of $p_C(w)$ cannot be computed for arbitrary pair-matched words. We provide some examples in Tables 10.1 and 10.2.

TABLE 10.1
$p_C(w)$ for monomials in H and T.

Monomial	Word	$p_C(w)$
TTHH	aabb	1
THTH	abab	2/3
TTTTHH	aabbcc	1
	abbacc	1
	ababcc	2/3
HHHHTT	aabbcc	1
	abbacc	1
	ababcc	0
TTHTTH	aabccb	1
	abcbac	1/2
	abcabc	1/2
HHTHHT	aabccb	1
	abcbac	1/2
	abcabc	0

As seen in Tables 10.1 and 10.2, $p_C(w)$ equals one for certain words. We now identify a class of such words. This has ramifications later in the study of freeness.

Just as the set of Catalan words of length m are in bijection with the set of non-crossing pair-partitions $NC_2(m)$, the colored Catalan words are in bijection with the following set of pair-partitions (denoted by $NC_2^{(p)}(m)$) and this correspondence will be useful in the proof of Theorem 10.3.1. For $p = (p(1), \ldots, p(m))$ integers (colors), denote

$$NC_2^{(p)}(m) = \{\pi \in NC_2(m) : p(\pi(r)) = p(r) \text{ for all } r = 1, \ldots, m\}.$$

Consider the following property of the number of matches between rows across all pairs of columns. All the five matrices satisfy this property.

TABLE 10.2
$p_C(w)$ for monomials in H, R and H, S.

Monomial	Word	$p_C(w)$	Monomial	Word	$p_C(w)$
RRHH	aabb	1	SSHH	aabb	1
RHRH	abab	0	SHSH	abab	2/3
RRRRHH	aabbcc	1	SSSSHH	aabbcc	1
	abbacc	1		abbacc	1
	ababcc	0		ababcc	1
HHHHRR	aabbcc	1	HHHHSS	aabbcc	1
	abbacc	1		abbacc	1
	ababcc	0		ababcc	0
RRHRRH	aabccb	1	HHHSHS	aabcbc	1/2
	abcbac	0		abbcac	1/2
	abcabc	2/3		abcabc	0
HHRHHR	aabccb	1	HHSHHS	aabccb	1
	abcbac	0		abcbac	1/2
	abcabc	1/2		abcabc	0

Property P: A link function L satisfies *Property P* if

$$M^* = \sup_n \sup_{i,j} \#\{1 \le k \le n : L(k,i) = L(k,j)\} < \infty. \qquad (10.10)$$

The next result extends Theorem 5.1.2 of Chapter 5 to multiple copies of colored matrices.

Theorem 10.2.3. (a) Suppose matrices X and Y satisfy Assumptions A1 and A2. Consider any monomial in X and Y of length $2k$. Then

$$\#\Pi_C^*(w) \ge n^{1+k} \text{ for any colored Catalan word } w.$$

As a consequence, $p_C(w) \ge 1$ for any colored Catalan word w.

(b) Suppose the link functions of X and Y satisfy *Property B* and *Property P* and the input satisfies Assumption A2. Then for any colored Catalan word, $p_C(w) = 1$. ◇

10.3 Freeness

We have seen in Chapter 9 that independent Wigner matrices are asymptotically free. It is an interesting question to identify situations where freeness is present and there are many results in this direction.

Voiculescu (1991)[102] showed that if we take k independent Hermitian random matrices $\{W_{i,n}\}_{1 \le i \le k}$ distributed as GUE then they are asymptotically free. He also showed the asymptotic freeness of GUE and diagonal constant matrices. Later, Voiculescu (1998)[103] improved the result to asymptotic freeness of GUE and general $n \times n$ deterministic matrices $\{D_{i,n}\}$ (having LSD) and satisfying

$$\sup_{n} \|D_{i,n}\| < \infty \text{ for each } i, \qquad (10.11)$$

where $\|\cdot\|$ denotes the operator norm. This inclusion of constant matrices had important implications in the factor theory of von Neumann algebras. Dykema (1993)[51] established a similar result for a family of independent Wigner matrices (symmetric matrices with i.i.d. real entries having uniformly bounded moments) and block-diagonal constant matrices with bounded block size. The results were also shown to hold with respect to ϕ_1 almost surely (see Hiai and Petz (2000)[64, 65] for details).

Various other extensions to Wishart ensembles, Gaussian orthogonal ensembles (GOE) and Gaussian symplectic ensembles (GSE) are also available. See Capitaine and Casalis (2004)[35], Capitaine and Martin (2007)[36], Collins, Guionnet and Segala (2009)[42], Schultz (2005)[91], Ryan (1998)[90] and Voiculescu (1998)[103].

By the results of Collins (2003)[41] and Collins and Sniady (2006)[43] the Wigner matrices are asymptotically free of general deterministic matrices which converge jointly. It appears that the existing results in the literature on freeness of Wigner and other random matrices need some conditions on the behavior of the trace of the matrices as pointed out in Remark 3.6 of Collins (2003)[41]. This condition (equation (3.4) therein) was studied in Capitaine and Casalis (2004)[35]. It was shown there that under the technical condition on the random matrices (see Condition C and C' there), there is asymptotic freeness between Wigner and other random matrices. In other available criteria for freeness, condition (10.11) appears (see Anderson, Guionnet and Zeitouni (2010)[1] and Theorem 22.2.4 of Speicher (2011)[98]).

Freeness is present elsewhere too and one important place is the Haar distributed matrices. It is well known that any unitary invariant matrix (in particular GUE) can be written as UDU^* where D is a diagonal matrix and U is Haar distributed on the space of unitary matrices and independent of D. Voiculescu (1991)[102] showed that $\{U, U^*\}$ and D are asymptotically free. Hiai and Petz (2000)[64] showed that the Haar unitaries and general deterministic matrices satisfying (10.11) are almost surely asymptotically free. Collins (2003)[41] showed that general deterministic matrices and Haar unitary matrices are asymptotically free almost surely, provided the deterministic matrices jointly converge. The case of orthogonal and symplectic groups were dealt with in Collins and Sniady (2006)[43].

The combinatorial properties of freeness have been extended to *operator valued freeness* (also called *freeness with amalgamation*) and in particular this is useful in describing the behavior of polynomials of random band matrices and rectangular random matrices (see Shlyakhtenko (1996)[95], Benaych-Georges (2009)[17, 18]).

Using the notions of circuits and words we are now able to show, in a relatively simple way, freeness of the Wigner with any of the other matrices, namely Toeplitz, Hankel, Reverse Circulant and Symmetric Circulant. This technique is similar in spirit to those in Chapter 22 of Nica and Speicher (2006)[79]. However, we bypass the detailed properties of the permutation group and the Weingarten functions that they use. It is quite plausible that the techniques in Collins (2003)[41] and Capitaine and Casalis (2004)[35] may be extended to prove Theorem 10.3.1. Incidentally, if we take the Wigner with complex entries then Theorem 10.3.1 holds for any patterned matrix satisfying Property B and having an LSD.

Theorem 10.3.1. Suppose $\{W_{i,n}, 1 \leq i \leq p, A_{i,n}, 1 \leq i \leq p\}$ are independent matrices which satisfy Assumption A2 where $W_{i,n}$ are Wigner matrices and $A_{i,n}$ are any of Toeplitz, Hankel, Symmetric Circulant or Reverse Circulant matrices. Then the collection $\{\frac{W_{i,n}}{\sqrt{n}}, 1 \leq i \leq p\}$ is asymptotically free of the collection $\{\frac{A_{i,n}}{\sqrt{n}}, 1 \leq i \leq p\}$. ◇

Remark 10.3.1. Incidentally, the freeness between the Gaussian unitary ensemble (GUE) and other patterned matrices is much easier to establish. Indeed, it can be shown that GUE and any patterned matrices (having Property B, satisfying (10.1) and having LSD) are asymptotically free. We provide a brief proof of this assertion after the proof of Theorem 10.3.1. ◇

10.4 Sum of independent patterned matrices

An important issue that has been studied in RMT is the spectrum (ESD) of $\frac{W_n}{\sqrt{n}} + P_n$ where W_n is a Wigner matrix and P_n is a suitable perturbation matrix which is independent of W_n. The spectrum of this perturbation has been of interest for a long time (see Fulton (1998)[57]). Suppose the ESD of P_n weakly converges to μ_P. Then the ESD of $\frac{W_n}{\sqrt{n}} + P_n$ converges weakly, almost surely and in expectation, to the *free convolution* of μ_P and the semi-circle law whenever μ_P has compact support or $\{P_n\}$ satisfy (10.11). These results were derived using results on the asymptotic freeness between deterministic (or random) matrices and the Wigner matrix. Pastur and Vasilchuk (2000)[82] extended these results for unbounded (possibly random) perturbations using the analytic machinery of the Stieltjes transform.

The special case where P_n has finite rank has received a considerable amount of interest. In this case, the LSD still the semi-circle law but the behavior at the edge has some interesting properties. See Féral and Péché(2007)[55], Capitaine, Donati-Martin, and Féral (2009)[37], Benaych-Georges, Guionnet, and Maida (2011)[19], Capitaine et al. (2011)[38] and Péché(2006)[83].

The following result on the sum of two patterned matrices essentially follows from Theorem 10.2.1. Figures 10.1 and 10.2 provide simulations in some special cases.

Corollary 10.4.1. Let A and B be two independent real symmetric patterned matrices which satisfy Assumptions A1 and A2. Suppose $p_C(w)$ exists for every q and every w. Then LSD for $\frac{A+B}{\sqrt{n}}$ exists in the almost sure sense, is symmetric and does not depend on the underlying distribution of the input sequences. Moreover, if either LSD of $\frac{A}{\sqrt{n}}$ or LSD of $\frac{B}{\sqrt{n}}$ has unbounded support then LSD of $\frac{A+B}{\sqrt{n}}$ also has unbounded support.   ◇

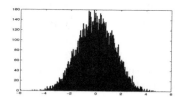 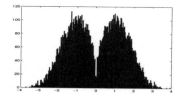

FIGURE 10.1
Histogram plots of ESD of (i) Reverse Circulant+ Symmetric Circulant (left) (ii) Circulant+Hankel (right). $n = 500$ with $N(0,1)$ entries.

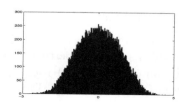 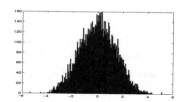

FIGURE 10.2
Histogram plots of ESD of Toeplitz+Hankel (left) (ii) Toeplitz+Symmetric Circulant (right). $n = 500$ with $N(0,1)$ entries.

Proof. The assumptions imply that LSD of $\frac{A}{\sqrt{n}}$ and $\frac{B}{\sqrt{n}}$ exist. By Theorem 10.2.1, $\{\frac{A}{\sqrt{n}}, \frac{B}{\sqrt{n}}\}$ converge jointly and hence $\lim_{n\to\infty} \frac{1}{n^{k/2+1}} E(\mathrm{Tr}(A + B)^k = \beta_k$ exists for all $k > 0$. Now let us fix k. Let Q_k be the set of monomials such that $(A + B)^k = \sum_{q \in Q_k} q(A, B)$. Hence

$$\frac{1}{n} \mathrm{Tr}(\frac{A + B}{\sqrt{n}})^k = \frac{1}{n^{1+k/2}} \sum_{q \in Q_k} \mathrm{Tr}(q(A, B)) = \sum_{q \in Q_k} \widetilde{\mu_n}(q)$$

where $\widetilde{\mu_n}(q)$ is as in Section 10.1. By Theorem 10.2.1(c), $\widetilde{\mu_n}(q) \to \alpha(q)$, almost surely and hence,

$$\beta_k = \lim_{n\to\infty} \frac{1}{n} \operatorname{Tr}\left(\frac{A+B}{\sqrt{n}}\right)^k = \sum_{q \in Q_k} \alpha(q) \quad \text{almost surely.}$$

Using Theorem 10.2.1(b), we have

$$\beta_{2k} = \sum_{q \in Q_{2k}} \alpha(q) \le |Q_{2k}| \frac{(2k)!}{k!2^k} \Delta^k = 2^{2k} \frac{(2k)!}{k!2^k} \Delta^k.$$

Hence $\{\beta_{2k}\}$ satisfies Riesz's condition, implying that the LSD exists.

To prove symmetry of the limit, let $q \in Q_{2k+1}$. Then from Theorem 10.2.1 (b), it follows that $\alpha(q) = 0$. Hence $\beta_{2k+1} = \sum_{q \in Q_{2k+1}} \alpha(q) = 0$ and the distribution is symmetric.

We leave the proof of unboundedness as an exercise. $\qquad\square$

When one of the matrix is Wigner, Theorem 10.3.1 implies that the limit is the free convolution of the semi-circle law and the other LSD. This result about the sum when one of them is Wigner also follows from the results of Pastur and Vasilchuk (2000)[82]. It also follows from the work of Biane (1997)[23] that any free convolution with the semi-circle law is continuous and the density can be expressed in terms of the Stieltjes transform of the LSD. Unfortunately, the Stieltjes transform of the LSD of the Toeplitz and Hankel are not known.

10.5   Proofs

To simplify the notational aspects we restrict ourselves to only $h = 2$.

Proof of Theorem 10.2.1. (a) We first show that

$$\Pi_C^*(w) = \Pi_{CI}^*(w) \quad \text{for all} \quad w \in CIW(2). \tag{10.12}$$

Let $\pi \in \Pi_{CI}^*(w)$. As q is fixed,

$$\psi(w)[i] = \psi(w)[j] \Rightarrow w[i] = w[j]$$
$$\Rightarrow (c_i, t_i, L_{c_i}(\pi(i-1), \pi(i))) = (c_j, t_j, L_{c_j}(\pi(j-1), \pi(j))) \quad (\text{as } \pi \in \Pi_{CI}^*(w)).$$

Hence, $L_{c_i}(\pi(i-1), \pi(i)) = L_{c_j}(\pi(j-1), \pi(j))$ and $\pi \in \Pi_C^*(\psi(w))$.

Now conversely, let $\pi \in \Pi_C^*(\psi(w))$. Then we have

$$w[i] = w[j]$$
$$\Rightarrow \psi(w)[i] = \psi(w)[j]$$
$$\Rightarrow L_{c_i}(\pi(i-1), \pi(i)) = L_{c_j}(\pi(j-1), \pi(j))$$
$$\Rightarrow Z_{c_i, t_i}(L_{c_i}(\pi(i-1), \pi(i))) = Z_{c_j, t_j}(L_{c_j}(\pi(j-1), \pi(j)))$$

as $w[i] = w[j] \Rightarrow c_i = c_j$ and $t_i = t_j$. Hence $\pi \in \Pi^*_{CI}(w)$. So (10.12) is established. As a consequence,

$$p_{CI}(w) = \lim_{n \to \infty} \frac{1}{n^{1+k/2}} \#\Pi^*_{CI}(w) = p_C(\psi(w)).$$

Hence by (10.12) $p_{CI}(w)$ exists for all $w \in CIW(2)$ and $p_{CI}(w) = p_C(\psi(w))$, proving (10.2.1).

(b) Recall that $\mathbf{Z}_\pi = \prod_{j=1}^k Z_{c_j, t_j}(L_{c_i}(\pi(j-1), \pi(j)))$ and using (10.5) and (10.6)

$$\bar{\mu}_n(q) = \frac{1}{n^{1+k/2}} \sum_{w:\ w \text{ matched}} \sum_{\pi \in \Pi_{CI}(w)} E(\mathbf{Z}_\pi). \qquad (10.13)$$

By using Assumption A2

$$\sup_\pi E |\mathbf{Z}_\pi| < K < \infty. \qquad (10.14)$$

By using arguments, which are by now familiar, for any colored and indexed word w which is matched but is not pair-matched,

$$\lim_{n \to \infty} \frac{1}{n^{1+k/2}} \Big| \sum_{\pi \in \Pi_{CI}(w)} |E(\mathbf{Z}_\pi)| \leq \frac{K}{n^{1+k/2}} |\Pi_{CI}(w)| \to 0. \qquad (10.15)$$

By using (10.15), and the fact that $E(\mathbf{Z}_\pi) = 1$ for every colored indexed pair-matched word (use Assumption A2), calculating the limit in (10.13) reduces to calculating $\lim \frac{1}{n^{1+k/2}} \sum_{w:\ w \in CIW(2)} \#\Pi_{CI}(w)$.

Now consider $w \in CIW(2)$. Any circuit in $\Pi^*_{CI}(w) - \Pi_{CI}(w)$ must have an edge of order four. Hence by (10.15),

$$\lim_{n \to \infty} \frac{\#(\Pi^*_{CI}(w) - \Pi_{CI}(w))}{n^{1+k/2}} = 0.$$

As a consequence, since there are finitely many words,

$$
\begin{aligned}
\lim_{n \to \infty} \bar{\mu}_n(q) &= \lim_{n \to \infty} \sum_{w \in CIW(2)} \frac{\#\Pi_{CI}(w)}{n^{1+k/2}} \\
&= \lim_{n \to \infty} \sum_{w \in CIW(2)} \frac{\#\Pi^*_{CI}(w)}{n^{1+k/2}} \\
&= \sum_{w \in CIW(2)} p_{CI}(w) = \alpha(q).
\end{aligned}
$$

To complete the proof of (b), we note that, if either k is odd or some index appears an odd number of times in q then for that q, $CIW(2)$ is empty and hence, $\alpha(q) = 0$. Now suppose that k is even and every index appears an even number of times. Then

$$\#CIW(2) \leq \#CW(2) \leq \frac{k!}{(k/2)!2^{k/2}}.$$

The first inequality above follows from the fact mentioned earlier that ψ is an injective map from $CIW(2)$ to $CW(2)$.

The second inequality follows by observing that the total number of *colored* pair-matched words of length k is less than the number of pair-matched words of length k.

Now note that $p_{CI}(w) \le \Delta^{k/2}$. Combining all these, $|\alpha(q)| \le \frac{k!\Delta^{k/2}}{(k/2)!2^{k/2}}$.

(c) Now we claim that

$$E[(\widetilde{\mu_n}(q) - \bar{\mu}_n(q))^4] = O(n^{-2}). \tag{10.16}$$

The proof of this claim is similar to the proof of Lemma 1.4.3 (b). We leave it as an exercise.

Now using the Borel-Cantelli lemma, $\widetilde{\mu_n}(q) - \bar{\mu}_n(q) \to 0$ almost surely as $n \to \infty$ and this completes the proof. $\qquad \square$

Proof of Theorem 10.2.2. Condition (10.9) which needs to be verified (only for even degree monomials), crucially depends on the type of the link function and hence we need to deal with every example differently. Since we are dealing with only two link functions, we simplify the notation. Let X and Y be patterned matrices with link function L_1 and L_2, respectively, with independent input sequences satisfying Assumptions A1 and A2. Let $q(X, Y)$ be any monomial such that both X and Y occur an even number of times in q. Let $deg(q) = 2k$ and let the number of times X and Y occurs in the monomial be k_1 and k_2 respectively. Note that we have $k = k_1 + k_2$. Then it is enough to show that (10.9) holds for every pair-matched colored word w of length $2k$ corresponding to q.

Let X and Y be any of the two following matrices: Wigner (W_n), Toeplitz (T_n), Hankel (H_n), Reverse Circulant (RC_n) and Symmetric Circulant (SC_n). The case when X and Y are of the same pattern was dealt with in Chapter 9.

Proof of Theorem 10.2.2 is immediate once we establish the following lemma.

Lemma 10.5.1. *Let X and Y be any of the matrices, W_n, T_n, H_n, RC_n and SC_n, satisfying Assumption A2. Let $w \in CW(2)$ corresponding to a monomial q of length $2k$. Then there exists a (finite) index set I independent of n and $\{\Pi^*_{C,l}(w) : l \in I\} \subset \Pi^*_C(w)$ such that*

(a) $\Pi^*_C(w) = \cup_{l \in I}\Pi^*_{C,l}(w)$, *and* $p_{C,l}(w) := \lim_{n \to \infty} \frac{\#\Pi^*_{C,l}(w)}{n^{1+k}}$ *exists for all $l \in I$,*

(b) *for $l \ne l'$ we have,*

$$\#(\Pi^*_{C,l}(w) \cap \Pi^*_{C,l'}(w)) = n^{1+k}. \tag{10.17}$$

$\diamond$

Assuming Lemma 10.5.1, $\#\Pi^*_C(w) = \#(\cup_{l \in I}\Pi^*_{C,l}(w))$ for some finite index set I and

$$p_C(w) \;=\; \lim_{n\to\infty} \frac{1}{n^{1+k}} \#\Pi_C^*(w)$$

$$\;=\; \sum_{l\in I} \lim_{n\to\infty} \frac{1}{n^{1+k}} \#\Pi_{C,l}^*(w) = \sum_{l\in I} p_{C,l}(w).$$

Proof of Lemma 10.5.1. We need to treat each pair of matrices separately. Since the arguments are similar for the different pairs, we provide only a selection of the arguments in most cases.

The set S of generating vertices of w is split into three classes $\{0\}\cup S_X\cup S_Y$ where

$$S_X = \{i\wedge j : c_i = c_j = X, w[i] = w[j]\}, \;\; S_Y = \{i\wedge j : c_i = c_j = Y, w[i] = w[j]\}.$$

For every $i \in S - \{0\}$, let j_i denote the index such that $w[j_i] = w[i]$. Let $\pi \in \Pi_C^*(w)$.

(i) **Toeplitz and Hankel**: Let X and Y be, respectively, the Toeplitz (T) and the Hankel (H) matrix. Observe that,

$$|\pi(i-1) - \pi(i)| = |\pi(j_i - 1) - \pi(j_i)| \text{ for all } i \in S_T$$

$$\pi(i-1) + \pi(i) = \pi(j_i - 1) + \pi(j_i) \text{ for all } i \in S_H.$$

Let I be $\{-1,1\}^{k_1}$ and $l = (l_1, ..., l_{k_1}) \in I$. Let $\Pi_{C,l}^*(w)$ be the subset of $\Pi_C^*(w)$ such that,

$$\pi(i-1) - \pi(i) = l_i(\pi(j_i - 1) - \pi(j_i)) \qquad \text{for all } i \in S_T,$$

$$\pi(i-1) + \pi(i) = \pi(j_i - 1) + \pi(j_i) \qquad \text{for all } i \in S_H.$$

Now clearly,

$$\Pi_C^*(w) = \bigcup_l \Pi_{C,l}^*(w) \text{ (not a disjoint union).}$$

Now let us define,

$$v_i = \frac{\pi(i)}{n} \qquad \text{and} \qquad U_n = \{0, \frac{1}{n}, ..., \frac{n-1}{n}\}. \tag{10.18}$$

Then,

$$\#\Pi_{C,l}^*(w) = \#\{(v_0, ..., v_{2k}) : v_i \in U_n \; \forall i, \; v_{i-1} - v_i = l_i(v_{j_i-1} - v_{j_i}) \; \forall i \in S_T$$
$$\text{and } v_{i-1} + v_i = v_{j_i-1} + v_{j_i} \;\; \forall i \in S_H, \;\; v_0 = v_{2k}\}.$$

Let us denote $\{v_i : i \in S\}$ by v_S. It can easily be seen from the above equations (other than $v_0 = v_{2k}$) that each of the $\{v_i : i \notin S\}$ can be written uniquely as an integer linear combination $L_i(v_S)$. Moreover, $L_i(v_S)$ only contains $\{v_j : j \in S, \; j < i\}$ with non-zero coefficients. Clearly,

$$\#\Pi_{C,l}^*(w) = \#\{(v_0, ..., v_{2k}) : v_i \in U_n \; \forall i, v_0 = v_{2k}, v_i = L_i(v_S) \;\; \forall i \notin S\}.$$

Any integer linear combinations of elements of U_n is again in U_n if and only if it is between 0 and 1. Hence,

$$\#\Pi^*_{C,l}(w) = \#\{v_S : v_i \in U_n \ \forall i \in S, v_0 = L_{2k}(v_S), 0 \le L_i(v_S) < 1 \ \forall i \notin S\}. \tag{10.19}$$

From (10.19) it follows that, $\frac{\#\Pi^*_{C,l}(w)}{n^{1+k}}$ is nothing but the Riemann sum for the function $I(0 \le L_i(v_S) < 1, i \notin S, v_0 = L_{2k}(v_S))$ over $[0,1]^{k+1}$ and converges to the integral and hence

$$\begin{aligned}
p_{C,l}(w) &= \lim_{n\to\infty} \frac{1}{n^{1+k}} \#\Pi^*_{C,l}(w) \\
&= \int_{[0,1]^{k+1}} I(0 \le L_i(v_S) < 1, i \notin S, v_0 = L_{2k}(v_S)) \, dv_S.
\end{aligned}$$

This shows part (a) of Lemma 10.5.1. For part (b) let $l \neq l'$. Without loss of generality, let us assume that, $l_{i_1} = -l'_{i_1}$. Let $\pi \in \Pi^*_{C,l}(w) \bigcap \Pi^*_{C,l'}(w)$. Then $\pi(i_1 - 1) = \pi(i_1)$ and hence $L^l_{i_1-1}(v_S) = v_{i_1}$. It now follows along the lines of the preceding arguments that

$$\lim_{n\to\infty} \frac{1}{n^{1+k}} \#(\Pi^*_{C,l}(w) \bigcap \Pi^*_{C,l'}(w)) \le \int_{[0,1]^{k+1}} \cdots \int I(v_i = L^l_{i_1-1}(v_S)) dv_S.$$

$L^l_{i_1-1}(v_S)$ contains $\{v_j : j \in S, \ j < i_1\}$ and hence $\{L^l_{i_1-1}(v_S) = v_i\}$ is a k-dimensional sub-space of $[0,1]^{k+1}$ and hence has Lebesgue measure 0.

(ii) **Hankel and Reverse Circulant**: Let X and Y be Hankel (H) and Reverse Circulant (RC) respectively. Then

$$\pi(i-1) + \pi(i) = \pi(j_i - 1) + \pi(j_i) \quad \text{for all } i \in S_H, \tag{10.20}$$

$$(\pi(i-1) + \pi(i)) \mod n = (\pi(j_i - 1) + \pi(j_i)) \mod n \quad \text{for all } i \in S_{RC}. \tag{10.21}$$

Clearly, as all the $\pi(i)$ are between 1 and n, relation (10.21) implies $(\pi(i-1) + \pi(i)) - (\pi(j_i - 1) + \pi(j_i)) = a_i n$ where $a_i \in \{0, 1, -1\}$

Let $a = (a_1, ..., a_{k_2}) \in I = \{-1, 0, 1\}^{k_2}$. Let $\Pi^*_{C,a}(w)$ be the subset of $\Pi^*_C(w)$ such that,

$$\pi(i-1) + \pi(i) = \pi(j_i - 1) + \pi(j_i) \ \forall i \in S_H \text{ and}$$

$$(\pi(i-1) + \pi(i)) - (\pi(j_i - 1) + \pi(j_i)) = a_i n \ \forall i \in S_{RC}.$$

Now clearly,

$$\Pi^*_C(w) = \bigcup_a \Pi^*_{C,a}(w) \text{ (a disjoint union)}.$$

Now we get that,

$$\begin{aligned}
&\#\Pi^*_{C,a}(w) \\
&= \#\{(v_0, ..., v_{2k}) : v_i \in U_n \ \forall i, \ v_{i-1} + v_i = v_{j_i-1} + v_{j_i} + a_i \ \forall i \in S_{RC} \\
&\qquad \text{and } v_{i-1} + v_i = v_{j_i-1} + v_{j_i} \ \forall i \in S_H, \ v_0 = v_{2k}\}.
\end{aligned}$$

Other than $v_0 = v_{2k}$, each $\{v_i : i \notin S\}$ can be written uniquely as an affine linear combination $L_i^a(v_S) + b_i^{(a)}$ for some integer $b_i^{(a)}$. Moreover, $L_i^a(v_S)$ only contains $\{v_j : j \in S, \ j < i\}$ with non-zero coefficients. Arguing as in the previous case,

$$\#\Pi_{C,a}^*(w) \tag{10.22}$$
$$= \#\{v_S : v_i \in U_n \ \forall i \in S, v_0 = L_{2k}^a(v_S) + b_{2k}^{(a)}, 0 \le L_i^a(v_S) + b_i^{(a)} < 1 \forall i \notin S\}.$$

As before,

$$p_{C,a}(w) = \lim_{n \to \infty} \frac{1}{n^{1+k}} \#\Pi_{C,a}^*(w)$$
$$= \int_{[0,1]^{k+1}} I\left(0 \le L_i^a(v_S) + b_i^{(a)} < 1, i \notin S, v_0 = L_{2k}^a(v_S) + b_{2k}^{(a)}\right) dv_S$$

and the proof of this case is complete.

(iii) **Hankel and Symmetric Circulant**: Let X and Y be Hankel (H) and Symmetric Circulant (SC) respectively. Note that

$$\pi(i-1) + \pi(i) = \pi(j_i - 1) + \pi(j_i) \ \ \forall i \in S_H \text{ and}$$

$$n/2 - |n/2 - |\pi(i-1) - \pi(i)|| = n/2 - |n/2 - |\pi(j_i-1) - \pi(j_i)|| \ \ \forall i \in S_S.$$

It can be easily seen from the second equation above that either $|\pi(i-1) - \pi(i)| = |\pi(j_i-1) - \pi(j_i)|$ or $|\pi(i-1) - \pi(i)| + |\pi(j_i-1) - \pi(j_i)| = n$. There are six cases for each Symmetric Circulant match $[i, j_i]$, and with $v_i = \pi(i)/n$, these are:

1. $v_{i-1} - v_i - v_{j_i-1} + v_{j_i} = 0.$
2. $v_{i-1} - v_i + v_{j_i-1} - v_{j_i} = 0.$
3. $v_{i-1} - v_i + v_{j_i-1} - v_{j_i} = 1.$
4. $v_{i-1} - v_i - v_{j_i-1} + v_{j_i} = 1.$
5. $v_i - v_{i-1} + v_{j_i-1} - v_{j_i} = 1.$
6. $v_i - v_{i-1} + v_{j_i} - v_{j_i-1} = 1.$

We can write $\Pi_C^*(w)$ as the (not disjoint) union of 6^{k_2} possible $\Pi_{C,l}^*(w)$ where l denotes the combination of cases (1)–(6) above that is satisfied in the k_2 matches of Symmetric Circulant. For each $\pi \in \Pi_{C,l}^*(w)$, each $\{v_i : i \notin S\}$ is a unique affine integer combination of v_S. As in the previous two pairs of matrices in (i) and (ii), $\lim_{n\to\infty} \frac{1}{n^{1+k}} \#\Pi_{C,l}^*(w)$ exists as an integral.

Now (10.17) can be checked case by case. As a typical case suppose Case 1 and Case 3 hold. Then $\pi(i-1) - \pi(i) = n/2$ and $v_{i-1} - v_i = 1/2$. Since i is generating and v_{i-1} is a linear combination of $\{v_j : j \in S, \ j < i\}$, this implies a non-trivial linear relation between the independent vertices v_S. This, in turn, implies that the number of circuits π satisfying the above conditions is $o(n^{1+k})$.

(iv) **Toeplitz and Symmetric Circulant**: Let X and Y be Toeplitz (T) and Symmetric Circulant (SC) respectively. Again note that,

$$|\pi(i-1) - \pi(i)| = |\pi(j_i - 1) - \pi(j_i)| \ \forall i \in S_T \text{ and}$$

$$n/2 - |n/2 - |\pi(i-1) - \pi(i)|| = n/2 - |n/2 - |\pi(j_i - 1) - \pi(j_i)|| \ \forall i \in S_{SC}. \quad (10.23)$$

Now, (10.23) implies either $|\pi(i-1) - \pi(i)| = |\pi(j_i - 1) - \pi(j_i)|$ or $|\pi(i-1) - \pi(i)| + |\pi(j_i - 1) - \pi(j_i)| = n$.

There are six cases for each Symmetric Circulant match as in Case (iii) above and two cases for each Toeplitz match.

As before we can write $\Pi_C^*(w)$ as the (not disjoint) union of $2^{k_1} \times 6^{k_2}$ possible $\Pi_{C,l}^*(w)$ where l denotes a combination of cases (1)–(6) for all SC matches (as in Case (iii)) and a combination of cases (1)–(2) for all T matches. As before, for each $\pi \in \Pi_{C,l}^*(w)$, each of the $\{v_i : i \notin S\}$ can be written uniquely as an affine integer combination of v_S. As earlier, $\lim_{n \to \infty} \frac{1}{n^{1+k}} \# \Pi_{C,l}^*(w)$ exists as an integral.

Now, (10.17) is again checked case by case. Suppose $l \neq l'$ and $\pi \in \Pi_{C,l}^*(w) \bigcap \Pi_{C,l'}^*(w)$. For $l \neq l'$, there must be one Toeplitz or Symmetric Circulant match such that two of the possible cases in (1)–(2) or in (1)–(6) occur simultaneously. Here we just deal with a typical pair, Case (1) and Case (2), for the Toeplitz match. Then we have $\pi(i-1) - \pi(i) = 0$ and hence $v_{i-1} - v_i = 0$. Since i is generating and v_{i-1} is a linear combination of $\{v_j : j \in S, j < i\}$, this implies there exists a non-trivial relation between the independent vertices v_S. This, in turn, implies that the number of circuits π satisfying the above conditions in $o(n^{1+k})$. Now suppose the Symmetric Circulant match happens for both Case (1) and Case (2). Then again we have $v_i = v_{i-1}$ and we can argue as before to conclude that (10.17) holds.

(v) **Toeplitz and Reverse Circulant**: Let X and Y be Toeplitz (T) and Reverse Circulant (RC) respectively. Note,

$$|\pi(i-1) - \pi(i)| = |\pi(j_i - 1) - \pi(j_i)| \quad \text{for all } i \in S_T,$$

$$(\pi(i-1) + \pi(i)) \mod n = (\pi(j_i - 1) + \pi(j_i)) \mod n \quad \text{for all } i \in S_{RC}.$$

Clearly, as all the $\pi(i)$ are between 1 and n, $(\pi(i-1) + \pi(i)) \mod n = (\pi(j_i - 1) + \pi(j_i)) \mod n$ implies $(\pi(i-1) + \pi(i)) - (\pi(j_i - 1) + \pi(j_i)) = a_i n$ where $a_i \in \{0, 1, -1\}$

Let the number of Toeplitz and Reverse Circulant matches be k_1 and k_2. respectively, and let $S_T = \{i_1, i_2, ..., i_{k_1}\}$, $S_{RC} = \{i_{k_1+1}, i_{k_1+2}, ..., i_{k_1+k_2}\}$. Let $l = (c, a) = (c_{i_1}, ..., c_{i_{k_1}}, a_{i_{k_1+1}}, ..., a_{i_{k_1+k_3}}) \in I = \{-1, 1\}^{k_1} \times \{-1, 0, 1\}^{k_3}$.

Let $\Pi_{C,l}^*(w)$ be the subset of $\Pi_C^*(w)$ such that,

$$\pi(i-1) - \pi(i) = c_i(\pi(j_i - 1) - \pi(j_i)) \ \forall i \in S_T$$

$$\pi(i-1) + \pi(i) = \pi(j_i - 1) + \pi(j_i) + a_i n \quad \forall i \in S_{RC}.$$

Now clearly,

$$\Pi_C^*(w) = \bigcup_{l \in I} \Pi_{C,l}^*(w),$$

and translating this into the language of v_i's, we get

$$\#\Pi_{C,l}^*(w)$$
$$= \#\{(v_0, ..., v_{2k}) : v_i \in U_n \ \forall i, \ v_{i-1} + v_i = (v_{j_i-1} + v_{j_i}) + a_i \ \forall i \in S_{RC}$$
$$\text{and} \quad v_{i-1} - v_i = c_i(v_{j_i-1} - v_{j_i}) \ \forall i \in S_T, \ v_0 = v_{2k}\}.$$

As in the previous cases, $\lim_{n\to\infty} \frac{\#\Pi_{C,l}^*(w)}{n^{1+k}}$ exists. It remains to show that, $\lim_{n\to\infty} \frac{\#(\Pi_{C,l}^*(w) \bigcap \Pi_{C,l'}^*(w))}{n^{1+k}} = 0$ for $l \neq l'$. If $l = (c,a) \neq l' = (c',a')$, then either $c \neq c'$ or $a \neq a'$. If $c = c'$, then clearly $\Pi_{C,l}^*(w)$ and $\Pi_{C,l'}^*(w)$ are disjoint. Let $c \neq c'$. Without loss of generality, we assume $c_{i_1} = -c_{i_1}$. Then clearly, for every $\pi \in \Pi_{C,l}^*(w) \bigcap \Pi_{C,l'}^*(w)$ we have $v_{i_1-1} = v_i$, which gives a non-trivial relation between $\{v_j : j \in S\}$. That in turn implies the required limit is 0.

(vi) **Reverse Circulant and Symmetric Circulant**: Let X and Y be Reverse Circulant (RC) and Symmetric Circulant (SC) respectively. Then

$$\pi(i-1) + \pi(i) \mod n = \pi(j_i - 1) + \pi(j_i) \mod n \ \forall i \in S_{RC} \text{ and}$$

$$n/2 - |n/2 - |\pi(i-1) - \pi(i)|| = n/2 - |n/2 - |\pi(j_i-1) - \pi(j_i)|| \ \forall i \in S_{SC}.$$

As before, the latter equation implies either $|\pi(i-1)-\pi(i)| = |\pi(j_i-1)-\pi(j_i)|$ or $|\pi(i-1) - \pi(i)| + |\pi(j_i-1) - \pi(j_i)| = n$.

There are now three cases for each Reverse Circulant match:

1. $v_{i-1} + v_i - v_{j_i-1} - v_{j_i} = 0.$
2. $v_{i-1} + v_i - v_{j_i-1} - v_{j_i} = 1.$
3. $v_{i-1} + v_i - v_{j_i-1} - v_{j_i} = -1.$

Also, there are six cases for each Symmetric Circulant match as in Case (iii).

As before we can write $\Pi_C^*(w)$ as the union of $3^{k_1} \times 6^{k_2}$ possible $\Pi_{C,l}^*(w)$. Hence arguing in a similar manner, $\lim_{n\to\infty} \frac{1}{n^{1+k}} \#\Pi_{C,l}^*(w)$ exists as an integral. Now, to check (10.17), case by case. Suppose $l \neq l'$ and $\pi \in \Pi_{C,l}^*(w) \bigcap \Pi_{C,l'}^*(w)$. Since $l \neq l'$, there must be one Reverse Circulant or Symmetric Circulant match such that two of the possible cases (1)–(3) or (1)–(6) (which appear in Case (iii)) occur simultaneously. It is easily seen that such an occurrence is impossible for a Reverse Circulant match. So we assume there is a Symmetric Circulant match and deal with one such typical match. Suppose then we have both Case (1) and Case (2). Then again we have $v_i = v_{i-1}$ and as a consequence (10.17) holds.

(vii) **Wigner and Hankel**: Let X and Y be Wigner (W) and Hankel (H) respectively. Observe that,

$$(\pi(i-1), \pi(i)) = \begin{cases} (\pi(j_i - 1), \pi(j_i)) & \text{(Constraint } C1) \\ (\pi(j_i), \pi(j_i - 1)) & \text{(Constraint } C2, \forall i \in S_W). \end{cases} \tag{10.24}$$

Also, $\pi(i-1) + \pi(i) = \pi(j_i - 1) + \pi(j_i)$ for all $i \in S_H$. So for each Wigner match there are two constraints and hence there are 2^{k_1} choices. Let λ be a typical choice of k_1 constraints and $\Pi^*_{C,\lambda}(w)$ be the subset of $\Pi^*_C(w)$ where the above relations hold. Hence

$$\Pi^*_C(w) = \bigcup_\lambda \Pi^*_{C,\lambda}(w) \quad \text{(not a disjoint union)}.$$

Now using equation (10.18) we have,

$$\#\Pi^*_{C,\lambda}(w)$$
$$= \#\{(v_0, \ldots, v_{2k}) : 0 \le v_i \le 1, v_0 = v_{2k}, v_{i-1} + v_i = v_{j_i - 1} + v_{j_i}, \ i \in S_H$$
$$v_{i-1} = v_{j_i - 1}, v_i = v_{j_i}, (C1), v_{i-1} = v_{j_i}, v_i = v_{j_i - 1}(C2), i \in S_W\}.$$

It can be seen from the above equations that each v_j, $j \notin S$ can be written (not uniquely) as a linear combination L_j^λ of elements in v_S. Hence as before,

$$\#\Pi^*_{C,\lambda}(w)$$
$$= \#\{v_S : v_i = L_i^\lambda(v_S), v_0 = v_{2k}, i \notin S, , v_{i-1} + v_i = v_{j_i - 1} + v_{j_i}, \ i \in S_H,$$
$$v_{i-1} = v_{j_i - 1}, v_i = v_{j_i}, (C1), v_{i-1} = v_{j_i}, v_i = v_{j_i - 1}(C2), i \in S_W\}.$$

So the limit of $\#\Pi^*_{C,\lambda}(w)/n^{1+k}$ exists and can be expressed as an appropriate Riemann integral.

Now we show (10.17). Without loss of generality assume λ_1 is a C_1 constraint and λ_2 is a C_2 constraint. For any $\pi \in \Pi^*_{C,\lambda_1}(w) \bigcap \Pi^*_{C,\lambda_2}(w)$ we note that for $i \in S$,

$$(\pi(j_i), \pi(j_i - 1)) = (\pi(i-1), \pi(i)) = (\pi(j_i - 1), \pi(j_i)),$$

which implies $\pi(i) = \pi(i-1)$. Now i is a generating vertex. But $\pi(i) = \pi(i-1)$ and hence is fixed, having chosen the first $i-1$ vertices. This lowers the order by a power of n and hence the claim follows.

(viii) **Wigner and other matrices**: These cases follow by similar and repetitive arguments. □

Proof of Theorem 10.2.3. Let w be a colored word of length $2k$ for a monomial $q = q(X, Y)$. Let w' be obtained from w by a cyclic permutation, that is, there exists l such that $w'[i] = w[(i+l) \bmod 2k]$. Note that w' is a colored word for the monomial q' obtained from q by the same cyclic permutation. We have the following lemma.

Lemma 10.5.2. $\#\Pi_C^*(w) = \#\Pi_C^*(w')$ and $p_C(w) = p_C(w')$.

Proof. Let $\pi \in \Pi_C^*(w)$. Let $\pi'(i) = \pi((i+l) \bmod 2k))$. Clearly, $\pi'(0) = \pi'(2k)$. Also

$$w'[i] = w'[j] \Rightarrow L^*(\pi'(i-1), \pi'(i)) = L^*(\pi'(j-1), \pi'(j))$$

where L^* is equal to L_1 or L_2 according when $w'[i] = w'[j]$ is an X match or a Y match. Hence, $\pi' \in \Pi_C^*(w')$.

As w can also be obtained from w' by another cyclic permutation, it follows that the map $\pi \to \pi'$ is a bijection between $\Pi_C^*(w)$ and $\Pi_C^*(w')$. Hence $\#\Pi_C^*(w) = \#\Pi_C^*(w')$ and $p_C(w) = p_C(w')$. □

Proof of Theorem 10.2.3. (a) We use induction on the length of the word.

If $k = 1$ then $q = XX$ or $q = YY$. The only colored Catalan word is aa (drop superscript for ease). In either case, $\pi(0) = i, \pi(1) = j, \pi(2) = i$ is a circuit in $\Pi_C^*(w)$ for $1 \leq i \leq n, 1 \leq j \leq n$. Hence, $\#\Pi_C^*(w) \geq n^2$ and the result is true for $k = 1$.

Now let us assume that the claim holds for all monomials q of length less than $2k$ and all Catalan words corresponding to q. By Lemma 10.5.2, without loss, we assume that $w = aaw_1$ where w_1 is a Catalan word of length $(2k-2)$. Now let $\pi' \in \Pi_C^*(w_1)$. For fixed j, $1 \leq j \leq n$, define π by

$$\pi(0) = \pi'(0) \tag{10.25}$$
$$\pi(1) = j \tag{10.26}$$
$$\pi(j) = \pi'(j-2), \qquad j \geq 2. \tag{10.27}$$

Clearly π is a circuit and $\pi(0) = \pi(2)$ implies $L(\pi(0), \pi(1)) = L(\pi(1), \pi(2))$. Hence $\pi \in \#\Pi_C^*(w)$ and so, $\#\Pi_C^*(w) \geq n\#\Pi_C^*(w_1) \geq n^{k+1}$ and hence (i) is proved.

(b) We shall now show that $p_C(w) \leq 1$ for matrices whose link functions satisfy *Property B* and *Property P*. The proof is the same as the proof given in Chapter 5 for the non-indexed and non-colored case with appropriate changes to add color and index. We indicate the changes while keeping the notation similar for easy comparison. The proof uses $(2k+1)$-tuple π which are not necessarily circuits, that is, $\pi(0) = \pi(2k)$ is not assumed. Let w be a colored Catalan word. Define

$$C'(w) =$$
$$\{\pi: w[i] = w[j] \Rightarrow c_i = c_j \text{ and } L_{c_i}(\pi(i-1), \pi(i)) = L_{c_j}(\pi(j-1), \pi(j))\}$$
$$\Gamma_{i,j}(w) =$$
$$\{\pi \in C(w): \pi(0) = i, \pi(2k) = j\}, \ (1 \leq i, j \leq n), \ \gamma_{i,j}(w) = \#\Gamma_{i,j}(w).$$

Clearly, $\#\Pi_C^*(w) = \sum_{i=1}^{n} \gamma_{i,i}(w)$. Now consider the following statement $\mathbf{S}_k'$ for all $k \geq 1$:

$\mathbf{S}'_k$: For any colored Catalan w of length $(2k)$, there exists $M_k > 0$ such that

$$\gamma_{i,j}(w) \le M_k n^{k-1} \text{ for all } i \ne j \text{ and } \frac{1}{n}\sum_{i=1}^{n}\left|\frac{\gamma_{i,i}(w)}{n^k} - 1\right| = O(n^{-1}).$$

The proof of $\mathbf{S}'_k$ easily follows by repeating the steps of the corresponding proof in Chapter 5 for non-colored and non-indexed words and changing the set $C(w)$ there by $C'(w)$ and using *Property B* and *Property P*. To avoid repetitive arguments we skip the details. Once the validity of $\mathbf{S}'_k$ is asserted, one gets $p_C(w) \le 1$ and the result now follows using part (a). $\qquad\square$

Proof of Theorem 10.3.1. We need the following development for describing freeness.

Let S_n be the group of permutations of $(1, \ldots n)$.

Definition 10.5.1. Let $\mathcal{A}$ be an algebra. Let $\psi_k : \mathcal{A}^k \longrightarrow C \ k > 0$ be multi'-linear functions. For $\alpha \in S_n$, let $c_1, \ldots, c_r$ be the cycles of α. Then define

$$\psi_\alpha[A_1, \ldots, A_n] = \psi_{c_1}[A_1, \ldots, A_n] \cdots \psi_{c_r}[A_1, \ldots, A_n]$$

where

$$\psi_c[A_1, \ldots, A_n] = \psi_p\left(A_{i_1} \cdots A_{i_p}\right) \text{ if } c = (i_1, i_2 \ldots, i_p).$$

$\diamond$

Freeness is intimately tied to non-crossing partitions. We describe the relevant portion of this relation in brief below. See Theorem 14.4 of Nica and Speicher (2006)[79] for more details. Let $NC_2(m)$ be the set of non-crossing pair-partitions of $\{1, \ldots, m\}$. A typical pair-partition π will be written in the form $\{(r, \pi(r)), \ r = 1, \ldots, m\}$. Recall that, $p = (p(1), \ldots, p(m))$ integers (also can be referred to as colors), let

$$NC_2^{(p)}(m) = \{\pi \in NC_2(m) : p(\pi(r)) = p(r) \text{ for all } r = 1, \ldots, m\}.$$

Suppose $d_1, \ldots, d_m, s_1, \ldots, s_m$ are elements in some non-commutative probability space $(\mathcal{B}, \phi)$. Suppose $\{s_1, \ldots, s_m\}$ are free and each s_i follows the semi-circle law. Then the collections $\{s_1, \ldots, s_m\}$ and $\{d_1, \ldots, d_m\}$ are free if and only if,

$$\phi(s_{p(1)}d_1 \cdots s_{p(m)}d_m) = \sum_{\pi \in NC(m)} k_\pi[s_{p(1)}, \ldots, s_{p(m)}] \cdot \phi_{\pi\gamma}[d_1, \ldots, d_m]$$

$$= \sum_{\pi \in NC_2^{(p)}(m)} \phi_{\pi\gamma}[d_1, \ldots, d_m], \qquad (10.28)$$

where $\gamma \in S_m$ is the cyclic permutation with one cycle and $\gamma = (1, \ldots, m -$

$1, m$). Here $\{k_n\}$ denote the free cumulants and k_π for a partition π is defined along the same lines as Definition 10.5.1.

We shall also drop the suffix C from $p_C(w)$, $\Pi_C(w)$, $\Pi_C^*(w)$ etc. for simplicity. Fix a monomial q of Wigner (W) and any other patterned matrix (A) of length $2k$. To prove freeness we show that the limiting variables satisfy the relation (10.28). We have already remarked that freeness is intimately tied to non-crossing partitions but freeness in the limit can also be roughly described in terms of colored words in the following manner.

1. If for a colored word the pair-partitions corresponding to the Wigner matrix cross, then $p(w) = 0$.

2. If the pair-partition corresponding to the letters of matrix A cross with any pair-partition of W then also $p(w) = 0$.

For example, $p(w_1 w_2 w_1 w_2 a_1 a_1) = 0$ and $p(w_1 a_1 w_1 a_1) = 0$. This is essentially the main content of Lemma 10.5.3 given below.

We will discuss in detail the proof of Theorem 10.3.1 for $p = 1$ and indicate how the results continue to hold for $p > 1$.

We now concentrate only on (colored) pair-matched words. For a word w, the pair $(i, j)\, 1 \le i < j \le 2k$ is said to be a match if $w[i] = w[j]$. A match (i, j) is said to be a W match or an A match according when $w[i] = w[j]$ is a Wigner or an A letter.

Define $w_{(i,j)}$ to be the word of length $j - i + 1$ as

$$w_{(i,j)}[k] = w[i - 1 + k] \text{ for all } 1 \le k \le j - i + 1.$$

Let $w_{(i,j)^c}$ be the word of length $t + i - j - 1$ obtained by removing $w_{(i,j)}$ from w, that is,

$$w_{(i,j)^c}[r] = \begin{cases} w[r] & \text{if } r < i, \\ w[r + j - i + 1] & \text{if } r \ge i. \end{cases}$$

Note that in general these sub-words may not be matched. If (i, j) is a W match, we will call $w_{(i,j)}$ a *Wigner string* of length $(j - i + 1)$. For instance, for the monomial $WAAAAWWW$, $w = abbccadd$ is a word and $abbcca$ and dd are Wigner strings of length six and two respectively. For any word w, we define the following two classes:

$$\Pi_{(C2)}^*(w) = \{\pi \in \Pi^*(w) : (i, j)\ W \text{ match} \Rightarrow (\pi(i - 1), \pi(i)) = (\pi(j), \pi(j - 1))\},$$
$$\Pi_{(i,j)}^*(w) = \{\pi \in \Pi^*(w) : (\pi(i - 1), \pi(i)) = (\pi(j), \pi(j - 1))\}.$$

The condition above involves a (C2) constraint defined in (10.24) and

$$\Pi_{(C2)}^*(w) = \bigcap_{(i,j):\, W\, match} \Pi_{(i,j)}^*(w). \tag{10.29}$$

We know that for a single sequence of Wigner matrices, $p(w) \ne 0$ if and only if all constraints are (C2). We need an extension now.

Lemma 10.5.3. Let w be of length $2k$, colored pair-matched and $p(w) \neq 0$. Then

(a) Every Wigner string is a colored pair-matched word.

(b) For any (i, j) which is a W match we have

$$\lim_{n \longrightarrow \infty} \frac{\#(\Pi^*(w) - \Pi^*_{(i,j)}(w))}{n^{1+k}} = 0. \tag{10.30}$$

(c)

$$\lim_{n \longrightarrow \infty} \frac{\#(\Pi^*(w) - \Pi^*_{(C2)}(w))}{n^{1+k}} = 0. \tag{10.31}$$

$\diamond$

By (10.29) as the number of pairs (i, j) is finite, (c) and (b) are equivalent.

Lemma 10.5.4. Suppose X_n has LSD and they satisfy Assumptions A1 and A2, then for any $l \geq 1$ and integers $(k_1, \ldots, k_l)$, we have

$$\mathrm{E}\Big[\prod_{i=1}^{l}\big(\frac{1}{n}\,\mathrm{Tr}\big((\frac{X_n}{\sqrt{n}})^{k_i}\big)\big)\Big] - \prod_{i=1}^{l}\mathrm{E}\Big[\frac{1}{n}\,\mathrm{Tr}\big((\frac{X_n}{\sqrt{n}})^{k_i}\big)\Big] \to 0 \text{ as } n \to \infty.$$

$\diamond$

We shall prove the lemmata after we prove Theorem 10.3.1.

We take a single copy of W and A to show the result. For multiple copies the proof essentially remains the same, modulo some notation. Let q be a typical monomial, $q = WA^{q(1)} \cdots WA^{q(m)}$ of length $2k$, where the $q(i)$'s may equal 0. So, $k = m/2 + (q(1) + \cdots + q(m))/2$. From Theorem 10.2.2, for every such monomial q, $\frac{1}{n^{k+1}}\,\mathrm{Tr}(q)$ converges to, say, $\phi(sa^{q(1)} \cdots sa^{q(m)})$, where s follows the semi-circle law and a is the marginal limit of A, and ϕ is the appropriate functional defined on the space of non-commutative polynomial algebra generated by a and s. It is enough to prove that ϕ satisfies (10.28).

Let us expand the expression for

$$\lim_{n\to\infty} \frac{1}{n^{1+k}}\,\mathrm{E}[\mathrm{Tr}(WA^{q(1)} \cdots WA^{q(m)})]$$

$$= \lim_{n\to\infty} \frac{1}{n^{1+k}} \sum_{\substack{i(1),\ldots,i(m) \\ j(1),\ldots,j(m)=1}}^{n} \mathrm{E}[w_{i(1)j(1)}a^{q(1)}_{j(1)i(2)}w_{i(2)j(2)}a^{q(2)}_{j(2)i(3)} \cdots w_{i(m)j(m)}a^{q(m)}_{j(m)i(1)}]$$

$$\tag{10.32}$$

$$= \lim_{n\to\infty} \frac{1}{n^{1+k}} \sum_{w\in CW(2)} \sum_{\pi\in\Pi^*(w)} \mathrm{E}[\mathbb{X}_\pi]$$

$$= \lim_{n\to\infty} \frac{1}{n^{1+k}} \sum_{w\in CW(2)} \sum_{\pi\in\Pi^*_{(C2)}(w)} \mathrm{E}[\mathbb{X}_\pi] \text{ (Lemma 10.5.3(c) and Assumption (10.1)).}$$

$$\tag{10.33}$$

Colored pair-matched words of length $2k$ are in bijection with the set of pair-partitions on $\{1, \ldots, 2k\}$ (denoted by $\mathcal{P}_2(2k)$). Now each such word w induces σ_w a pair-partition of $\{1, \ldots, m\}$ that is induced by only the Wigner matches (i.e. $(a, b) \in \sigma_w$ iff (a, b) is a Wigner match). So given any pair-partition σ of $\{1, \ldots, m\}$, we denote by $[\sigma]_W$ the class of all w which induce the partition σ. So the sum in (10.33) can be written as

$$\lim_{n \to \infty} \frac{1}{n^{1+k}} \sum_{\sigma \in \mathcal{P}_2(m)} \sum_{w \in [\sigma]_W} \sum_{\pi \in \Pi^*_{(C2)}(w)} E[\mathbb{X}_\pi]. \tag{10.34}$$

By the (C2) constraint imposed on the class $\Pi^*_{(C2)}(w)$, if (r, s) is a W match then $(i(r), j(r)) = (j(s), i(s))$ (or, equivalently in terms of π we have, $(\pi(r - 1), \pi(r)) = (\pi(s), \pi(s - 1)))$.

Therefore, we have the following string of equalities. The equality in (10.35) follows from (10.32) and (10.33). The steps in (10.36), (10.37) and (10.38) follow easily from calculations similar to Proposition 22.32 of Nica and Speicher (2006)[79]. The last step follows from the fact that the number of cycles of $\sigma\gamma$ is equal to $1 + m/2$ if and only if $\sigma \in NC_2(m)$. The notation $\mathrm{Tr}_{\sigma\gamma}$ is defined via Definition 10.5.1 by noting that Tr is multilinear.

$$\lim_{n \to \infty} \frac{1}{n^{k+1}} \mathrm{E}[\mathrm{Tr}(WA^{q(1)} \cdots WA^{q(m)})]$$

$$= \lim_{n \to \infty} \frac{1}{n^{k+1}} \sum_{\sigma \in \mathcal{P}_2(m)} \sum_{\substack{i(1), \ldots, i(m) \\ j(1), \ldots, j(m)=1}}^{n} \prod_{(r,s) \in \sigma} \delta_{i(r)j(s)} \delta_{i(s)j(r)} \, \mathrm{E}[a^{q(1)}_{j(1)i(2)} \cdots a^{q(m)}_{j(m)i(1)}] \tag{10.35}$$

$$= \lim_{n \to \infty} \frac{1}{n^{k+1}} \sum_{\sigma \in \mathcal{P}_2(m)} \sum_{\substack{i(1), \ldots, i(m) \\ j(1), \ldots, j(m)=1}}^{n} \prod_{(r,s) \in \sigma} \delta_{i(r)j(s)} \delta_{i(s)j(r)} \, \mathrm{E}[\prod_{k=1}^{m} a^{q(k)}_{j(k)i(\gamma(k))}] \tag{10.36}$$

$$= \lim_{n \to \infty} \frac{1}{n^{k+1}} \sum_{\sigma \in \mathcal{P}_2(m)} \sum_{\substack{i(1), \ldots, i(m) \\ j(1), \ldots, j(m)=1}}^{n} \prod_{r=1}^{m} \delta_{i(r)j(\sigma(r))} \, \mathrm{E}[a^{q(1)}_{j(1)i(\gamma(1))} \cdots a^{q(m)}_{j(m)i(\gamma(m))}] \tag{10.37}$$

$$= \lim_{n \to \infty} \frac{1}{n^{k+1}} \sum_{\sigma \in \mathcal{P}_2(m)} \sum_{j(1), \ldots, j(m)=1}^{n} \mathrm{E}[a^{q(1)}_{j(1)j(\sigma\gamma(1))} \cdots a^{q(m)}_{j(m)j(\sigma\gamma(m))}] \tag{10.38}$$

$$= \sum_{\sigma \in NC_2(m)} \lim_{n \to \infty} \mathrm{E}\left(\mathrm{Tr}_{\sigma\gamma}[A^{(q_1)}, \ldots, A^{(q_m)}]\right).$$

Now it follows from Lemma 10.5.4 that,

$$\sum_{\sigma \in NC_2(m)} \lim_{n \to \infty} \mathrm{E} \left(\mathrm{Tr}_{\sigma\gamma}[A^{(q_1)}, \ldots, A^{(q_m)}] \right) = \sum_{\sigma \in NC_2(m)} \lim_{n \to \infty} (\mathrm{E}\,\mathrm{Tr})_{\sigma\gamma}[A^{(q_1)}, \ldots, A^{(q_m)})]$$

$$= \sum_{\sigma \in NC_2(m)} \phi_{\sigma\gamma}[a^{(q_1)}, a^{(q_2)}, \ldots, a^{(q_m)})].$$

This establishes (10.28) and hence freeness in the limit.

The above method can be easily extended to plug in more independent copies of W and A. The following details will be necessary.

(1) The extension of Lemmata 10.5.3 and 10.5.4. Note that these extensions can be easily obtained using the injective mapping ψ described in Section 10.2 and used in Theorem 10.2.1.

(2) When we consider several independent copies of the Wigner matrix the product in (10.37) gets replaced by

$$\prod_{r=1}^{m} \delta_{i(r)j(\sigma(r))} \delta_{p(r)p(\sigma(r))}.$$

Here $(p(1), \ldots, p(m))$ denotes the colors corresponding to the independent Wigner matrices. Then the calculations are similar to the proof of Theorem 22.35 of Nica and Speicher (2006)[79].

The rest are some algebraic details, which we skip.

Having proved the theorem we now come back to the proofs of Lemma 10.5.3 and 10.5.4. The next lemma turns out to be the most essential ingredient in proving Lemma 10.5.3 and it points out the behavior of a colored pair-matched word which contains a Wigner string inside it.

Lemma 10.5.5. For any colored pair-matched word w and a Wigner string $w_{(i,j)}$ which is a pair-matched word and satisfies equation (10.30)

$$p(w) = p(w_{(i,j)})p(w_{(i,j)^c}). \tag{10.39}$$

Further, if $w_{(i+1,j-1)}$ and $w_{(i,j)^c}$ satisfy (10.31) then so does w. ◇

Proof. Given any $\pi_1 \in \Pi^*(w_{(i+1,j-1)})$ and $\pi_2 \in \Pi^*(w_{(i,j)^c})$ construct π as: $(\pi_2(0), \ldots, \pi_2(i-1), \pi_1(0), \ldots, \pi_1(j-i-1)) = (\pi_1(0), \pi_2(i-1), \ldots \pi_2(2k-j+i-1)) \in \Pi^*_{(i,j)}(w)$. Conversely, from any $\pi \in \Pi^*_{(i,j)}(w)$ one can construct π_1 and π_2 by reversing the above construction.

So we have

$$\#\Pi^*_{(i,j)}(w) = \#\Pi^*(w_{(i+1,j-1)})\#\Pi^*(w_{(i,j)^c}). \tag{10.40}$$

Let $|w_{(i+1,j-1)}| = 2l_1$ and $|w_{(i,j)^c}| = 2l_2$ and note that $(1+l_1)+(1+l_2) = k+1$.

Now using the fact that $w_{(i,j)}$ satisfies (10.30) and dividing equation (10.40) by n^{k+1} we get in the limit,

$$p(w) = p(w_{(i+1,j-1)})p(w_{(i,j)}^c).$$

Now we claim that

$$\#\Pi^*(w_{(i,j)}) = n\#\Pi^*(w_{(i+1,j-1)}). \tag{10.41}$$

Now given $\pi \in \Pi^*(w_{(i,j)})$, one can always get a $\pi' \in \Pi^*(w_{(i+1,j-1)})$, where the $\pi(i-1)$ is arbitrary and hence $\frac{\#\Pi^*(w_{(i,j)})}{n} \le \#\Pi^*(w_{(i+1,j-1)})$. Also given a $\pi' \in \Pi^*(w_{(i+1,j-1)})$ one can choose $\pi(i-1)$ in n ways and also assign $\pi(j) = \pi(i-1)$ or $\pi(i)$, making j a dependent vertex. So we get that, $\#\Pi^*(w_{(i,j)}) \ge n\#\Pi^*(w_{(i+1,j-1)})$. This shows (10.41). So from (10.41) it follows that

$$p(w_{(i,j)}) = p(w_{(i+1,j-1)}),$$

whenever $w_{(i,j)}$ is a Wigner string.

Also note that from the first construction,

$$\#\Pi^*_{(C2)}(w) = \#\Pi^*_{(C2)}(w_{(i+1,j-1)})\#\Pi^*_{(C2)}(w_{(i,j)^c}).$$

Now suppose $w_{(i+1,j-1)}$ and $w_{(i,j)^c}$ satisfy (10.31). So we have that

$$\#\Pi^*(w_{(i+1,j-1)}) = \#\Pi^*_{(C2)}(w_{(i+1,j-1)}) + o(n^{l_1+1}) \text{ and}$$
$$\#\Pi^*(w_{(i,j)^c}) = \#\Pi^*_{(C2)}(w_{(i,j)^c}) + o(n^{l_2+1}).$$

Multiplying these and using the fact (from (10.40))

$$\#\Pi^*(w) = \#\Pi^*(w_{(i+1,j-1)})\#\Pi^*(w_{(i,j)^c}) + o(n^{k+1}),$$

the result follows. $\qquad\square$

Proof of Lemma 10.5.3. We use induction on the length l of the Wigner string. Let w be a pair-matched colored word of length $2k$ with $p(w) \ne 0$. First suppose the Wigner string is of length 2, that is, $l = 2$. We may without loss of generality assume them in the starting position. So for any $\pi \in \Pi^*(w)$ with above property we have

$$(\pi(0), \pi(1)) = \begin{cases} (\pi(1), \pi(2)) \\ (\pi(2), \pi(1)). \end{cases}$$

In the first case $\pi(0) = \pi(1) = \pi(2)$ and so $\pi(1)$ is not a generating vertex and this lowers the number of generating vertices (which is not possible as $p(w) \ne 0$). Hence, the only possibility is $(\pi(0), \pi(1)) = (\pi(2), \pi(1))$ and the circuit is complete for the Wigner string and so it is a pair-matched word, proving part (a). Also, as a result of the above arguments only (C2) constraints survive, which shows (b).

Now suppose the result holds for all Wigner strings of length strictly less than l. Consider a Wigner string of length l, say $w_{(1,l)}$ (we assume it to start from the first position). We break the proof into two Cases I and II.

Case I: Here we suppose that the Wigner string has another one of smaller order and then use induction hypothesis and Lemma 10.5.5. Suppose that $w_{(1,l)}$ contains a Wigner string of length less than l at the position (p,q) with $1 \le p < q \le l$. Since $w_{(p,q)}$ is a Wigner string, by Lemma 10.5.5 we have,

$$p(w) = p(w_{(p,q)})p(w_{(p,q)^c}) \neq 0.$$

So by induction hypothesis and the fact that both $p(w_{(p,q)})$ and $p(w_{(p,q)^c})$ are not equal to zero we have, $w_{(p,q)}$ and $w_{(p,q)^c}$ are pair-matched words and they also satisfy (10.30). So $w_{(1,l)}$ is a pair-matched word, as it is made up of $w_{(p,q)}$ and $w_{(p,q)^c}$ which are pair-matched. Also from the second part of Lemma 10.5.5, we have $w_{(1,l)}$ satisfies part (b) and (c).

Case II: We now assume that there is no Wigner string inside. So there is a string of letters coming from matrix A after a Wigner letter. We show that this string is pair-matched and the last Wigner letter before the l-th position is essentially at the first position. This also implies that the string within a Wigner string does not cross a Wigner letter.

Suppose there is no Wigner string in the first l positions. Consider the last Wigner letter in the first $l-1$ positions, say at position j_0. Since there is no Wigner string of smaller length, $\pi(j_0)$ is a generating vertex. Also, as j_0 is the last Wigner letter, the positions from j_0 to $l-1$ are all letters coming from the matrix A.

Now we use the structure of the matrix A.

Subcase II(i): Suppose A is a Toeplitz matrix. For $i = 1, 2, \ldots, l-1-j_0$, let $s_i = (\pi(j_0 + i) - \pi(j_0 + i - 1))$. Now consider the following equation

$$s_1 + \cdots + s_{l-1-j_0} = (\pi(l-1) - \pi(j_0)). \tag{10.42}$$

If for any j, $w[j]$ is the first appearance of that letter, then consider s_j to be an independent variable (can be chosen freely). Then due to the Toeplitz link function, if $w[k] = w[j]$, where $k > j$, then $s_k = \pm s_j$. Since $(1,l)$ is a W match, $\pi(l-1)$ is either $\pi(0)$ or $\pi(1)$ and hence $\pi(l-1)$ is not a generating vertex. Note that (10.42) is a constraint on the independent variables unless $s_1 + \cdots + s_{l-1-j_0} = 0$. If this is non-zero, this non-trivial constraint lowers the number of independent variables and hence the limit contribution will be zero, which is not possible as $p(w) \neq 0$. So we must have,

$$\pi(l-1) = \pi(j_0) \quad \text{and} \quad j_0 = 1.$$

This also shows $(\pi(l), \pi(l-1)) = (\pi(0), \pi(1))$ and hence $w_{(1,l)}$ is a colored word. As $s_1 + \cdots + s_{l-1-j_0} = 0$, all the independent variables occur twice with different signs in the left side, since otherwise it would again mean a non-trivial relation among them and thus would lower the order. Hence we conclude that the Toeplitz letters inside the first l positions are also pair-matched. Since the (C2) constraint is satisfied at the position $(1, l)$, part (b) also holds.

Subcase II(ii): Suppose A is a Hankel matrix. We write, $t_i = (\pi(j_0 + i) + \pi(j_0 + i - 1))$ and consider

$$-t_1 + t_2 - t_3 + \cdots (-1)^{l-j_0-1} t_{l-j_0-1} = (-1)^{l-j_0-1}(\pi(l-1) - \pi(j_0)). \quad (10.43)$$

Now, again as earlier, the t_i's are independent variables, and so this implies that again to avoid a non-trivial constraint which would lower the order, both sides of the equation (10.43) have to vanish, which automatically leads to the conclusion that $\pi(l-1) = \pi(j_0) = \pi(1)$. So $j_0 = 1$ and again the Wigner paired string of length l is pair-matched. Part (b) also follows as the (C2) constraint holds.

Subcase II(iii): A is Symmetric or Reverse Circulant. Note that they have link functions which are quite similar to Toeplitz and Hankel respectively, the proofs are very similar to the above two cases and hence we skip them. $\quad\square$

Proof of Lemma 10.5.4. We first show that,

$$\mathrm{E}\Big[\prod_{i=1}^{l}\Big(\mathrm{Tr}\,\frac{X_n^{k_i}}{n^{k_i/2}} - \mathrm{E}\Big[\mathrm{Tr}\,\frac{X_n^{k_i}}{n^{k_i/2}}\Big]\Big)\Big] = O(n^{-1}) \text{ as } n \to \infty, \quad (10.44)$$

where Tr denotes the normalized trace. To prove (10.44), we see that,

$$\mathrm{E}\Big[\prod_{i=1}^{l}\Big(\mathrm{Tr}\,\frac{X_n^{k_i}}{n^{k_i/2}} - \mathrm{E}\Big[\mathrm{Tr}\,\frac{X_n^{k_i}}{n^{k_i/2}}\Big]\Big)\Big] = \frac{1}{n^{\sum_{i=1}^{l}k_i/2+l}} \sum_{\pi_1,\dots,\pi_l} \mathrm{E}[(\prod_{j=1}^{l}(X_{\pi_i} - \mathrm{E}(X_{\pi_i})))].$$

$$(10.45)$$

If the circuit π_i is not jointly-matched with the other circuits then $\mathrm{E}\,X_{\pi_i} = 0$ and

$$\mathrm{E}[(\prod_{j=1}^{l}(X_{\pi_i} - \mathrm{E}(X_{\pi_i})))] = \mathrm{E}[X_{\pi_i}(\prod_{j\neq i}(X_{\pi_i} - \mathrm{E}(X_{\pi_i})))] = 0.$$

If any of the circuits is self-matched, i.e., it has no cross-matched edge then

$$\mathrm{E}[(\prod_{j=1}^{l}(X_{\pi_i} - \mathrm{E}(X_{\pi_i})))] = \mathrm{E}[X_{\pi_i} - \mathrm{E}(X_{\pi_i})]\,\mathrm{E}[(\prod_{j\neq i}(X_{\pi_i} - \mathrm{E}(X_{\pi_i})))] = 0.$$

The total number of circuits $\{\pi_1, \dots, \pi_l\}$ where each edge appears at least twice and one edge at least thrice is bounded above by $\leq Cn^{\sum_{i=1}^{l}k_i/2+l-1}$ due to Property B. Hence using Assumption A2 such terms in (10.45) are of the order $O(n^{-1})$. Now consider the rest of the terms where all the edges appear exactly twice. As a consequence $\sum_{i=1}^{l}k_i$ is even. Also the number of partitions of $\frac{1}{2}\sum_{i=1}^{l} k_i$ into l circuits is independent of n. We need to consider only $\{\pi_1, \ldots, \pi_l\}$ which are jointly-matched but not self-matched.

If we prove that for such a partition the number of circuits is less than $Cn^{\sum_{i=1}^{l} k_i + l - 1}$ we are done since the number of such partitions is independent of n and due to the relation (10.14).

Since π_1 is not self-matched we can without loss of generality assume that the edge value for $(\pi(0), \pi(1))$ occurs exactly once in π_1. So construct π_1 as follows. First choose $\pi_1(0) = \pi_1(k_1)$ and then choose the remaining vertices in the order $\pi_1(k_1), \pi_1(k_1 - 1), \ldots, \pi_1(1)$. One sees that we lose one degree of freedom as in this way the edge $(\pi(0), \pi(1))$ is determined and we cannot choose it arbitrarily.

The result now follows from (10.44) by using induction. For $l = 2$, expanding and using the fact that expected normalized trace of the powers of $X_n/\sqrt{n}$ converges we get,

$$\mathrm{E}\left[\prod_{i=1}^{2}\left(\mathrm{Tr}\,\frac{X_n^{k_i}}{n^{k_i/2}} - \mathrm{E}\left[\mathrm{Tr}\,\frac{X_n^{k_i}}{n^{k_i/2}}\right]\right)\right]$$

$$= \mathrm{E}\left[\left(\mathrm{Tr}\,\frac{X_n^{k_1}}{n^{k_1/2}} - \mathrm{E}\left[\mathrm{Tr}\,\frac{X_n^{k_1}}{n^{k_1/2}}\right]\right)\left(\mathrm{Tr}\,\frac{X_n^{k_2}}{n^{k_2/2}} - \mathrm{E}\left[\mathrm{Tr}\,\frac{X_n^{k_2}}{n^{k_2/2}}\right]\right)\right]$$

$$= \mathrm{E}\left[\mathrm{Tr}\,\frac{X_n^{k_1}}{n^{k_1/2}}\,\mathrm{Tr}\,\frac{X_n^{k_2}}{n^{k_2/2}}\right] - \mathrm{E}\left[\mathrm{Tr}\,\frac{X_n^{k_1}}{n^{k_1}}\right]\mathrm{E}\left[\mathrm{Tr}\,\frac{X_n^{k_2}}{n^{k_2}}\right] \to 0 \text{ as } n \to \infty.$$

So the result holds for $l = 2$. Now suppose it is true for all $2 \le m < l$. We expand

$$\lim_{n\to\infty} \mathrm{E}\left[\prod_{i=1}^{l}\left(\mathrm{Tr}\left(\left(\frac{X_n}{\sqrt{n}}\right)^{k_i}\right) - \mathrm{E}\left(\mathrm{Tr}\left(\left(\frac{X_n}{\sqrt{n}}\right)^{k_i}\right)\right)\right)\right] = 0$$

to get

$$\lim_{n\to\infty} \sum_{m=1}^{l} (-1)^m \sum_{i_1 < \cdots < i_m} \mathrm{E}\left[\prod_{j=1}^{m} \mathrm{Tr}\left(\left(\frac{X_n}{\sqrt{n}}\right)^{k_{i_j}}\right)\right] \prod_{i \notin \{i_1, \ldots, i_m\}} \mathrm{E}\left[\mathrm{Tr}\left(\left(\frac{X_n}{\sqrt{n}}\right)^{k_i}\right)\right] = 0.$$

Now using the result for products of smaller order successively,

$$\lim_{n\to\infty} (-1)^l\,\mathrm{E}\left[\prod_{j=1}^{l} \mathrm{Tr}\left(\left(\frac{X_n}{\sqrt{n}}\right)^{k_j}\right)\right] = \lim_{n\to\infty} \sum_{m<l} (-1)^m \sum_{i_1 < \cdots < i_m} \mathrm{E}\left[\prod_{j=1}^{m} \mathrm{Tr}\left(\left(\frac{X_n}{\sqrt{n}}\right)^{k_{i_j}}\right)\right]$$

$$\times \prod_{i \notin \{i_1, \ldots, i_m\}} \mathrm{E}\left[\mathrm{Tr}\left(\left(\frac{X_n}{\sqrt{n}}\right)^{k_i}\right)\right].$$

Now every term in the right side equals, by induction hypothesis, $\lim_{n\to\infty} \prod_{i=1}^{l} \mathrm{E}[\mathrm{Tr}((\frac{X_n}{\sqrt{n}})^{k_i})]$ and from this the lemma follows. $\square$

Proof of Remark 10.3.1. This is similar to the proof of Theorem 10.3.1 but is much easier. We provide a sketch. If W is a centered GUE with variance $1/n$,

$$E[W_{ij}W_{kl}] = \frac{1}{n}\delta_{il}\delta_{jk}. \tag{10.46}$$

We follow the steps (10.34) onward in the proof of Theorem 10.3.1. The above equation (10.46) provides the necessary (C2) constraint.

$$\lim_{n\to\infty} \frac{1}{n^{k+1}} E[\mathrm{Tr}(WA^{q(1)}WA^{q(2)}\cdots WA^{q(m)})]$$

$$= \lim_{n\to\infty} \frac{1}{n^{k+1}} \sum_{\substack{\sigma\in\mathcal{P}_2(m)}} \sum_{\substack{i(1),\dots,i(m)\\j(1),\dots,j(m)=1}}^{n} \prod_{(r,s)\in\sigma} \delta_{i(r)j(s)}\delta_{i(s)j(r)} E[a^{q(1)}_{j(1)i(2)}\cdots a^{q(m)}_{j(m)i(1)}]$$

$$= \sum_{\sigma\in NC_2(m)} \lim_{n\to\infty} E\left(\mathrm{Tr}_{\sigma\gamma}[A^{(q_1)}, A^{(q_2)},\dots, A^{(q_m)})]\right).$$

The result follows using Lemma 10.5.4 which holds due to Property B and the existence of the LSD. $\qquad\qquad\qquad\qquad\qquad\qquad\qquad\qquad\qquad\qquad$ □

10.6   Exercises

1. Prove Remark 10.2.1(b).

2. Complete the proof of Theorem 10.2.1.

3. Show that to prove the existence of the LSD for sums of two independent patterned matrices whose i.i.d. inputs satisfy Property B, it is enough to assume the finiteness of the second moments.

4. Show that if $\mathbb{P}$ is a symmetric polynomial in any two of the matrices Wigner, Toeplitz, Hankel, Symmetric Circulant and Reverse Circulant, where the input sequence satisfies Assumption (A2), then the LSD of $n^{-1/2}\mathbb{P}$ converges almost surely.

5. Establish the unboundedness claim in the proof of Corollary 10.4.1.

6. Show that all conclusions in Corollary 10.4.1 hold when A and B are any two of the Toeplitz, Hankel, Reverse Circulant and Symmetric Circulant matrices.

7. Show that Corollary 10.4.1 continues to hold if the two input sequences are i.i.d. with finite second moments.

8. Work out the detailed proof of (10.16).

9. Simulate the ESD of sums of independent patterned random matrices considered in this chapter.

10. Show that $W_n^u/\sqrt{n}$ and $W_n^l/\sqrt{n}$ considered in Chapter 8 jointly converge if they are independent and the limit is not free.

11

Autocovariance matrix

Let $X = \{X_t\}$ be a *stationary* process with $E(X_t) = 0$ and $E(X_t^2) < \infty$. The *autocovariance function* (ACVF) $\gamma_X(\cdot)$ and the *autocovariance matrix* (ACVM) $\Sigma_n(X)$ of order n are defined as:

$$\gamma_X(k) = cov(X_0, X_k), \ k = 0, 1, \ldots \ \text{and} \ \Sigma_n(X) = ((\gamma_X(i-j)))_{1 \le i, j \le n}.$$

To every ACVF, there corresponds a unique distribution, called the *spectral distribution*, $F_X(\cdot)$ which satisfies

$$\gamma_X(h) = \int_{(0, \, 1]} \exp(2\pi i h x) dF_X(x) \ \text{for all} \ h. \tag{11.1}$$

We shall assume that

$$\sum_{k=1}^{\infty} |\gamma_X(k)| < \infty. \tag{11.2}$$

Then $F_X(\cdot)$ has a density, known as the *spectral density of X or of $\gamma_X(\cdot)$*, which equals

$$f_X(t) = \sum_{k=-\infty}^{\infty} \exp(-2\pi i t k) \gamma_X(k), \ t \in (0, \, 1]. \tag{11.3}$$

The *non-negative definite* estimate of $\Sigma_n(X)$ is the *sample ACVM*

$$\Gamma_n(X) = ((\hat{\gamma}_X(i-j)))_{1 \le i, j \le n} \ \text{where} \ \hat{\gamma}_X(k) = n^{-1} \sum_{i=1}^{n-|k|} X_i X_{i+|k|}.$$

The matrix $\Gamma_n(X)$ is a random Toeplitz matrix with a triangular input sequence. The autocovariances are of course crucial objects in time series analysis. They are used in estimation, prediction, model fitting and white noise tests. Under suitable assumptions on $\{X_t\}$, for every fixed k, $\hat{\gamma}_X(k) \to \gamma_X(k)$ almost surely. There are also results on the asymptotic distribution of specific functionals of the autocovariances. See Brockwell and Davis (1991)[33].

Recently there has been growing interest in the matrix $\Gamma_n(X)$ itself. For instance, the largest eigenvalue of $\Sigma_n(X) - \Gamma_n(X)$ does not converge to zero, even under reasonable assumptions (see Wu and Pourahmadi (2009)[108], Mc-Murry and Politis (2010)[75] and Xiao and Wen (2011)[110]).

If T_n is a symmetric Toeplitz matrix with entrie $\{t_k\}$, from Szegö's theory of Toeplitz operators (see Bottcher and Silberman (1999)[31]), we know that if $\sum |t_k| < \infty$, then the LSD of T_n equals $f(U)$ where U is uniformly distributed on $(0, 1]$ and $f(x) = \sum_{k=-\infty}^{\infty} t_k \exp(-2\pi i x k)$, $x \in (0, 1]$. In particular if (11.2) holds, then the LSD of $\Sigma_n(X)$ equals $f_X(U)$ where $f_X(\cdot)$ is as defined in (11.3).

The question is, does the LSD of $\Gamma_n(X)$ exist, and if so, does it coincide with the LSD of $\Sigma_n(X)$? In this chapter we study the behavior of $\Gamma_n(X)$, and a few other natural estimators of $\Sigma_n(X)$, as $n \to \infty$.

11.1   Preliminaries

We assume that the stationary process $\{X_t\}$ is a linear process, so that

$$X_t = \sum_{k=0}^{\infty} \theta_k \varepsilon_{t-k} \qquad (11.4)$$

where $\{\theta_k\}$ satisfies a weak condition and $\{\varepsilon_t, \ t \in \mathbb{Z}\}$ is a sequence of independent random variables with appropriate conditions. The simulations of Sen (2006)[92] suggested that the LSD of $\Gamma_n(X)$ exists and is independent of the distribution of $\{\varepsilon_t\}$ as long as they are i.i.d. with mean zero and variance one. Basak (2009)[11] and Sen (2010)[94] initially studied, respectively, the special cases where $\{X_t\}$ is an i.i.d. process or it is an MA(1) process, $X_t = \varepsilon_t + \theta_1 \varepsilon_{t-1}$.

Theorem 11.2.1 says that, if $\{X_t\}$ satisfies (11.4) and $\sum_{k=0}^{\infty} |\theta_k| < \infty$ then the LSD of $\Gamma_n(X)$ exists, and it is *universal* when $\{\varepsilon_t\}$ are independent with mean zero and variance 1 and are either uniformly bounded or identically distributed. We show that the LSD is unbounded when $\theta_i \geq 0$ for all i.

When $\{X_t\}$ is a finite-order process, that is, only finitely many θ_i are nonzero, the moments of the LSD can be written as multinomial type sums of the autocovariances (see (11.10)). When X is of infinite order, the limit moments are the limits of these sums as the order tends to infinity. Additional properties of the limit moments are available in Basak, Bose and Sen (2014)[14].

Incidentally, $\Gamma_n(X)$ reminds us of the Sample covariance matrix, S, for the i.i.d. setup, whose spectral properties we have seen in Chapter 3. In particular recall that the LSD of S (with i.i.d. entries) under suitable conditions is the Marčenko-Pastur law and is supported on the interval $[0, (1 + \sqrt{y})^2]$. Thus, the LSD of $\Gamma_n(X)$ is in sharp contrast.

Since the LSD of $\Sigma_n(X)$ is $f_X(U)$ where U is uniformly distributed on $[0, 1]$, it is reasonable to call a sequence of estimators $\{E_n\}$ of $\Sigma_n(X)$ *consistent* if its LSD is also $f_X(U)$. Since $f_X(U)$ has bounded support, $\{\Gamma_n(X)\}$ is inconsistent. This phenomenon is mainly due to the estimation of a large number of autocovariances by $\Gamma_n(X)$.

$\Gamma_n(X)$ can be modified by suitable *tapering* or *banding* to achieve consistency. Consider a sequence of integers $m := m_n \to \infty$, and a kernel function $K(\cdot)$. Define

$$\hat{f}_X(t) = \sum_{k=-m}^{m} K(k/m) \exp\left(-2\pi itk\right)\hat{\gamma}_X(k), \ t \in (0, 1] \qquad (11.5)$$

as the kernel density estimate of $f_X(\cdot)$. Viewing this function as a spectral density, the corresponding ACVF is given by (for $-m \le h \le m$):

$$\begin{aligned}
\gamma_K(h) &= \int_{(0, 1]} \exp(2\pi ihx)\hat{f}_X(x)dx \\
&= \sum_{k=-m}^{m} K(k/m) \int_{(0, 1]} \exp\{2\pi ihx - 2\pi ixk\}\hat{\gamma}_X(k)dx \\
&= K(h/m)\hat{\gamma}_X(h)
\end{aligned}$$

and is 0 otherwise. This motivates the consideration of the *tapered sample ACVM*

$$\Gamma_{n,K}(X) = ((K((i-j)/m)\hat{\gamma}_X(i-j)))_{1 \le i,j \le n}. \qquad (11.6)$$

If K is a non-negative definite function then $\Gamma_{n,K}(X)$ is also non-negative definite. Theorem 11.2.3(c) states that under the minimal condition $m_n/n \to 0$, if K is bounded, symmetric and continuous at 0 and $K(0) = 1$, then $\Gamma_{n,K}(X)$ is consistent.

The second approach is to use banding as in McMurry and Politis (2010)[75] who used it to develop their bootstrap procedures. We study two such banded matrices. Let $\{m_n\}_{n\in\mathbb{N}} \to \infty$ be such that $\alpha_n := m_n/n \to \alpha \in [0,1]$. Then the *Type I banded sample autocovariance matrix* $\Gamma_n^{\alpha,I}(X)$ is the same as $\Gamma_n(X)$ except that we substitute 0 for $\hat{\gamma}_X(k)$ whenever $|k| \ge m_n$. This is the same as $\Gamma_{n,K}$ with $K(x) = I_{\{|x|\le 1\}}$. The *Type II banded ACVM* $\Gamma_n^{\alpha,II}(X)$ is the $m_n \times m_n$ principal sub-matrix of $\Gamma_n(X)$. Theorem 11.2.3 (a)–(b) gives our results on these banded ACVMs. In particular, the LSD exists for all α and is unbounded when $\alpha \ne 0$. When $\alpha = 0$, the LSD is $f_X(U)$ and thus those estimate matrices are consistent.

A related matrix, which may be of interest, especially to probabilists, is,

$$\Gamma_n^*(X) = ((\gamma_X^*(|i-j|)))_{1 \le i,j \le n} \text{ where } \gamma_X^*(k) = n^{-1}\sum_{i=1}^{n} X_i X_{i+k} \ , \ k = 0, 1, \dots.$$

$\Gamma_n^*(X)$ does not have a "data" interpretation unless one assumes we have $2n-1$ observations $X_1, \dots, X_{2n-1}$. It is not non-negative definite and hence many of the techniques applied to $\Gamma_n(X)$ are not available for it. Theorem 11.2.2 states that its LSD also exists but under stricter conditions on $\{X_t\}$. Its moments dominate those of the LSD of $\Gamma_n(X)$ when $\theta_i \ge 0$ for all i. Simulations show that the LSD of $\Gamma_n^*(X)$ has significant positive mass on the negative axis.

11.2   Main results

We shall assume that $X = \{X_t\}_{t \in \mathbb{Z}}$ is a linear $(MA(\infty))$ process

$$X_t = \sum_{k=0}^{\infty} \theta_k \varepsilon_{t-k} \tag{11.7}$$

where $\{\varepsilon_t, \ t \in \mathbb{Z}\}$ is a sequence of independent random variables. A special case of this process is the so-called $MA(d)$ where $\theta_k = 0$ for all $k > d$. We denote this process by

$$X^{(d)} = \{X_{t,d} \equiv \theta_0 \varepsilon_t + \theta_1 \varepsilon_{t-1} + \cdots + \theta_d \varepsilon_{t-d}, t \in \mathbb{Z}\} \quad (\theta_0 \neq 0).$$

Note that working with two-sided moving average entails no difference. The conditions on $\{\varepsilon_t\}$ and on $\{\theta_k\}$ that will be used are:

Assumption A. (a) $\{\varepsilon_t\}$ are i.i.d. with $\mathrm{E}[\varepsilon_t] = 0$ and $\mathrm{E}[\varepsilon_t^2] = 1$.

(b) $\{\varepsilon_t\}$ are independent, uniformly bounded with $\mathrm{E}[\varepsilon_t] = 0$ and $\mathrm{E}[\varepsilon_t^2] = 1$.

Assumption B. (a) $\theta_j \geq 0$ for all j.

(b) $\sum_{j=0}^{\infty} |\theta_j| < \infty$.

The series in (11.7) converges a.s. under Assumptions A(a) (or A(b)) and B(b). Further, X and $X^{(d)}$ are strongly stationary and ergodic under Assumption A(a) and weakly (second-order) stationary under Assumptions A(b) and B(b).

The ACVF of $X^{(d)}$ and X are given by

$$\gamma_{X^{(d)}}(j) = \sum_{k=0}^{d-j} \theta_k \theta_{j+k} \quad \text{and} \quad \gamma_X(j) = \sum_{k=0}^{\infty} \theta_k \theta_{j+k}. \tag{11.8}$$

Let $\{k_i\}$ stand for suitable integers and let

$$\mathbf{k} = (k_0 \ldots k_d), \quad S_{h,d} = \{\mathbf{k} : k_0, \ldots, k_d \geq 0, k_0 + \cdots + k_d = h\}. \tag{11.9}$$

Theorem 11.2.1. Suppose Assumption A(a) or A(b) holds.

(a) Then $F^{\Gamma_n(X^{(d)})}$ converges almost surely to some F_d which is non-random and does not depend on the distribution of $\{\varepsilon_t\}$. Further,

$$\beta_{h,d} = \int x^h dF_d(x) = \sum_{S_{h,d}} p_{\mathbf{k}}^{(d)} \prod_{i=0}^{d} [\gamma_{X^{(d)}}(i)]^{k_i}, \tag{11.10}$$

where $\{p_{\mathbf{k}}^{(d)}\}$ are universal constants independent of the θ_i and the $\{\epsilon_i\}$. They are defined by a limiting process given in (11.19) and (11.31).

(b) Under Assumption B(b), a.s., $F^{\Gamma_n(X)} \overset{w}{\to} F$ which is non-random and independent of the distribution of $\{\varepsilon_t\}$. Further, for every fixed h, as $d \to \infty$,

$$F_d \overset{w}{\to} F \quad \text{and} \quad \beta_{h,d} \to \beta_h = \int x^h \, dF(x).$$

(c) Under Assumption B(a), F_d has unbounded support and $\beta_{h,d-1} \le \beta_{h,d}$ if $d \ge 1$. Consequently, if Assumptions B(a) and B(b) hold, then F has unbounded support. Therefore $\{\Gamma_n(X)\}$ is inconsistent. ⬦

Theorem 11.2.2. Suppose Assumption A(b) holds. Then conclusions of Theorem 11.2.1 continue to hold for $\Gamma_n^*(X)$, $d \le \infty$, and (11.10) holds with modified universal constants $\{p_\mathbf{k}^{*(d)}\}$. ⬦

Remark 11.2.1. (i) From the proofs, it will follow that the limit moments $\{\beta_{h,d}\}$ and $\{\beta_h\}$ of the above LSD are dominated by $\frac{4^h(2h)!}{h!}(\sum_{k=0}^{\infty} |\theta_k|)^{2h}$ which are the $(2h)$th moment of a Gaussian variable with mean zero and variance $4((\sum_{k=0}^{\infty} |\theta_k|)^2)$. Hence the limit moments satisfy Carleman's/Riesz's condition and uniquely identify the LSD.

All the above LSDs have unbounded support while $f_X(U)$ has support contained in $[-\sum_{-\infty}^{\infty} |\gamma_X(k)|, \sum_{-\infty}^{\infty} |\gamma_X(k)|]$. Simulations show that the LSD of $\Gamma_n^*(X)$ has positive mass on the negative real axis.

(iii) Since $\Gamma_n^*(X)$ is not non-negative definite, the proof of Theorem 11.2.2 for $d = \infty$ is different from the proof of Theorem 11.2.1 and needs Assumption A(b).

(iv) Unfortunately, the moments of the LSD of $\Gamma_n(X)$ has no easy description. There is no easy description of the constants $\{p_k^{(d)}\}$ either. Recall that the LSD of the $n^{-1/2}T_{n,\varepsilon} = n^{-1/2}((\varepsilon_{|i-j|}))$ where $\{\varepsilon_t\}$ is i.i.d. with mean zero variance 1 also does not have any easy description and the limit moments were expressed as sums of certain integrals. The quantities $p_k^{(d)}$ can be expressed in terms of similar but more complicated integrals. ⬦

Theorem 11.2.3. Suppose Assumption A(b) holds.

(a) Let $0 < \alpha \le 1$. Then all the conclusions of Theorem 1 hold for $\Gamma_n^{\alpha,I}(X^{(d)})$ and $\Gamma_n^{\alpha,II}(X^{(d)})$ with modified universal constants $\{p_\mathbf{k}^{\alpha,I,(d)}\}$ and $\{p_\mathbf{k}^{\alpha,II,(d)}\}$, respectively, in (11.10). The same conclusions continue to hold also for $d = \infty$.

(b) If $\alpha = 0$, and Assumption B(b) holds, the LSD of $\Gamma_n^{\alpha,I}(X)$ and $\Gamma_n^{\alpha,II}(X)$ are $f_X(U)$.

(a) and (b) remain true for $\Gamma_n^{\alpha,II}(X^{(d)})$ and $\Gamma_n^{\alpha,II}(X)$ under Assumption A(a).

(c) Suppose Assumption B(b) holds. Let K be bounded, symmetric and continuous at 0, $K(0) = 1$, $K(x) = 0$ for $|x| > 1$. Suppose $m_n \to \infty$ such that $m_n/n \to 0$. Then the LSD of $\Gamma_{n,K}(X)$ is $f_X(U)$ for $d \le \infty$. If K is non-negative definite, the above holds under Assumption A(a).   ◇

11.3   Proofs

The proof of Theorem 11.2.1 is challenging, mainly because of the non-linear dependence between X_t and the Toeplitz structure of $\Gamma_n(X)$. None of the earlier results on Toeplitz matrices seem to be helpful.

When $\alpha = 1$, then without loss of generality for asymptotic purposes, we assume that $m_n = n$. We visualize the full ACVM $\Gamma_n(X)$ as the case with $\alpha = 1$. The longest and hardest part of the proof is to verify (M1). We first assume that the variables are bounded and d is finite. The W_2 metric is used to remove the boundedness assumption as well as to deal with the infinite-order case. Easy modifications of these arguments yield the existence of the LSD when $0 \le \alpha \le 1$ in Theorem 11.2.3(a) and (b). The proof of Theorem 11.2.2 is a by-product of the arguments in the proof of Theorem 11.2.1. However, due to the matrix now not being non-negative definite, we impose Assumption A(b). The proof of Theorem 11.2.1(a) is given in detail. All other proofs are sketched and details are available in Basak, Bose and Sen (2014)[14].

Proof of Theorem 11.2.1. The method of proof of the following lemma is by now standard for us and we omit it. For convenience, we will write

$$\Gamma_n(X^{(d)}) = \Gamma_{n,d}.$$

Lemma 11.3.1. If for every $\{\varepsilon_t\}$ which satisfies Assumption A(b), $\Gamma_n(X^{(d)})$ has the same LSD a.s., then this LSD continues to hold if it satisfies Assumption A(a).   ◇

Thus, from now on we assume that Assumption A(b) holds.

$$\Gamma_{n,d} = \frac{1}{n}((Y_{i,j}^{(n)}))_{i,j=1,\dots,n}, \quad \text{where} \;\; Y_{i,j}^{(n)} = \sum_{t=1}^{n} X_{t,d} X_{t+|i-j|,d} \mathbb{I}_{(t+|i-j| \le n)},$$

$$\beta_h(\Gamma_{n,d}) = \frac{1}{n} \operatorname{Tr}(\Gamma_{n,d}^h) = n^{-(h+1)} \sum_{1 \le \pi_0 = \pi_h, \pi_1,\dots,\pi_{h-1} \le n} Y_{\pi_0,\pi_1}^{(n)} \cdots Y_{\pi_{h-1},\pi_h}^{(n)}$$

$$= \sum_{\substack{1 \le \pi_0,\dots,\pi_h \le n \\ \pi_h = \pi_0}} \prod_{j=1}^{h} \sum_{t_j=1}^{n} \frac{X_{t_j,d} X_{t_j+|\pi_{j-1}-\pi_j|,d} \mathbb{I}_{(t_j+|\pi_{j-1}-\pi_j| \le n)}}{n^{h+1}}. \quad (11.11)$$

To express the above in a neater and more amenable form, define

$$
\begin{aligned}
\mathbf{t} &= (t_1, \ldots, t_h), \quad \pi = (\pi_0, \ldots, \pi_{h-1}), \\
\mathcal{A} &= \big\{ (\mathbf{t}, \pi) : 1 \le t_1, \ldots, t_h, \pi_0, \ldots, \pi_{h-1} \le n, \pi_h = \pi_0 \big\}, \\
\mathbf{a}(\mathbf{t}, \pi) &= \big(t_1, \ldots, t_h, t_1 + |\pi_0 - \pi_1|, \ldots, t_h + |\pi_{h-1} - \pi_h| \big), \\
\mathbf{a} &= (a_1, \ldots, a_{2h}) \in \{1, 2, \ldots, 2n\}^{2h}, \\
X_{\mathbf{a}} &= \prod_{j=1}^{2h} (X_{a_j, d}) \quad \text{and} \quad \mathbb{I}_{\mathbf{a}(\mathbf{t}, \pi)} = \prod_{j=1}^{h} \mathbb{I}_{(t_j + |\pi_{j-1} - \pi_j| \le n)}.
\end{aligned}
$$

Then using (11.11) we can write the trace formula,

$$
\mathrm{E}[\beta_h(\Gamma_{n,d})] = \frac{1}{n^{h+1}} \mathrm{E}\Big[\sum_{(\mathbf{t}, \pi) \in \mathcal{A}} X_{\mathbf{a}(\mathbf{t}, \pi)} \mathbb{I}_{\mathbf{a}(\mathbf{t}, \pi)} \Big]. \tag{11.12}
$$

Matches and negligibility. By independence of $\{\varepsilon_t\}$, $\mathrm{E}[X_{\mathbf{a}(\mathbf{t}, \pi)}] = 0$ if there is at least one component of the product that has no ε_t common with any other component. Motivated by this, we introduce the following notion of a match and show that certain higher-order terms can be asymptotically neglected in (11.12). We say

• $\mathbf{a}$ is *d-matched* (in short *matched*) if $\forall\, i \le 2h, \exists j \ne i$ such that $|a_i - a_j| \le d$. When $d = 0$ this means $a_i = a_j$.

• $\mathbf{a}$ is *minimal d-matched* (in short *minimal matched*) if there is a partition $\mathcal{P}$ of $\{1, \ldots, 2h\}$,

$$
\{1, \ldots, 2h\} = \cup_{k=1}^{h} \{i_k, j_k\}, \quad i_k < j_k \tag{11.13}
$$

such that $\{i_k\}$ are in ascending order and

$$
|a_x - a_y| \le d \Leftrightarrow \{x, y\} = \{i_k, j_k\} \text{ for some } k.
$$

For example, for $d = 1$, $h = 3$, $(1, 2, 3, 8, 9, 10)$ is matched but not minimal matched and $(1, 2, 5, 6, 9, 10)$ is both matched and minimal matched.

Lemma 11.3.2. $\#\{\mathbf{a} : \mathbf{a}$ is matched but not minimally$\} = \mathrm{O}(n^{h-1})$. ◇

Proof. Consider the graph with vertices $\{1, 2, \ldots, 2h\}$. Vertices i and j have an edge if $|a_i - a_j| \le d$. Let $k = \#$ connected components. Consider a typical $\mathbf{a}$. Let l_j be the number of vertices in the j-th component. Since $\mathbf{a}$ is matched, $l_j \ge 2$ for all j and $l_j > 2$ for at least one j. Hence $2h = \sum_{j=1}^{k} l_j > 2k$. That implies $k \le h - 1$. Also if i and j are in the same connected component then $|a_i - a_j| \le 2dh$. Hence the number of a_i's such that i belongs to any given component is $\mathrm{O}(n)$ and the result follows. □

Now we can rewrite (11.12) as

$$
\mathrm{E}[\beta_h(\Gamma_{n,d})] \;=\; \frac{1}{n^{h+1}}\,\mathrm{E}[\sum_1 X_{\mathbf{a}(\mathbf{t},\pi)}\mathbb{I}_{\mathbf{a}(\mathbf{t},\pi)}] + \frac{1}{n^{h+1}}\,\mathrm{E}[\sum_2 X_{\mathbf{a}(\mathbf{t},\pi)}\mathbb{I}_{\mathbf{a}(\mathbf{t},\pi)}]
$$
$$
+ \frac{1}{n^{h+1}}\,\mathrm{E}[\sum_3 X_{\mathbf{a}(\mathbf{t},\pi)}\mathbb{I}_{\mathbf{a}(\mathbf{t},\pi)}] = T_1 + T_2 + T_3 \quad \text{(say)},
$$

where the three summations are over $(\mathbf{t},\pi) \in \mathcal{A}$ such that $\mathbf{a}(\mathbf{t},\pi)$ is respectively, (i) minimal matched, (ii) matched but not minimal matched and (iii) not matched.

By mean zero assumption, $T_3 = 0$. Since X_i's are uniformly bounded, by Lemma 11.3.2, $T_2 \le \frac{C}{n}$ for some constant C. So provided the limit exists,

$$
\lim_{n\to\infty} \mathrm{E}[\beta_h(\Gamma_{n,d})] = \lim_{n\to\infty} \frac{1}{n^{h+1}}\,\mathrm{E}[\sum_{\substack{(\mathbf{t},\pi)\in\mathcal{A}:\mathbf{a}(\mathbf{t},\pi)\text{ is}\\ \text{minimal matched}}} X_{\mathbf{a}(\mathbf{t},\pi)}\mathbb{I}_{\mathbf{a}(\mathbf{t},\pi)}]. \qquad (11.14)
$$

Hence, from now our focus will be only on minimal matched words.

Verification of (M1) for Theorem 11.2.1(a): This is the hardest and lengthiest part of the entire proof. Our starting point is equation (11.14). We first define an *equivalence relation* on the set of minimal matched $\mathbf{a} = \mathbf{a}(\mathbf{t},\pi)$. To define the equivalence relation, consider the collection of $(2d+1)h$ symbols (letters)

$$
\mathcal{W}_h = \{w^k_{-d},\ldots,w^k_0,\ldots,w^k_d : k = 1,\ldots,h\}.
$$

Any minimal d−matched $\mathbf{a} = (a_1,\ldots,a_{2h})$ induces a partition as in (11.13). With this $\mathbf{a}$, associate the *word* $w = w[1]w[2]\ldots w[2h]$ of length $2h$ where

$$
w[i_k] = w^k_0, \; w[j_k] = w^k_l \; \text{if} \; a_{i_k} - a_{j_k} = l, \; 1 \le k \le h. \qquad (11.15)
$$

For example, consider $d = 1, h = 3$ and $\mathbf{a} = (a_1,\ldots,a_6) = (1,21,1,20,39,40)$. Then the unique partition of $\{1,2,\ldots,6\}$ and the unique word associated with $\mathbf{a}$ are $\{\{1,3\},\{2,4\},\{5,6\}\}$ and $[w^1_0 w^2_0 w^1_0 w^2_1 w^3_0 w^3_{-1}]$ respectively.

To any partition $\mathcal{P} = \{\{i_k,j_k\}, 1 \le k \le h\}$, there are several $\mathbf{a}$ associated with it and there are exactly $(2d+1)^h$ words that arise from it. For example with $d = 1, h = 2$ consider the partition $\mathcal{P} = \{\{1,2\},\{3,4\}\}$. Then the nine words corresponding to $\mathcal{P}$ are $w^1_0 w^1_i w^2_0 w^2_j$ where $i,j = -1,0,1$.

By a slight abuse of notation we write $w \in \mathcal{P}$ if the partition corresponding to w is the same as $\mathcal{P}$. We will say that

• $w[x]$ *matches* with $w[y]$ (say $w[x] \approx w[y]$) iff $w[x] = w^k_l$ and $w[y] = w^k_{l'}$ for some k, l, l'.

• w is d-*pair-matched* if it is *induced* by a minimal d-matched $\mathbf{a}$ (so $w[x]$ matches with $w[y]$ iff $|a_x - a_y| \le d$).

This induces an *equivalence relation* on all d minimal matched $\mathbf{a}$ and the equivalence classes can be indexed by d pair matched w. Given such a w, the corresponding equivalence class is given by

$$\Pi(w) \quad = \quad \{(\mathbf{t}, \pi) \in \mathcal{A} : \ w[i_k] = w_0^k, w[j_k] = w_l^k \Leftrightarrow$$
$$\mathbf{a}(\mathbf{t}, \pi)_{i_k} - \mathbf{a}(\mathbf{t}, \pi)_{j_k} = l \ \text{ and } \ \mathbb{I}_{\mathbf{a}(\mathbf{t}, \pi)} = 1\}. \quad (11.16)$$

Then we rewrite (11.14) as (*provided the second limit exists*)

$$\lim_{n \to \infty} \mathrm{E}[\beta_h(\Gamma_{n,d})] = \sum_{\mathcal{P}} \sum_{w \in \mathcal{P}} \lim_{n \to \infty} \frac{1}{n^{h+1}} \sum_{(t,\pi) \in \Pi(w)} \mathrm{E}[X_{\mathbf{a}(\mathbf{t},\pi)} \mathbb{I}_{\mathbf{a}(\mathbf{t},\pi)}]. \quad (11.17)$$

Let
$$\mathcal{W}(\mathbf{k}) = \{w : \ \#\{s : |w[i_s] - w[j_s]| = i\} = k_i, \ i = 0, 1, \ldots, d\}.$$

Using the definitions of $\gamma_{X^{(d)}}(\cdot)$ and $S_{h,d}$ given in (11.9), rewrite (11.17) as

$$\lim_{n \to \infty} \mathrm{E}[\beta_h(\Gamma_{n,d})] = \sum_{\mathcal{P}} \sum_{S_{h,d}} \sum_{w \in \mathcal{P} \cap \mathcal{W}(\mathbf{k})} \lim_{n \to \infty} \frac{1}{n^{h+1}} \#\Pi(w) \prod_{i=0}^{d} [\gamma_{X^{(d)}}(i)]^{k_i}$$
$$(11.18)$$

provided the following limit exists for every word w of length $2h$.

$$p_w^{(d)} \equiv \lim_{n \to \infty} \frac{1}{n^{h+1}} \#\Pi(w). \quad (11.19)$$

To show that this limit exists, it is convenient to work with $\Pi^*(w) \supseteq \Pi(w)$ defined as

$$\Pi^*(w) \quad = \quad \{(\mathbf{t}, \pi) \in \mathcal{A} : \ w[i_k] = w_0^k, \ w[j_k] = w_l^k \Rightarrow$$
$$a(\mathbf{t}, \pi)_{i_k} - a(\mathbf{t}, \pi)_{j_k} = l \ \text{ and } \ \mathbb{I}_{\mathbf{a}(\mathbf{t},\pi)} = 1\}. \quad (11.20)$$

By Lemma 11.3.2 for every w, $n^{-(h+1)} \#(\Pi^*(w) - \Pi(w)) \to 0$. Thus, it is enough to show that $\lim_{n \to \infty} \frac{1}{n^{h+1}} \#\Pi^*(w)$ exists.

For a pair matched w, we divide its coordinates according to the position of the matches as follows. For $1 \leq i < j \leq h$, let the sets S_i be defined as

$$S_1(w) \quad = \quad \{i : w[i] \approx w[j]\}, \qquad S_2(w) = \{j : w[i] \approx w[j]\},$$
$$S_3(w) \quad = \quad \{i : w[i] \approx w[j+h]\}, \qquad S_4(w) = \{j : w[i] \approx w[j+h]\},$$
$$S_5(w) \quad = \quad \{i : w[i+h] \approx w[j+h]\}, \quad S_6(w) = \{j : w[i+h] \approx w[j+h]\}.$$

Let E, and $G \subset E$, be defined as

$$E \quad = \quad \{t_1, \ldots, t_h, \pi_0, \ldots, \pi_h\},$$
$$G \quad = \quad \{t_i | i \in S_1(w) \cup S_3(w)\} \cup \{\pi_0\} \cup \{\pi_i | i + h \in S_5(w)\}.$$

Elements in G are the indices where any matched letter appears for the *first generating vertices*. G has $(h+1)$ elements, say $u_1^n, \ldots, u_{h+1}^n$, and we will write

$$G \equiv U_n = (u_1^n, \ldots, u_{h+1}^n) \ \text{ and } \ \mathcal{N}_n = \{1, 2, \ldots, n\}.$$

Claim 1: Each element of E is a linear expression, (say $\lambda_{\mathbf{i}}$) of the generating vertices that are all to the *left* of the element.

Proof. Let the constants in the proposed linear expressions be $\{m_j\}$. Then observe the following.

(a) For elements of E which are generating vertices, choose $m_j = 0$ and the linear combination as the identity mapping so that

$$\text{for all } i \in S_1(w) \cup S_3(w), \ \lambda_i \equiv t_i,$$

$$\lambda_{h+1} \equiv \pi_0,$$

$$\text{and for all } i+h \in S_5(w), \ \lambda_{i+h+1} \equiv \pi_i.$$

(b) Using the relations between $S_1(w)$ and $S_2(w)$ induced by w, we can write

$$\text{for all } j \in S_2(w), \ t_j = \lambda_j + n_j,$$

for some n_j such that $|n_j| \leq d$ and define $m_j = n_j$ for $j \in S_2(w)$ and $\lambda_j \equiv \lambda_i$.

(c) Note that for every π we can write

$$|\pi_{i-1} - \pi_i| = b_i(\pi_{i-1} - \pi_i) \ \text{ for some } \ b_i \in \{-1, 1\}.$$

Consider the vector $\mathbf{b} = (b_1, b_2, \ldots, b_h) \in \{-1, 1\}^h$. It will be a valid choice if we have

$$b_i(\pi_{i-1} - \pi_i) \geq 0 \ \text{ for all } \ i. \tag{11.21}$$

We then have the following two cases:

Case 1: $w[i]$ matches with $w[j+h]$, $j+h \in S_4(w)$ and $i \in S_3(w)$. Then we get

$$t_i = t_j + b_j(\pi_{j-1} - \pi_j) + n_{j+h} \ \text{ for some integer } \ n_{j+h} \in \{-d, \ldots, 0 \ldots, d\}. \tag{11.22}$$

Case 2: $w[i+h]$ matches with $w[j+h]$, $j+h \in S_6(w)$ and $i+h \in S_5(w)$. Then we have

$$t_i + |\pi_{i-1} - \pi_i| = t_j + |\pi_{j-1} - \pi_j| + n_{j+h} \text{ where } n_{j+h} \in \{-d, \ldots, 0, \ldots, d\}. \tag{11.23}$$

So we note that inductively from left to right we can write

$$\pi_j = \lambda^{\mathbf{b}}_{j+1+h} + m_{j+1+h}, \ j+h \in S_4(w) \cup S_6(w). \tag{11.24}$$

Hence, inductively, π_j is a linear combination $\{\lambda^{\mathbf{b}}_j\}$ of the generating vertices up to an appropriate constant. The superscript $\mathbf{b}$ emphasizes the dependence on $\mathbf{b}$. Further, $\{\lambda^{\mathbf{b}}_j\}$ depends *only* on the vertices present to the left of it. □

Now we are almost ready to write down an expression for the limit. If λ_i were unique for each $\mathbf{b}$, then we could write $\#\Pi^*(w)$ as a sum of all possible choices of $\mathbf{b}$ and we could tackle the expression for each $\mathbf{b}$ separately. However, λ_i's may be the same for several choices $b_i \in \{-1, 1\}$. For example, for $w_0^1 w_0^2 w_0^1 w_0^2$, we can choose any $\mathbf{b}$. We circumvent this problem as follows: Let

$$\mathcal{T} = \{j + h \in S_4(w) \cup S_6(w) : \ \lambda^{\mathbf{b}}_{j+h} - \lambda^{\mathbf{b}}_{j+h-1} \equiv 0 \ \forall \ b_j\}.$$

Note that the definition of $\mathcal{T}$ depends on w only through the partition $\mathcal{P}$ it generates.

Suppose $j + h \in \mathcal{T}$. Define

$$
\begin{aligned}
L_j(U_n) &:= b_j(\lambda^{\mathbf{b}}_{\mathbf{j+h-1}}(U^n) - \lambda^{\mathbf{b}}_{\mathbf{j+h}}(U^n)) + m_{j+h-1} - m_{j+h} \quad (11.25) \\
&:= \tilde{L}_j(U_n) + m_{j+h-1} - m_{j+h}. \quad (11.26)
\end{aligned}
$$

Then from (11.22) and (11.23) the region given by (11.21) is

$$
\{L_j(U_n) \geq 0\} \equiv \{\tilde{L}_j(U_n) + m_{j+h-1} - m_{j+h} \geq 0\}. \quad (11.27)
$$

Claim 2: Expression (11.27) is the same for all choices of $\{b_j\}$, for $j + h \in \mathcal{T}$.

Proof. First we show that if $j + h \in \mathcal{T}$ then we must have

$$
t_j = t_j + |\pi_{j-1} - \pi_j| + n_j \text{ for some integer } |n_j| \leq d. \quad (11.28)
$$

Suppose this is not true. So first assume that $j + h \in S_6(w)$. Then we will have a relation

$$
t_i + b_i(\pi_{i-1} - \pi_i) = t_j + b_j(\pi_{j-1} - \pi_j) + n_j, \text{ where } i + h \in S_5(w). \quad (11.29)
$$

Since $\lambda^{\mathbf{b}}_{\mathbf{j}}$ depends only on the vertices to the left of it, in (11.29), the coefficient of π_i would be non-zero and hence we must have $\lambda^{\mathbf{b}}_{\mathbf{j+h-1}} - \lambda^{\mathbf{b}}_{\mathbf{j+h}} \neq 0$.

Now assume $j + h \in S_4(w)$ and $w[i]$ matches with $w[j + h]$ for $i \neq j$. Then we can repeat the argument above to arrive at a similar contradiction. This shows that if $j + h \in \mathcal{T}$ then our relation must be like (11.28). Now a simple calculation shows that for such relations,

$$
b_j(\lambda^{\mathbf{b}}_{\mathbf{j+h-1}}(U_n) - \lambda^{\mathbf{b}}_{\mathbf{j+h}}(U_n)) + m_{j+h-1} - m_{j+h} = -n_j
$$

which is of course the same across all choices of **b**. This proves our claim. □

Now note that if $j + h \in \mathcal{T}$ and if $n_{j+h} \neq 0$ then as we change b_j it changes the value of m_{2h+1}. Further, we can have at most two choices for π_j for every choice of π_{j-1} if $n_{j+h} \neq 0$ depending on b_j.

However for $j + h \in \mathcal{T}$ and $n_j = 0$ we have only one choice for π_j given the choice for π_{j-1} for every choice of b_j. On the other hand we know $\mathbf{b} \in \{-1, 1\}^h$ must satisfy (11.21). Keeping the above in view, let

$$
\mathcal{B}(w) = \{\mathbf{b} \in \{-1, 1\}^h \mid b_j = 1 \text{ if } n_j = 0 \text{ for } j \in \mathcal{T}\}
$$

where $\{n_j\}$ is as in Claim 2. For ease of writing we introduce a few more notations:

$$\mathbb{I}_{m,h}(U_n) \quad := \quad \mathbb{I}(\lambda^{\mathbf{b}}{}_{2h+1}(U_n) + m_{2h+1} = \lambda^{\mathbf{b}}{}_{h+1}(U_n) + m_{h+1}),$$

$$\mathbb{I}_{\lambda^{\mathbf{b}},L}(U_n) \quad := \quad \prod_{j=1}^{h} \mathbb{I}(\lambda^{\mathbf{b}}_{\mathbf{j}}(U_n) + L_j(U_n) \le n),$$

$$\mathbb{I}_{\lambda^{\mathbf{b}},m}(U_n) : \quad = \quad \prod_{j=1}^{2h} \mathbb{I}(\lambda^{\mathbf{b}}_{\mathbf{j}}(U_n) + m_j \in \mathcal{N}_n), \text{ and}$$

$$\mathbb{I}_{\mathcal{T}}(U_n) \quad := \quad \prod_{1 \le j \le h, \ j \notin \mathcal{T}} \mathbb{I}(L_j(U_n) \ge 0) \times \prod_{j \in \mathcal{T}} \mathbb{I}(n_j \le 0). \qquad (11.30)$$

Now we note that,

$$p_w^{(d)} \quad := \quad \lim_n \frac{1}{n^{h+1}} \#\Pi^*(w)$$

$$= \quad \lim_n \frac{1}{n^{h+1}} \sum_{\mathbf{b} \in \mathcal{B}(w)} \sum_{U_n \in \mathcal{N}_n^{h+1}} \mathbb{I}_{m,h}(U_n) \times \mathbb{I}_{\lambda^{\mathbf{b}},m}(U_n) \times \mathbb{I}_{\lambda^{\mathbf{b}},L}(U_n) \times \mathbb{I}_{\mathcal{T}}(U_n)$$

$$= \quad \lim_n \sum_{\mathbf{b} \in \mathcal{B}(w)} \mathrm{E}_{U_n} \left[\mathbb{I}_{m,h}(U_n) \times \mathbb{I}_{\lambda^{\mathbf{b}},m}(U_n) \times \mathbb{I}_{\lambda^{\mathbf{b}},L}(U_n) \times \mathbb{I}_{\mathcal{T}}(U_n) \right].$$

Now it only remains to identify the limit. First fix a partition $\mathcal{P}$ and $\mathbf{b} \in \{-1,1\}^h$. If $d = 0$, then there is one and only one word corresponding to it. However, across any d and any fixed $k_0, k_1, \ldots, k_d$, the linear functions λ_j's continue to remain the same. The only possible changes will be in the values of m_j's. We now identify the cases where the above limit is zero.

Claim 3: Suppose w is such that $\mathcal{R} := \{\lambda^{\mathbf{b}}_{2h+1}(U_n) + m_{2h+1} = \lambda^{\mathbf{b}}_{h+1}(U_n) + m_{h+1}\}$ is a lower-dimensional subset of $\mathcal{N}_n^{h+1}$. Then the above limit is zero.

Proof. First assume that $d = 0$. Then $m_j = 0$, $\forall\, j$. Note that $\mathcal{R}$ lies in a hypercube. Hence the result follows by convergence of the Riemann sum to the corresponding Riemann integral. For general d the corresponding region is just a translate of the region considered for $m_j = 0$. Hence the result follows. $\qquad \square$

Hence for a fixed $w \in \mathcal{P}$, a positive limit contribution is possible only when $\mathcal{R} = \mathcal{N}_n^{h+1}$. This implies that we must have

$$\lambda^{\mathbf{b}}_{2h+1}(U_n) - \lambda^{\mathbf{b}}_{h+1}(U_n) \quad \equiv \quad 0 \ \ (\text{for } d = 0)$$
$$\lambda^{\mathbf{b}}_{2h+1}(U_n) - \lambda^{\mathbf{b}}_{h+1}(U_n) \quad \equiv \quad 0 \ \text{ and } \ m_{2h+1} - m_{h+1} = 0 \ \ (\text{for general } d).$$

The first relation depends only on the partition $\mathcal{P}$ but the second relation is determined by the word w. Now $\lambda^{\mathbf{b}}_{\mathbf{j}}$ being linear forms with integer coefficients,

$$\lambda^{\mathbf{b}}_{\mathbf{j}}(U_n) + m_j \in \{1, \ldots, n\} \iff \lambda^{\mathbf{b}}_{\mathbf{j}}\left(\frac{U_n}{n}\right) + \frac{m_j}{n} \in (0, 1].$$

Define $\mathbb{I}_{m,h}(U)$, $\mathbb{I}_{\lambda^\flat,\tilde{L}}(U)$, $\mathbb{I}_{\lambda^\flat}(U)$ and $\tilde{\mathbb{I}}_{\mathcal{T}}(U)$ as in (11.30) with U_n replaced by U, L replaced by $\tilde{L}$, $\mathcal{N}_n$ replaced by the interval $(0,1)$, n replaced by 1, and dropping m_j's in $\mathbb{I}_{\lambda^\flat,m}$. Note that $\frac{U_n}{n} \overset{w}{\Rightarrow} U$ and U follows the uniform distribution on $[0,1]^{h+1}$. Hence $\frac{1}{n^{h+1}}\lim \#\Pi^*(w)$ equals

$$p_w^{(d)} = \sum_{\mathbf{b}\in\mathcal{B}(w)} \mathrm{E}_U\left[\mathbb{I}_{m,h}(U) \times \mathbb{I}_{\lambda^\flat,\tilde{L}}(U) \times \mathbb{I}_{\lambda^\flat}(U) \times \tilde{\mathbb{I}}_{\mathcal{T}}(U)\right].$$

Now the verification of (M1) is complete by observing that (11.18) becomes

$$\lim_{n\to\infty} \mathrm{E}[\beta_h(\Gamma_{n,d})] = \sum_{\mathcal{P}}\sum_{\mathbf{k}\in S_{h,d}} p_{\mathbf{k}}^{\mathcal{P},d}\prod_{i=0}^{d}[\gamma_{X^{(d)}}(i)]^{k_i}$$

$$= \sum_{\mathbf{k}\in S_{h,d}} p_{\mathbf{k}}^{(d)}\prod_{i=0}^{d}[\gamma_{X^{(d)}}(i)]^{k_i}$$

where

$$p_{\mathbf{k}}^{\mathcal{P},d} = \sum_{w\in\mathcal{P}\cap\mathcal{W}(\mathbf{k})} p_w^{(d)} \quad \text{and} \quad p_{\mathbf{k}}^{(d)} = \sum_{\mathcal{P}} p_{\mathbf{k}}^{\mathcal{P},d}. \tag{11.31}$$

There is no explicit expression known for the moments of the LSD. Table 11.3 provides the first three moments when the input sequence is i.i.d. and MA(1). To calculate these moments we need to find the contributions $p_w^{(d)}$ for words w. These contributions of different relevant words, are provided in Table 11.1, and in Table 11.2, for the i.i.d. case. Using these values, the contributions in the MA(1) case in Table 11.3 can be worked out.

TABLE 11.1
Contribution from words of length 4 for the i.i.d. case.

Word w	Contribution $p_w^{(0)}$
aabb	2/3
abab	1
abba	0

Verification of (M2) and (M4) for Theorem 11.2.1(a):

Lemma 11.3.3. (a) $\mathrm{E}\left[n^{-1}\mathrm{Tr}(\Gamma_{n,d}^h) - n^{-1}\mathrm{E}[\mathrm{Tr}(\Gamma_{n,d}^h)]\right]^4 = \mathrm{O}\left(n^{-2}\right)$. Hence $\frac{1}{n}\mathrm{Tr}(\Gamma_{n,d}^h)$ converges to $\beta_{h,d}$ a.s..

(b) $\{\beta_{h,d}\}_{h\geq 0}$ satisfies Riesz's condition (R). $\qquad\qquad \diamond$

Proof. (a) Here we use ideas that we have seen many times though some modifications are needed. Details are left as an exercise.

TABLE 11.2
Contributions from words of length 6 for the i.i.d. case.

Word w	Contribution $p_w^{(0)}$	Word w	Contribution $p_w^{(0)}$
aabccb	2/3	abbcac	1/6
aabbcc	1/6	abcabc	1
aabcbc	1/6	abcacb	0
ababcc	1/6	abcbac	0
abacbc	2/3	abcbca	0
abaccb	1/6	abccab	0
abbacc	2/3	abccba	0
abbcca	1/6		

TABLE 11.3
First three moments for the i.i.d. and MA(1) input sequence.

	IID	MA(1)
Mean	θ_0^2	$\theta_0^2 + \theta_1^2$
Second moment	$\frac{5}{3}\theta_0^4$	$\frac{5}{3}(\theta_0^2 + \theta_1^2)^2 + \frac{20}{3}\theta_0^2\theta_1^2$
Third moment	$4\theta_0^6$	$4(\theta_0^2 + \theta_1^2)^3 + 24(\theta_0^2 + \theta_1^2)(2\theta_0\theta_1)^2$

(b) Using (11.31) and (11.10) and noting that the number of ways of choosing the partition $\{1,...,2h\} = \cup_{l=1}^{h}\{i_l, j_l\}$ for $\mathbf{a}(\mathbf{t}, \pi)$ is $\frac{(2h)!}{2^h h!}$, it easily follows that

$$
\begin{aligned}
|\beta_{h,d}| &\leq \sum_{S_{h,d}} \frac{4^h(2h)!}{h!} \frac{h!}{k_0!\dots k_d!} \prod_{i=0}^{d} |\gamma_{X^{(d)}}(i)|^{k_i} \\
&\leq \frac{4^h(2h)!}{h!} \left(\sum_{j=0}^{d}\sum_{k=0}^{d-j} |\theta_k\theta_{k+j}|\right)^h \leq \frac{4^h(2h)!}{h!} \left(\sum_{k=0}^{d} |\theta_k|\right)^{2h}. \quad (11.32)
\end{aligned}
$$

Proof of Theorem 11.2.1(a) is now complete. $\square$

Proof of Theorem 11.2.1(b) (infinite-order case). *First we assume* $\{\varepsilon_t\}$ *is i.i.d.*. Fix $\varepsilon > 0$. Choose d such that $\sum_{k\geq d+1} |\theta_k| \leq \varepsilon$. For convenience we will write $\Gamma_n(X) = \Gamma_n$. Clearly, $\Gamma_n = A_n A_n^T$ where

$$
(A_n)_{i,j} = \begin{cases} X_{j-i}, & \text{if } 1 \leq j - i \leq n \\ 0, & \text{otherwise.} \end{cases}
$$

By *ergodic theorem*, a.s., we have the following two relations:

$$\frac{1}{n}\left[\operatorname{Tr}(\Gamma_{n,d}+\Gamma_n)\right] = \frac{1}{n}\left[\sum_{t=1}^{n}X_{t,d}^2+\sum_{t=1}^{n}X_t^2\right]$$

$$\rightarrow \operatorname{E}[X_{t,d}^2+X_t^2]\le 2\sum_{k=0}^{\infty}\theta_k^2.$$

$$\frac{1}{n}\operatorname{Tr}[(A_{n,d}-A_n)(A_{n,d}-A_n)^T] = \frac{1}{n}\sum_{t=1}^{n}(X_{t,d}-X_t)^2$$

$$\rightarrow \operatorname{E}[X_{t,d}-X_t]^2\le \sum_{k=d+1}^{\infty}\theta_k^2\le\varepsilon^2.$$

Hence using Lemma 3.2.1(a), a.s.

$$\limsup_n W_2^2(F^{\Gamma_{n,d}},F^{\Gamma_n})\le 2(\sum_{k=0}^{\infty}|\theta_k|)^2\varepsilon^2. \tag{11.33}$$

Now note that $F^{\Gamma_{n,d}}\xrightarrow{\mathcal{D}}F_d$ a.s. and there is almost sure moment convergence also. Now recall that W_2 metrizes weak convergence of probability measures with finite second moment. Hence, as $n\to\infty$, $W_2(F^{\Gamma_{n,d}},F_d)\longrightarrow 0$, a.s. and moreover, $\{F^{\Gamma_{n,d}}\}_{n\ge 1}$ is Cauchy with respect to W_2 almost surely.

Hence by triangle inequality, and (11.33),

$$\limsup_{m,n} W_2(F^{\Gamma_n},F^{\Gamma_m})\le 2\sqrt{2}(\sum_{k=0}^{\infty}|\theta_k|)\varepsilon$$

so that $\{F^{\Gamma_n}\}_{n\ge 1}$ is Cauchy with respect to W_2 a.s.

As a consequence, since W_2 is complete, there exists a probability measure F on $\mathbb{R}$ such that $F^{\Gamma_n}\to F$ weakly almost surely. Further,

$$W_2(F_d,F)=\lim_n W_2(F^{\Gamma_{n,d}},F^{\Gamma_n})\le\sqrt{2}(\sum_{k=0}^{\infty}|\theta_k|)\varepsilon,$$

and hence $F_d\to F$ weakly. Since $\{F_d\}$ are non-random, so is F.

Now if $\{\varepsilon_t\}$ is not i.i.d. but independent and uniformly bounded by some $C>0$ then the above proof is even simpler. We omit the details.

Now we show the convergence of $\{\beta_{h,d}\}$. Assumption B(b), (11.32) yields

$$\sup_d|\beta_{h,d}|\le c_h:=\frac{4^h(2h)!}{h!}(\sum_{k=0}^{\infty}|\theta_k|)^{2h}<\infty,\ \forall\ h\ge 0. \tag{11.34}$$

Hence for every fixed h, $\{A_d^h\}$ is uniformly integrable where $A_d \sim F_d$. Since $F_d \to F$ weakly,

$$\beta_h = \int x^h \, dF = \lim_d \int x^h \, dF_d = \lim_{d \to \infty} \beta_{h,d}.$$

This completes the proof of (b). Since $|\beta_h| \le c_h$, it satisfies Riesz's condition. This completes the proof of Theorem 11.2.1(b). $\qquad \square$

Proof of Theorem 11.2.1(c).

(i) $\beta_{h-1,d} \le \beta_{h,d}$: We first claim that for $d \ge 0$ $p_{k_0,\ldots,k_d}^{(d)} = p_{k_0,\ldots,k_d,0}^{(d+1)}$. To see this, consider a graph G with $2h$ vertices with h connected components and two vertices in each component. Let

$$\mathcal{M} = \{ \mathbf{a} : \quad \mathbf{a} \text{ is minimal } d \text{ matched, induces } G \text{ and } |a_x - a_y| = d + 1$$
$$\text{for some } x, y \text{ belonging to distinct components of } G \}.$$

Then one can easily argue that $\#\mathcal{M} = \mathrm{O}(n^{h-1})$ and consequently

$$\#\{ (\mathbf{t}, \pi) \in \mathcal{A} | \ \mathbf{a}(\mathbf{t}, \pi) \in \mathcal{M} \} = \mathrm{O}(n^h).$$

Hence

$$p_{k_0,\ldots,k_d}^{(d)}$$
$$= \lim_{n \to \infty} \frac{1}{n^{h+1}} \#\{ (\mathbf{t}, \pi) \in \mathcal{A} \ | \mathbf{a}(\mathbf{t}, \pi) \text{ is minimal } d \text{ matched with partition}$$
$$\{1, \ldots, 2h\} = \cup_{l=1}^{h} \{i_l, j_l\} \text{ and there are exactly } k_s \text{many } l\text{'s for which}$$
$$|\mathbf{a}(\mathbf{t}, \pi)(i_l) - \mathbf{a}(\mathbf{t}, \pi)(j_l)| = s, s = 0, \ldots, d, \mathbb{I}_{\mathbf{a}(\mathbf{t}, \pi)} = 1 \text{ and}$$
$$|\mathbf{a}(\mathbf{t}, \pi)(x) - \mathbf{a}(\mathbf{t}, \pi)(y)| \ge d + 2 \text{ if } x, y \text{ are in different partition blocks}\}$$
$$= p_{k_0,\ldots,k_d,0}^{(d+1)}.$$

Thus, for $\theta_0, \ldots, \theta_d \ge 0$ and $d \ge 1$,

$$\beta_{h,d} \ge \sum_{S_{h,d-1}} p_{k_0,\ldots,k_{d-1},0}^{(d)} \prod_{i=0}^{d-1} [\gamma_{X^{(d)}}(i)]^{k_i}$$

$$\ge \sum_{S_{h,d-1}} p_{k_0,\ldots,k_{d-1}}^{(d-1)} \prod_{i=0}^{d-1} [\gamma_{X^{(d-1)}}(i)]^{k_i} = \beta_{h,d-1},$$

proving the result.

Incidentally, if Assumption B(a) is violated, then the ordering need not hold.

(ii) Unbounded support of F_d and F: For any word w, let $|w|$ denote the length of the word. Let

$$\mathcal{W} = \{ w = w_1 w_2 : |w_1| = 2h = |w_2|; \ w, w_1, w_2 \text{ are zero pair-matched};$$
$$w_1[x] \text{ matches with } w_1[y] \text{ iff } w_2[x] \text{ matches with } w_2[y] \}.$$

Then

$$\beta_{2h,d} \geq [\gamma_{X^{(d)}}(0)]^{2h} p_{2h,0,\dots,0} \geq [\gamma_{X^{(d)}}(0)]^{2h} \sum_{w \in \mathcal{W}} \lim_n n^{-(2h+1)} \#\Pi^*(w).$$

(11.35)

For $w = w_1 w_2 \in \mathcal{W}$, let $\{1, \dots, 2h\} = \cup_{i=1}^h (i_s, j_s)$ be the partition corresponding to w_1. Then

$$\lim_n \frac{\#\Pi^*(w)}{n^{2h+1}} \geq \lim_n \frac{1}{n^{2h+1}} \#\{(\mathbf{t}, \pi): \ t_{i_s} = t_{j_s} \text{ and } \pi_{i_s} - \pi_{i_s-1} = \pi_{j_s-1} - \pi_{j_s}$$
$$\text{for } 1 \leq s \leq h; t_j + |\pi_j - \pi_{j-1}| \leq n, \text{ for } 1 \leq j \leq 2h\}. \ (11.36)$$

Now using the ideas in the proof of unboundedness of the LSD of the Toeplitz matrix with i.i.d. inputs (Chapter 2), for each d finite F_d has unbounded support. Since $\{\beta_{h,d}\}$ increases to β_h, the same conclusion is true for F. Proof of Theorem 11.2.1(c) is now complete. □

Proof of Theorem 11.2.2. Proceeding as earlier it is easy to see that the limit exists, and for each word w, the limiting contribution is given by,

$$p_w^{*,(d)} = \sum_{b \in \mathcal{B}(w)} \mathrm{E}_U \left[\mathbb{I}_{m,h}(U) \times \mathbb{I}_{\lambda^\flat}(U) \times \tilde{I}_\mathcal{T}(U) \right].$$

Comparing the above expression with the corresponding expression for the sequence $\Gamma_{n,d}$,

$$\beta_{h,d} \leq \beta_{h,d}^* \quad \text{if } \theta_j \geq 0, \ 0 \leq j \leq d.$$

Relation (11.32) holds if $\beta_{h,d}$ is replaced by $\beta_{h,d}^*$. We can use this to prove tightness of $\{F_d^*\}$ under Assumption B(a). Clearly Carleman's/Riesz's condition are also satisfied.

Since Γ_n^* and $\Gamma_{n,d}^*$ are no longer positive definite matrices, many of the ideas used earlier cannot be adapted here. We proceed as follows instead: Note that

$$\mathrm{E}[\beta_h(\Gamma_n^*)] = \frac{1}{n^{h+1}} \mathrm{E} \left[\sum_{(\mathbf{t}, \pi) \in \mathcal{A}} \prod_{j=1}^h X_{t_j} \prod_{j=1}^h X_{t_j + |\pi_{j-1} - \pi_j|} \right].$$

Write

$$X_{t_j} = \sum_{k_j \geq 0} \theta_{k_j} \varepsilon_{t_j - k_j} \quad \text{and} \quad X_{t_j + |\pi_{j-1} - \pi_j|} = \sum_{k_j' \geq 0} \theta_{k_j'} \varepsilon_{t_j + |\pi_{j-1} - \pi_j| - k_j'}.$$

Then using Assumption B(b) and applying DCT we get

$$\mathrm{E}[\beta_h(\Gamma_n^*)] = \sum_{\substack{k_j, k_j' \geq 0 \\ j=1,\dots,h}} \prod_{j=1}^h (\theta_{k_j} \theta_{k_j'}) \frac{1}{n^{h+1}} \mathrm{E} \left[\sum_{(\mathbf{t}, \pi) \in \mathcal{A}} \prod_{j=1}^h \varepsilon_{t_j - k_j} \varepsilon_{t_j + |\pi_{j-1} - \pi_{j-1}| - k_j'} \right].$$

Using the fact that $\{\varepsilon_t\}$ are uniformly bounded and $\sum_{k=1}^{\infty} |\theta_k| < \infty$ we note that it is enough to show that the limit below exists.

$$\lim_{n\to\infty} n^{-(h+1)} \, \mathrm{E}\Big[\sum_{(\mathbf{t},\pi)\in\mathcal{A}} \prod_{j=1}^{h} (\varepsilon_{t_j - k_j} \varepsilon_{t_j + |\pi_j - \pi_{j-1}| - k'_j}) \Big].$$

One can proceed as in the proof of Theorem 11.2.1 to show that only pair matched words contribute and hence enough to argue that as $n \to \infty$, the following expression converges.

$$\frac{1}{n^{h+1}} \#\{(\mathbf{t},\pi) \in \mathcal{A} : \{t_j - k_j, \ t_j + |\pi_j - \pi_{j-1}| - k'_j, \ 1 \le j \le h\} \text{ is pair-matched}\}.$$

This follows by adapting the ideas used in the proof of Theorem 11.2.1. Note that appropriate compatibility is needed among $\{k_j, k'_j, \ j = 1, \ldots, h\}$, the word w and the signs $b_i \ (= \pm 1)$ to ensure that the condition $\pi_0 = \pi_h$ is satisfied. So the above limit will depend on $\{k_j, k'_j, \ j = 1, \ldots, h\}$.

We also note that

$$\lim_n \frac{1}{n^{h+1}} \sum_{w \in \mathcal{W}_{2h}(2)} \#\{(\mathbf{t},\pi) \in \mathcal{A} : (t_j - k_j, \ t_j + |\pi_j - \pi_{j-1}| - k'_j)_{1 \le j \le h} \in \Pi(w)\}$$

$$\le \frac{4^h (2h)!}{h!}.$$

Hence F^* is uniquely determined by its moments and using DCT, $\beta^*_{h,d} \to \beta^*_h$. Whence it also follows that $F^*_d \xrightarrow{w} F^*$. Proof of part (c) is similar to the proof of Theorem 11.2.1(c). □

Proof of Theorem 11.2.3(a), (b) when $0 < \alpha < 1$. Let $\beta_h(\Gamma^{\alpha,I}_{n,d})$ and $\beta_h(\Gamma^{\alpha,II}_{n,d})$ be the hth moments, respectively, of the ESD of Type I and Type II ACVM with parameter α. We begin by noting that the expression for these contain an extra indicator term

$$\mathbb{I}_1 = \prod_{i=1}^{h} \mathbb{I}(|\pi_{i-1} - \pi_i| \le m_n) \text{ and } \mathbb{I}_2 = \prod_{i=1}^{h} \mathbb{I}(1 \le \pi_i \le m_n)$$

respectively.

For Type II ACVM, since there are m_n eigenvalues instead of n, the normalizing denominator is now m_n. Hence

$$\beta_h(\Gamma^{\alpha,I}_{n,d}) \tag{11.37}$$

$$= \frac{1}{n^{h+1}} \sum_{\substack{1 \le \pi_0, \ldots, \pi_h \le n \\ \pi_h = \pi_0}} \prod_{j=1}^{h} \Big(\sum_{t;j=1}^{n} X_{t_j,d} X_{t_j + |\pi_j - \pi_{j-1}|, d} \mathbb{I}_{(t_j + |\pi_j - \pi_{j-1}| \le n)} \Big) \mathbb{I}_1,$$

and

$$\frac{m_n}{n} \beta_h(\Gamma_{n,d}^{\alpha,II}) \tag{11.38}$$

$$= \frac{1}{n^{h+1}} \sum_{\substack{1 \leq \pi_0, \ldots, \pi_h \leq n \\ \pi_h = \pi_0}} \left[\prod_{j=1}^{h} \left(\sum_{t_j=1}^{n} X_{t_j,d} X_{t_j+|\pi_j-\pi_{j-1}|,d} \mathbb{I}_{(t_j+|\pi_j-\pi_{j-1}|\leq n)} \right) \right] \mathbb{I}_2.$$

It is thus enough to establish the limits on the right side of the above expressions and we can follow similar steps as in the proof of Theorem 11.2.1.

Since there are only the extra indicator terms, the negligibility of higher-order edges and verification of (M2) and (M4) needs no new arguments. Likewise, verification of (M1) is also similar except that there is now an extra indicator term in the expression for $p_w^{(d)}$. This takes care of the finite d case.

For $d = \infty$, note that the Type II ACVMs are $m_n \times m_n$ principal submi-nors of the original sample ACVMs and hence are automatically non-negative definite. We can write $\Gamma_n^{\alpha,II}(X^{(d)}) = (A_{n,d}^{\alpha,II})(A_{n,d}^{\alpha,II})^T$ where $A_{n,d}^{\alpha,II}$ is the first m_n rows of $A_{n,d}$. Thus, imitating the proof of Theorem 11.2.1, we can move from finite d to $d = \infty$.

However, for Type I ACVM, we cannot apply these arguments, as these matrices are not necessarily non-negative definite. Rather we proceed as in the proof of Theorem 11.2.2. Previous proof of unbounded support now needs only minor changes. We omit the details.

Since $\Gamma_{n,d}^{\alpha,II}$ is non-negative definite, the technique of proof of Theorem 11.2.1 can be adopted to prove the result under Assumption A(a). $\qquad\square$

Proof of Theorem 11.2.3(b) for Type I band ACVM.

Existence: Let $p_w^{(d),0,I}$ be the limiting contribution of the word w for the Type I ACVM with band parameter $\alpha = 0$. Then

$$p_w^{(d),0,I} := \lim_n \frac{1}{n^{h+1}} \sum_{b \in \mathcal{B}(w)} E_{U_n} \left[\mathbb{I}_{m,h}(U_n) \times \mathbb{I}_{\lambda^b,m}(U_n) \times \mathbb{I}_{\lambda^b,L}(U_n) \times \mathbb{I}_{\mathcal{T}}^{I}(U_n) \right].$$

where

$$\mathbb{I}_{\mathcal{T}}^{I}(U_n) = \mathbb{I}_{\mathcal{T},L}(U_n) \times \mathbb{I}_{\mathcal{T},m} := \prod_{\substack{j=1 \\ j \notin \mathcal{T}}}^{h} \mathbb{I}(0 \leq L_j(U_n) \leq m_n) \times \prod_{j \in \mathcal{T}} \mathbb{I}(-m_n \leq n_j \leq 0).$$

If w, $\lambda_{j+h-1}^b \neq \lambda_{j+h}^b$ for some j then $\mathbb{I}_{\mathcal{T},L}(U_n) \to 0$ as $n \to \infty$ and thus the limiting contribution from that word will be 0. Thus, only those words w for which $\lambda_{h+1}^b = \lambda_{j+h}^b$ for all $j \in \{1, 2, \ldots, h+1\}$ may contribute a non-zero quantity in the limit. This condition also implies that, for such words no π_i belongs to the generating set except π_0. This observation together with Lemma 6 of Basak, Bose and Sen (2014)[14], and the expression for limiting moments for $\Gamma_n(X)$ shows that $w \in \mathcal{W}_0^h$ may contribute a non-zero quantity, where

$$\mathcal{W}_0^h = \{w : \ |w| = 2h, \ w[i] \text{ matches with } w[i+h], \ n_i \leq 0, \ i = 1, 2, \ldots, h\}.$$

Further if $w \in \mathcal{W}_0^h$ then $\mathcal{T} = \{h+1, h+2, \ldots, 2h\}$, and thus $\mathbb{I}_{\mathcal{T},L} \equiv 1$. For $d = 0$ note that $\#\mathcal{W}_0^h = 1$ for every h and one can easily check that the contribution from that word is 1. Thus, $\beta_{h,0}^0 = \theta_0^{2h}$ and as a consequence, the LSD is $\delta_{\theta_0^2}$.

Now let $0 < d < \infty$. Note that if $m_n \geq d$, then

$$\mathbb{I}_{\lambda\flat,m} \times \mathbb{I}_{\lambda\flat,L} \times \mathbb{I}_{\mathcal{T},m} \to \prod_{j=1}^{h} \mathbb{I}(n_j \leq 0), \text{ as } n \to \infty.$$

Combining the above arguments we get that for any $w \in \mathcal{W}_0^h$, $p_w^{(d),0,I}$ is the number of choices of $\mathbf{b} \in \mathcal{B}(w)$, and $\{n_1, \ldots, n_h; \ n_i \leq 0\}$, such that $\sum_i n_i b_i = 0$.

Type I ACVMs are not necessarily non-negative definite, Hence we need to adapt the proof of Theorem 11.2.2. Details are omitted.   □

Identification of the LSD: Now it remains to argue that the limit we obtained is the same as $f_X(U)$. For $d = 0$ the LSD is $\delta_{\theta_0^2}$ and it is trivial to check that it is the same as $f_X(U)$.

For $0 < d < \infty$, note that the proof does not use the fact that $m_n \to \infty$ and we further note that for any sequence $\{m_n\}$ the limit we obtained above will be the same whenever $\liminf_{n \to \infty} m_n \geq d$. So in particular the limit will be the same if we choose another sequence $\{m_n'\}$ such that $m_n' = d$ for all n. Let $\Gamma_{n',d}^I$ denote the Type I ACVM where we put 0 instead of $\hat{\gamma}_{X^{(d)}}(k)$ whenever $k > m_n'$ and let $\Sigma_{n,d}$ be the $n \times n$ matrix whose (i,j)th entry is the population autocovariance $\gamma_{X^{(d)}}(|i-j|)$. Now from Lemma 3.2.1(a) we get

$$W_2^2(F^{\Gamma_{n',d}^I}, F^{\Sigma_{n,d}}) \leq \frac{1}{n} \text{Tr}(\Gamma_{n',d}^I - \Sigma_{n,d})^2$$
$$\leq 2[(\hat{\gamma}_{X^{(d)}}(0) - \gamma_{X^{(d)}}(0))^2 + \cdots + (\hat{\gamma}_{X^{(d)}}(d) - \gamma_{X^{(d)}}(d))^2].$$

For any j as $n \to \infty$, $\hat{\gamma}_{X^{(d)}}(j) \to \gamma_{X^{(d)}}(j)$ a.s.. Since d is finite, the right side of the above expression goes to 0 a.s.. This proves the claim for d finite.

To prove the result for the case $d = \infty$, first note that we already have

$$LSD(\Gamma_{n,d}^{0,I}) = LSD(\Sigma_{n,d}) := G_d \text{ and } LSD(\Gamma_{n,d}^{0,I}) \xrightarrow{w} LSD(\Gamma_n^{0,I}) \text{ as } d \to \infty.$$

Thus, it is enough to prove that $G_d \xrightarrow{\mathcal{D}} G(= LSD(\Sigma_n))$ as $d \to \infty$ where Σ_n is the $n \times n$ matrix whose (i,j)th entry is $\gamma_X(|i-j|)$. Define a sequence of $n \times n$ matrices $\bar{\Sigma}_{n,d}$ whose (i,j)th entry is $\gamma_X(|i-j|)$ if $|i-j| \leq d$ and otherwise 0. By triangle inequality,

$$W_2^2(F^{\Sigma_{n,d}}, F^{\Sigma_n}) \leq 2W_2^2(F^{\Sigma_{n,d}}, F^{\bar{\Sigma}_{n,d}}) + 2W_2^2(F^{\bar{\Sigma}_{n,d}}, F^{\Sigma_n}).$$

Fix any $\varepsilon > 0$. Fix d_0 such that $\left(\sum_{j=0}^{\infty} |\theta_j|\right)^2 \left(\sum_{l=d+1}^{\infty} |\theta_l|\right)^2 \leq \frac{\varepsilon^2}{32}$ for all $d \geq d_0$. Now again using Lemma 3.2.1(a) we get the following two relations:

$$\limsup_n W_2^2(F^{\Sigma_{n,d}}, F^{\bar{\Sigma}_{n,d}}) \leq 2\sum_{j=1}^{d}(\gamma_{X^{(d)}}(j) - \gamma_X(j))^2$$

$$= 2\sum_{j=0}^{d}\left(\sum_{k=d-j+1}^{\infty} \theta_k\theta_{j+k}\right)^2 \leq \frac{\varepsilon^2}{16},$$

$$W_2^2(F^{\bar{\Sigma}_{n,d}}, F^{\Sigma_n}) \leq \limsup_n \frac{1}{n}\mathrm{Tr}(\bar{\Sigma}_{n,d} - \Sigma_n)^2 \leq \frac{\varepsilon^2}{16}.$$

Thus $\limsup_n W_2^2(F^{\Sigma_{n,d}}, F^{\Sigma_n}) \leq \varepsilon/2$, for any $d \geq d_0$, and therefore by triangle inequality, $W_2^2(F^{G_d}, F^G) \leq \varepsilon$. This completes the proof. $\square$

Proof of Theorem 11.2.3(b) for Type II band autocovariance matrix. By Lemma 11.3.2 we need to consider only minimal matched terms. Let

$$G_t = \{t_i : t_i \in G\} \text{ and } G_\pi = \{\pi_i : \pi_i \in G\}.$$

Since $1 \leq \pi_i \leq m_n$ for all i, by similar arguments as in Lemma 11.3.2 we get

$$\text{the number of choices of } \mathbf{a}(\mathbf{t}, \pi) = \mathrm{O}(n^{\#G_t} m_n^{\#G_\pi}).$$

Thus, for any word w for which $\#G_t < h$, the limiting contribution will be 0. Hence the only contributing words in this case are those for which $\#S_3(w) = \#S_4(w) = h$, and from Lemma 6 of Basak, Bose and Sen (2014)[14], the only contributing words are those belonging to $\mathcal{W}_0^h$. Therefore using the same arguments as in the proof of Theorem 11.2.3, for Type I ACVM, for $\alpha = 0$ we obtain the same limit. All the remaining conclusions here follow from the proof for Type I ACVM with parameter $\alpha = 0$.

Since Type II ACVMs are non-negative definite, the connection between the LSD for finite d and $d = \infty$ is proved by adapting the ideas from the proof of Theorem 11.2.1.

Proof of Theorem 11.2.3(c). Since K is bounded, negligibility of higher-order edges and verification of $(M2)$ and (M4) is the same as before. Verification of (M1) is also the same, with an extra indicator in the limiting expression. Denoting $p_w^{(d),K}$ to be the limiting contribution from a word w, we have,

$$p_w^{(d),K} = \lim_n \mathbb{E}_{U_n}\left[\mathbb{I}_{m,h}(U_n) \times \mathbb{I}_{\lambda^\flat,m}(U_n) \times \mathbb{I}_{\lambda^\flat,L}(U_n) \times \mathbb{I}_{\mathcal{T}}^{\mathbb{I}}(U_n) \times \mathbb{I}_K(U_n)\right],$$

where

$$\mathbb{I}_K(U_n) := \prod_{j=1}^{h} K\left(\frac{L_j(U_n)}{m_n}\right).$$

Since $m_n \to \infty$, and $K(\cdot)$ is continuous at 0, $K(0) = 1$, note that $\mathbb{I}_K \to 1$. Arguing as earlier, we get $p_w^{(d),0,I} = p_w^{(d),K}$ for every word w and thus the limiting distributions are the same in both cases. When $d = \infty$ the arguments are also similar to earlier and the details are omitted. $\qquad\square$

11.4 Exercises

1. Write down a separate and easier proof of Theorem 11.2.1(a) for the case $d = 0$.

2. Simulate $\Gamma_n^*(X)$ for different models and verify that the ESD has significant positive mass on the negative real axis.

3. Show that each of the LSDs in this chapter are identical for the combinations $(\theta_0, \theta_1, \theta_2, \ldots)$, $(\theta_0, -\theta_1, \theta_2, \ldots)$ and $(-\theta_0, \theta_1, -\theta_2, \ldots)$. See Basak, Bose and Sen (2014)([14] for a proof which is based on properties of the limit moments.

4. Show that the distribution of $f_X(U)$ and of the LSD of $\Sigma_n(X)$ are identical for processes with autocovariances $(\gamma_0, \gamma_1, \ldots, \gamma_d)$ and $(\gamma_0, -\gamma_1, \ldots, (-1)^d \gamma_d)$.

5. Verify the contents of Tables 11.1 and 11.2.

6. Prove Lemma 11.3.1.

7. Prove Lemma 11.3.3.

8. In the context of proof of Theorem 11.2.1(c), consider an MA(2) and an MA(1) process with parameters θ_0, θ_1, θ_2 and where $\theta_2 = -\kappa\theta_0$, $\theta_0, \theta_1 > 0$. Show that $\beta_{2,2} < \beta_{2,1}$ if we choose $\kappa > 0$ sufficiently small.

9. Verify unboundedness of the support of F_d.

Bibliography

[1] G. Anderson, A. Guionnet, and O. Zeitouni. *An Introduction to Random Matrices*. Cambridge University Press, Cambridge, UK, 2009.

[2] T. W. Anderson. *An Introduction to Multivariate Statistical Analysis*. Wiley Series in Probability and Mathematical Statistics. John Wiley & Sons, Inc., New York, second edition, 1984.

[3] T. M. Apostol. *Introduction to Analytic Number Theory*. Springer-Verlag, New York-Heidelberg, 1976. Undergraduate Texts in Mathematics.

[4] L. Arnold. On the asymptotic distribution of the eigenvalues of random matrices. *J. Math. Anal. Appl.*, 20:262–268, 1967.

[5] Z. D. Bai. Methodologies in spectral analysis of large-dimensional random matrices, a review. *Statist. Sinica*, 9(3):611–677, 1999. With comments by G. J. Rodgers and Jack W. Silverstein; and a rejoinder by the author.

[6] Z. D. Bai and Y. Q. Yin. Convergence to the semicircle law. *Ann. Probab.*, 16(2):863–875, 1988.

[7] Z. D. Bai and W. Zhou. Large sample covariance matrices without independence structures in columns. *Statist. Sinica*, 18(2):425–442, 2008.

[8] Z.D. Bai and J. W. Silverstein. *Spectral Analysis of Large Dimensional Random Matrices*. Springer Series in Statistics. Springer, New York, second edition, 2010.

[9] S. Banerjee and A. Bose. Noncrossing partitions, Catalan words, and the semicircle law. *J. Theoret. Probab.*, 26(2):386–409, 2013.

[10] T. Banica, S. Curran, and R. Speicher. de Finetti theorems for easy quantum groups. *Ann. Probab.*, 40(1):401–435, 2012.

[11] A. Basak. *Large Dimensional Random Matrices*. Master's dissertation. Indian Statistical Institute, Kolkata, 2009.

[12] A. Basak and A. Bose. Balanced random Toeplitz and Hankel matrices. *Electron. Commun. Probab.*, 15:134–148, 2010.

[13] A. Basak and A. Bose. Limiting spectral distribution of some band matrices. *Periodica Hungarica*, 43(1):113–150, 2011.

[14] A. Basak, A. Bose, and S. Sen. Limiting spectral distribution of sample autocovariance matrices. *Bernoulli*, 20(3):1234–1259, 2014.

[15] R. Basu, A. Bose, S. Ganguly, and R. S. Hazra. Joint convergence of several copies of different patterned random matrices. *Electron. J. Probab.*, 17:no. 82, 33, 2012.

[16] R. Basu, A. Bose, S. Ganguly, and R. S. Hazra. Limiting spectral distribution of block matrices with Toeplitz block structure. *Statist. Probab. Lett.*, 82(7):1430–1438, 2012.

[17] F. Benaych-Georges. Rectangular random matrices, entropy, and Fisher's information. *J. Operator Theory*, 62(2):371–419, 2009.

[18] F. Benaych-Georges. Rectangular random matrices, related convolution. *Probab. Theory Related Fields*, 144(3-4):471–515, 2009.

[19] F. Benaych-Georges, A. Guionnet, and M. Maida. Fluctuations of the extreme eigenvalues of finite rank deformations of random matrices. *Electron. J. Probab.*, 16:no. 60, 1621–1662, 2011.

[20] S. Bhamidi, S. N. Evans, and A. Sen. Spectra of large random trees. *J. Theoret. Probab.*, 25(3):613–654, 2012.

[21] R. Bhatia. *Matrix Analysis*, volume 169 of *Graduate Texts in Mathematics*. Springer-Verlag, New York, 1997.

[22] R. N. Bhattacharya and R. Ranga Rao. *Normal Approximation and Asymptotic Expansions*. John Wiley & Sons, New York-London-Sydney, 1976. Wiley Series in Probability and Mathematical Statistics.

[23] P. Biane. On the free convolution with a semi-circular distribution. *Indiana Univ. Math. J.*, 46(3):705–718, 1997.

[24] A. Bose, S. Gangopadhyay, and K. Saha. Convergence of a class of Toeplitz type matrices. *Random Matrices Theory Appl.*, 2(3):1350006, 21, 2013.

[25] A. Bose, S. Gangopadhyay, and A. Sen. Limiting spectral distribution of XX' matrices. *Ann. Inst. Henri Poincaré Probab. Stat.*, 46(3):677–707, 2010.

[26] A. Bose, R. S. Hazra, and K. Saha. Convergence of joint moments for independent random patterned matrices. *Ann. Probab.*, 39(4):1607–1620, 2011.

[27] A. Bose and J. Mitra. Limiting spectral distribution of a special circulant. *Statist. Probab. Lett.*, 60(1):111–120, 2002.

[28] A. Bose, J. Mitra, and A. Sen. Large dimensional random k circulants. *J. Theoret. Probab.*, 25(3):771–797, 2012.

[29] A. Bose and A. Sen. Another look at the moment method for large dimensional random matrices. *Electron. J. Probab.*, 13:no. 21, 588–628, 2008.

[30] A. Bose and S. Sen. Finite diagonal random matrices. *J. Theoret. Probab.*, 26(3):819–835, 2013.

[31] A. Böttcher and B. Silbermann. *Introduction to Large Truncated Toeplitz Matrices*. Universitext. Springer-Verlag, New York, 1999.

[32] M. Bożejko and R. Speicher. Interpolations between Bosonic and Fermionic relations given by generalized Brownian motions. *Math. Z.*, 222(1):135–159, 1996.

[33] P. J. Brockwell and R. A. Davis. *Time Series: Theory and Methods*. Springer Series in Statistics. Springer-Verlag, New York, second edition, 1991.

[34] W. Bryc, A. Dembo, and T. Jiang. Spectral measure of large random Hankel, Markov and Toeplitz matrices. *Ann. Probab.*, 34(1):1–38, 2006.

[35] M. Capitaine and M. Casalis. Asymptotic freeness by generalized moments for Gaussian and Wishart matrices. Application to beta random matrices. *Indiana Univ. Math. J.*, 53(2):397–431, 2004.

[36] M. Capitaine and C. Donati-Martin. Strong asymptotic freeness for Wigner and Wishart matrices. *Indiana Univ. Math. J.*, 56(2):767–803, 2007.

[37] M. Capitaine, C. Donati-Martin, and D. Féral. The largest eigenvalues of finite rank deformation of large Wigner matrices: Convergence and nonuniversality of the fluctuations. *Ann. Probab.*, 37(1):1–47, 2009.

[38] M. Capitaine, C. Donati-Martin, D. Féral, and M. Février. Free convolution with a semicircular distribution and eigenvalues of spiked deformations of Wigner matrices. *Electron. J. Probab.*, 16:no. 64, 1750–1792, 2011.

[39] T. Carleman. *Les Fonctions Quasi Analytiques (in French)*. Paris: Gauthier-Villars, 1926.

[40] S. Chatterjee. A simple invariance theorem. *arXiv:math/0508213 [math.PR]*, 2005.

[41] B. Collins. Moments and cumulants of polynomial random variables on unitary groups, the Itzykson-Zuber integral, and free probability. *Int. Math. Res. Not.*, 2003(17):953–982, 2003.

[42] B. Collins, A. Guionnet, and E. Maurel-Segala. Asymptotics of unitary and orthogonal matrix integrals. *Adv. Math.*, 222(1):172–215, 2009.

[43] B. Collins and P. Śniady. Integration with respect to the Haar measure on unitary, orthogonal and symplectic group. *Comm. Math. Phys.*, 264(3):773–795, 2006.

[44] R. M. Corless, G. H. Gonnet, D. E. G. Hare, D. J. Jeffrey, and D. E. Knuth. On the Lambert W function. *Adv. Comput. Math.*, 5(4):329–359, 1996.

[45] R. Couillet and M. Debbah. *Random Matrix Methods for Wireless Communications*. Cambridge University Press, Cambridge, UK, 2011.

[46] R. Couillet, M. Debbah, and J. W. Silverstein. A deterministic equivalent for the analysis of correlated MIMO multiple access channels. *IEEE Trans. Inform. Theory*, 57(6):3493–3514, 2011.

[47] P. J. Davis. *Circulant Matrices*. John Wiley & Sons, New York-Chichester-Brisbane, 1979. A Wiley-Interscience Publication, Pure and Applied Mathematics.

[48] P. Deift. *Orthogonal Polynomials and Random Matrices: A Riemann-Hilbert Approach*, volume 3 of *Courant Lecture Notes*. American Mathematical Society and Courant Institute of Mathematical Sciences, 1999.

[49] P. Deift, J. Baik, and T. Suidan. *Combinatorics and Random Matrix Theory*, volume 172 of *Graduate Studies in Mathematics*. American Mathematical Society and Courant Institute of Mathematical Sciences, 2016.

[50] P. Deift and D. Gioev. *Random Matrix Theory: Invariant Ensembles and Universality*, volume 18 of *Courant Lecture Notes*. American Mathematical Society and Courant Institute of Mathematical Sciences, 2009.

[51] K. Dykema. On certain free product factors via an extended matrix model. *J. Funct. Anal.*, 112(1):31–60, 1993.

[52] K. Dykema and U. Haagerup. DT-operators and decomposability of Voiculescu's circular operator. *Amer. J. Math.*, 126(1):121–189, 2004.

[53] L. Erdos and Horng-Tzer Yau. *A Dynamical Approach to Random Matrix Theory*, volume 28 of *Courant Lecture Notes*. American Mathematical Society and Courant Institute of Mathematical Sciences, 2017.

[54] J. Fan and Q. Yao. *Nonlinear Time Series*. Springer Series in Statistics. Springer-Verlag, New York, 2003.

[55] D. Féral and S. Péché. The largest eigenvalue of rank one deformation of large Wigner matrices. *Comm. Math. Phys.*, 272(1):185–228, 2007.

[56] P. J. Forrester. *Log-gases and Random Matrices*, volume 34 of *London Mathematical Society Monographs Series*. Princeton University Press, Princeton, NJ, 2010.

[57] W. Fulton. Eigenvalues of sums of Hermitian matrices (after A. Klyachko). *Astérisque*, 40:Exp. No. 845, 5, 255–269, 1998. Séminaire Bourbaki.

[58] S. Georgiou and C. Koukouvinos. Multi-level k-circulant supersaturated designs. *Metrika*, 64(2):209–220, 2006.

[59] V. L. Girko and K. Yu. Repin. The quarter-of-circumference law (English. Russian original). *Theory of Probab. Math. Stat*, 50:66–69, 1995.

[60] R.M. Gray. *Toeplitz and Circulant Matrices: A Review*. Now Publishers, Inc., 2006.

[61] U. Grenander. *Probabilities on Algebraic Structures*. John Wiley & Sons, Inc., New York-London; Almqvist & Wiksell, Stockholm-Göteborg-Uppsala, 1963.

[62] U. Grenander and J. W. Silverstein. Spectral analysis of networks with random topologies. *SIAM J. Appl. Math.*, 32(2):499–519, 1977.

[63] C. Hammond and S. J. Miller. Distribution of eigenvalues for the ensemble of real symmetric Toeplitz matrices. *J. Theoret. Probab.*, 18(3):537–566, 2005.

[64] F. Hiai and D. Petz. Asymptotic freeness almost everywhere for random matrices. *Acta Sci. Math. (Szeged)*, 66(3-4):809–834, 2000.

[65] F. Hiai and D. Petz. *The Semicircle Law, Free Random Variables and Entropy*, volume 77 of *Mathematical Surveys and Monographs*. American Mathematical Society, Providence, RI, 2000.

[66] A. J. Hoffman and H. W. Wielandt. The variation of the spectrum of a normal matrix. *Duke Math. J.*, 20:37–39, 1953.

[67] D. Jonsson. Some limit theorems for the eigenvalues of a sample covariance matrix. *J. Multivariate Anal.*, 12(1):1–38, 1982.

[68] E. Kaltofen. Analysis of Coppersmith's block Wiedemann algorithm for the parallel solution of sparse linear systems. *Math. Comp.*, 64(210):777–806, 1995.

[69] V. Kargin. Spectrum of random Toeplitz matrices with band structure. *Electron. Commun. Probab.*, 14:412–421, 2009.

[70] A. M. Khorunzhy, B. A. Khoruzhenko, and L. A. Pastur. Asymptotic properties of large random matrices with independent entries. *J. Math. Phys.*, 37(10):5033–5060, 1996.

[71] D. Z. Liu and Z. D. Wang. Limit distributions for random Hankel, Toeplitz matrices and independent products. *arXiv preprint arXiv:0904.2958*, 2009.

[72] C. Male. The norm of polynomials in large random and deterministic matrices. *Probab. Theory Related Fields*, 154(3-4):477–532, 2012. With an appendix by Dimitri Shlyakhtenko.

[73] V. A. Marčenko and L. A. Pastur. Distribution of eigenvalues in certain sets of random matrices. *Mat. Sb. (N.S.)*, 72 (114):507–536, 1967.

[74] A. Massey, S. J. Miller, and J. Sinsheimer. Distribution of eigenvalues of real symmetric palindromic Toeplitz matrices and circulant matrices. *J. Theoret. Probab.*, 20(3):637–662, 2007.

[75] T. L. McMurry and D. N. Politis. Banded and tapered estimates for autocovariance matrices and the linear process bootstrap. *Journal of Time Series Analysis*, 1:471–482, 2010.

[76] M. W. Meckes. Some results on random circulant matrices. In *High dimensional probability V: The Luminy Volume*, volume 5 of *Inst. Math. Stat. (IMS) Collect.*, pages 213–223. Inst. Math. Statist., Beachwood, OH, 2009.

[77] M. L. Mehta. *Random Matrices and the Statistical Theory of Energy Levels*. Academic Press, New York-London, 1967.

[78] M. L. Mehta. *Random Matrices*, volume 142 of *Pure and Applied Mathematics (Amsterdam)*. Elsevier/Academic Press, Amsterdam, third edition, 2004.

[79] A. Nica and R. Speicher. *Lectures on the Combinatorics of Free Probability*, volume 335 of *London Mathematical Society Lecture Note Series*. Cambridge University Press, Cambridge, 2006.

[80] T. Oraby. The spectral laws of Hermitian block-matrices with large random blocks. *Electron. Comm. Probab.*, 12:465–476, 2007.

[81] L. A. Pastur. The spectrum of random matrices. *Teoret. Mat. Fiz.*, 10(1):102–112, 1972.

[82] L. A. Pastur and V. Vasilchuk. On the law of addition of random matrices. *Comm. Math. Phys.*, 214(2):249–286, 2000.

[83] S. Péché. The largest eigenvalue of small rank perturbations of Hermitian random matrices. *Probab. Theory Related Fields*, 134(1):127–173, 2006.

[84] D. S. G. Pollock. Circulant matrices and time-series analysis. *Internat. J. Math. Ed. Sci. Tech.*, 33(2):213–230, 2002.

[85] I. Popescu. General tridiagonal random matrix models, limiting distributions and fluctuations. *Probab. Theory Related Fields*, 144(1-2):179–220, 2009.

[86] C. E. Porter. *Statistical Theories of Spectra: Fluctuations: A Collection of Reprints and Original Papers, with an Introductory Review*. Academic Press, 1965.

[87] R. Rashidi Far, T. Oraby, W. Bryc, and R. Speicher. On slow-fading MIMO systems with nonseparable correlation. *IEEE Trans. Inform. Theory*, 54(2):544–553, 2008.

[88] M. Riesz. Surle probleme des moments. Troisieme note (in French). *Ark.for matematik. Astronomi och Fysik*, 16:1–52, 1923.

[89] J. Riordan. *Combinatorial Identities*. Robert E. Krieger Publishing Co., Huntington, N.Y., 1979. Reprint of the 1968 original.

[90] Øyvind Ryan. On the limit distributions of random matrices with independent or free entries. *Comm. Math. Phys.*, 193(3):595–626, 1998.

[91] H. Schultz. Non-commutative polynomials of independent Gaussian random matrices. The real and symplectic cases. *Probab. Theory Related Fields*, 131(2):261–309, 2005.

[92] A. Sen. *Large Dimensional Random Matrices*. Masters Dissertation. Indian Statistical Institute, Kolkata, 2006.

[93] A. Sen and B. Virág. Absolute continuity of the limiting eigenvalue distribution of the random Toeplitz matrix. *Electron. Commun. Probab.*, 16:706–711, 2011.

[94] S. Sen. *Limiting Spectral Distribution of Random Matrices*. Master's dissertation. Indian Statistical Institute, Kolkata, 2010.

[95] D. Shlyakhtenko. Random Gaussian band matrices and freeness with amalgamation. *Internat. Math. Res. Notices*, 1996(20):1013–1025, 1996.

[96] J. W Silverstein and A. M. Tulino. Theory of large dimensional random matrices for engineers. In *Spread Spectrum Techniques and Applications, 2006 IEEE Ninth International Symposium on*, pages 458–464. IEEE, 2006.

[97] R. Speicher. On universal products. In *Free Probability Theory (Waterloo, ON, 1995)*, volume 12 of *Fields Inst. Commun.*, pages 257–266. Amer. Math. Soc., Providence, RI, 1997.

[98] R. Speicher. Free probability theory. In *The Oxford Handbook of Random Matrix Theory*, pages 452–470. Oxford Univ. Press, Oxford, 2011.

[99] V. V. Strok. Circulant matrices and spectra of de Bruijn graphs. *Ukraïn. Mat. Zh.*, 44(11):1571–1579, 1992.

[100] T. Tao. *Topics in Random Matrix Theory*, volume 132 of *Graduate Studies in Mathematics*. American Mathematical Society, Providence, RI, 2012.

[101] A. M. Tulino and S. Verdú. Random matrix theory and wireless communications. *Foundations and Trends® in Communications and Information Theory*, 1(1):1–182, 2004.

[102] D. Voiculescu. Limit laws for random matrices and free products. *Invent. Math.*, 104(1):201–220, 1991.

[103] D. Voiculescu. A strengthened asymptotic freeness result for random matrices with applications to free entropy. *Internat. Math. Res. Notices*, 1998(1):41–63, 1998.

[104] K. W. Wachter. The strong limits of random matrix spectra for sample matrices of independent elements. *Ann. Probability*, 6(1):1–18, 1978.

[105] E. P. Wigner. Characteristic vectors of bordered matrices with infinite dimensions. *Ann. of Math.*, 62:548–564, 1955.

[106] E. P. Wigner. On the distribution of the roots of certain symmetric matrices. *Ann. of Math. (2)*, 67:325–327, 1958.

[107] J. Wishart. The generalised product moment distribution in samples from a normal multivariate population. *Biometrika*, pages 32–52, 1928.

[108] W. B. Wu and M. Pourahmadi. Nonparametric estimation of large covariance matrices of longitudinal data. *Biometrica*, 90:831–844, 2003.

[109] Y.K. Wu, R. Z. Jia, and Q. Li. g-Circulant solutions to the (0, 1) matrix equation Am= Jn* 1. *Linear Algebra and Its Applications*, 345(1-3):195–224, 2002.

[110] H. Xiao and W. B. Wu. Covariance matrix estimation for stationary time series. *Ann. Statist.*, 40(1):466–493, 2012.

[111] Khintchine A. Y. On unimodal distributions (in Russian). *Izv. Nauchno-Issled. Inst. Mat. Mekh., Tomsk. Gos. Univ.*, 2:1–7, 1938.

[112] Y. Q. Yin. Limiting spectral distribution for a class of random matrices. *J. Multivariate Anal.*, 20(1):50–68, 1986.

[113] Y. Q. Yin and P. R. Krishnaiah. Limit theorem for the eigenvalues of the sample covariance matrix when the underlying distribution is isotropic. *Teor. Veroyatnost. i Primenen.*, 30(4):810–816, 1985.

[114] J. T. Zhou. A formula solution for the eigenvalues of g-circulant matrices. *Math. Appl. (Wuhan)*, 9(1):53–57, 1996.

Index